Studies in Computational Intelligence

Volume 971

Series Editor

Janusz Kacprzyk, Polish Academy of Sciences, Warsaw, Poland

The series "Studies in Computational Intelligence" (SCI) publishes new developments and advances in the various areas of computational intelligence—quickly and with a high quality. The intent is to cover the theory, applications, and design methods of computational intelligence, as embedded in the fields of engineering, computer science, physics and life sciences, as well as the methodologies behind them. The series contains monographs, lecture notes and edited volumes in computational intelligence spanning the areas of neural networks, connectionist systems, genetic algorithms, evolutionary computation, artificial intelligence, cellular automata, self-organizing systems, soft computing, fuzzy systems, and hybrid intelligent systems. Of particular value to both the contributors and the readership are the short publication timeframe and the world-wide distribution, which enable both wide and rapid dissemination of research output.

Indexed by SCOPUS, DBLP, WTI Frankfurt eG, zbMATH, SCImago.

All books published in the series are submitted for consideration in Web of Science.

More information about this series at http://www.springer.com/series/7092

Uttam Ghosh · Yassine Maleh · Mamoun Alazab ·
Al-Sakib Khan Pathan
Editors

Machine Intelligence and Data Analytics for Sustainable Future Smart Cities

 Springer

Editors
Uttam Ghosh 🆔
Department of EECS
Vanderbilt University
Nashville, TN, USA

Yassine Maleh 🆔
ENSAK
Sultan Moulay Slimane University
Beni-Mellal, Morocco

Mamoun Alazab 🆔
Charles Darwin University
Darwin, Northern Territory, Australia

Al-Sakib Khan Pathan 🆔
Department of Computer Science
and Engineering
Independent University
Dhaka, Bangladesh

ISSN 1860-949X   ISSN 1860-9503 (electronic)
Studies in Computational Intelligence
ISBN 978-3-030-72064-3   ISBN 978-3-030-72065-0 (eBook)
https://doi.org/10.1007/978-3-030-72065-0

This Springer imprint is published by the registered company Springer Nature Switzerland AG
The registered company address is: Gewerbestrasse 11, 6330 Cham, Switzerland

Preface

The subjects of the smart city and machine intelligence are the sources of many claims, hopes and fears. In concrete terms, it is a set of challenges, made up of threats and opportunities, relating to the production of cities and the management of urban services. Dealing with these challenges requires a better balance between the performance of the artificial and real expectations. Smart and sustainable cities need an intelligent infrastructure that is stable, secure, reliable and interoperable, capable of supporting a huge volume of ICT-enabled applications and services. In recent years, machine intelligence approaches have emerged as powerful computational models and have shown significant success to deal with a massive amount of data in unsupervised settings. Machine intelligence techniques influence various technologies because they offer an effective learning representation and allow the system to learn features from data without explicitly designing them. With the emerging technologies, the Internet of Things (IoT), wearable devices, cloud computing and data analytics offer the potential to acquire and process a tremendous amount of data from the physical world. Machine intelligence algorithms help efficiently leverage IoT and Big Data aspects to develop personalized services in sustainable smart cities.

Machine Intelligence and Data Analytics for Sustainable Future Smart Cities comprise a number of state-of-the-art contributions from both scientists and practitioners working in machine intelligence and smart cities. It aspires to provide a relevant reference for students, researchers, engineers, and professionals working in this area or those interested in grasping its diverse facets and exploring the latest advances on machine intelligence and data analytics for sustainable future smart city applications. More specifically, the book contains 21 chapters classified into three pivotal sections: the first part presents the role and importance of machine intelligence and data analytics techniques for sustainable future smart cities. The second part describes and analyzes various applications of machine intelligence and data analytics in smart cities, such as Internet of Things, vision systems, and healthcare to mitigate the impact of COVID-19 (or similar pandemic situation) in sustainable smart cities. The third part introduces the machine and deep learning techniques in smart cities' surveillance and cybersecurity.

We want to take this opportunity and express our thanks to the contributors to this volume and the reviewers for their tremendous efforts in reviewing and providing interesting feedback to the authors of the chapters. The editors would like to thank Dr. Thomas Ditzinger (Springer, Editorial Director, Interdisciplinary Applied Sciences) and Prof. Janusz Kacprzyk (Series Editor-in-Chief), and Ms. Jennifer Sweety Johnson (Springer Project Coordinator), for the editorial assistance and support to produce this important scientific work. Without this collective effort, this book would not have been possible to be completed.

Nashville, TN, USA   Prof. Uttam Ghosh
Khouribga, Morocco   Prof. Yassine Maleh
Darwin, Australia   Prof. Mamoun Alazab
Dhaka, Bangladesh   Prof. Al-Sakib Khan Pathan

Contents

About the Editors

Uttam Ghosh is working as an Assistant Professor of the Practice in the Department of Electrical Engineering and Computer Science, Vanderbilt University, Nashville, TN, USA. He obtained his Ph.D. in Electronics and Electrical Engineering from the Indian Institute of Technology Kharagpur, India in 2013, and has Post-doctoral experience at the University of Illinois in Urbana-Champaign, Fordham University, and Tennessee State University. He has been awarded the 2018–2019 Junior Faculty Teaching Fellow (JFTF) and has been promoted to a Graduate Faculty position at Vanderbilt University. He has published 50 papers at reputed international journals including IEEE Transaction, Elsevier, Springer, IET, Wiley, InderScience, and IETE, and also in top international conferences sponsored by IEEE, ACM, and Springer. He has conducted several sessions and workshops related to Cyber-Physical Systems (CPS), SDN, IoT, and smart cities as co-chair at top international conferences including IEEE MASS, SECON, CPSCOM, IEMCON, ICDCS, and so on. He has served as a Technical Program Committee (TPC) member at renowned international conferences including ACM SIGCSE, IEEE LCN, IEMCON, STPSA, SCS SpringSim, and IEEE Compsac. He is serving as an Associate Editor of the International Journal of Computers and Applications, Taylor & Francis, and also a reviewer for international journals including IEEE Transactions, Elsevier, Springer, and Wiley. He is contributing as Guest Editor for special issues with ACM Transactions on Internet Technology (TOIT), Springer MTAP, MDPI Future Internet, and Wiley ITL. He is a senior member of the IEEE and a member of AAAS, ASEE, ACM, and Sigma Xi.

Dr. Ghosh is actively working for editing two edited volumes on Emerging CPS, Security, Deep/Machine Learning with CRC Press, and Chapman Hall Big Data Series. His main research interests include Cybersecurity, Computer Networks, Wireless Networks, Information Centric Networking, and Software-Defined Networking.

Yassine Maleh is an Associate Professor at the National School of Applied Sciences at Sultan Moulay Slimane University, Morocco. He received his Ph.D. degree in Computer Science from Hassan 1st University, Morocco. He is a cybersecurity and information technology researcher and practitioner with industry and academic experience. He worked for the National Ports Agency in Morocco as an IT manager from 2012 to 2019. He is a senior member of IEEE, member of the International Association of Engineers IAENG and The Machine Intelligence Research Labs. He has made contributions in the fields of Information Security and Privacy, Internet of Things Security, and Wireless and Constrained Networks Security. His research interests include Information Security and Privacy, Internet of Things, Networks Security, Information system, and IT Governance. He has published more than 50 papers (book chapters, international journals, and conferences/workshops), 7 edited books, and 3 authored books. He is the Editor-in-Chief of the International Journal of Smart Security Technologies (IJSST). He serves as an Associate Editor for IEEE Access (2019 Impact Factor 4.098), the International Journal of Digital Crime and Forensics (IJDCF), and the International Journal of Information Security and Privacy (IJISP). He was also a Guest Editor of a special issue on Recent Advances on Cyber Security and Privacy for Cloud-of-Things of the International Journal of Digital Crime and Forensics (IJDCF), Volume 10, Issue 3, July–September 2019. He has served and continues to serve on executive and technical program committees and as a reviewer of numerous international conference and journals such as Elsevier Ad Hoc Networks, IEEE Network Magazine, IEEE Sensor Journal, ICT Express, and Springer Cluster Computing. He was the Publicity Chair of BCCA 2019 and the General Chair of the MLBDACP 19 symposium.

Mamoun Alazab is the Associate Professor in the College of Engineering, IT, and Environment at Charles Darwin University, Australia. He received his Ph.D. degree is in Computer Science from the Federation University of Australia, School of Science, Information Technology and Engineering. He is a cybersecurity researcher and practitioner with industry and academic experience. His research is multidisciplinary that focuses on cybersecurity and digital forensics of computer systems including current and emerging issues in the cyberenvironment like cyber-physical systems and the Internet of things, by taking into consideration the unique challenges present in these environments, with a focus on cybercrime detection and prevention. He looks into the intersection use of machine learning as an essential tool for cybersecurity, for example, for detecting attacks, analyzing malicious code, or uncovering vulnerabilities in software. He has more than 100 research papers. He is the recipient of short fellowship from Japan Society for the Promotion of Science (JSPS) based on his nomination from the Australian Academy of Science. He delivered many invited and keynote speeches, 27 events in 2019 alone. He convened and chaired more than 50 conferences and workshops. He is the Founding Chair of the IEEE Northern Territory Subsection: (Feb 2019—current). He is a senior member of the IEEE, Cybersecurity Academic Ambassador for Oman's Information Technology Authority (ITA); member of the IEEE Computer Society's Technical Committee on Security and Privacy (TCSP); and has worked closely with government and industry on many projects, including IBM, Trend Micro, the Australian Federal Police (AFP), the Australian Communications and Media Authority (ACMA), Westpac, UNODC, and the Attorney General's Department.

Al-Sakib Khan Pathan is a Professor of Computer Science and Engineering. Currently, he is with the Independent University, Bangladesh as an Adjunct Professor. He received Ph.D. degree in Computer Engineering in 2009 from Kyung Hee University, South Korea and B.Sc. degree in Computer Science and Information Technology from Islamic University of Technology (IUT), Bangladesh in 2003. In his academic career so far, he worked as a faculty member at the CSE Department of Southeast University, Bangladesh during 2015–2020; at Computer Science Department, International Islamic University Malaysia (IIUM), Malaysia during 2010–2015; at BRACU, Bangladesh during 2009–2010; and at NSU, Bangladesh during 2004–2005. He was a Guest Lecturer for the STEP project at the Department of Technical and Vocational Education, Islamic University of Technology, Bangladesh in 2018. He also worked as a Researcher at Networking Lab, Kyung Hee University, South Korea from September 2005 to August 2009 where he completed his MS leading to Ph.D. His research interests include wireless sensor networks, network security, cloud computing, and e-services technologies. Currently, he is also working on some multidisciplinary issues. He is a recipient of several awards/best paper awards and has several notable publications in these areas. So far, he has delivered over 20 Keynotes and Invited speeches at various international conferences and events. He was awarded the IEEE Outstanding Leadership Award for his role in IEEE GreenCom'13 Conference. He is currently serving as the Editor-in-Chief of International Journal of Computers and Applications, Taylor & Francis, UK, Editor of Ad Hoc and Sensor Wireless Networks, Old City Publishing, International Journal of Sensor Networks, Inderscience Publishers, and Malaysian Journal of Computer Science; Associate Editor of Connection Science, Taylor & Francis, UK, International Journal of Computational Science and Engineering, Inderscience; Area Editor of International Journal of Communication Networks and Information Security; Guest Editor of many special issues of top-ranked journals; and Editor/Author of 21 books. One of his books has been included twice in Intel Corporation's Recommended Reading List for Developers, second half of 2013 and first half of 2014; three books were included in IEEE Communications Society's (IEEE ComSoc) Best Readings in Communications and Information Systems Security, 2013; two other books were indexed with all the titles (chapters) in Elsevier's acclaimed abstract and citation database, Scopus, in February 2015; and a seventh book is translated to simplified Chinese language from English version. Also, two of his journal papers and one conference paper were included under different categories in IEEE Communications Society's (IEEE ComSoc) Best Readings Topics on Communications and Information Systems Security, 2013. He also serves as a referee of many prestigious journals. He received some awards for his reviewing activities like: one of the most active reviewers of IAJIT several times; Elsevier Outstanding Reviewer for Computer Networks, Ad Hoc Networks, FGCS, and JNCA in multiple years. He is a Senior Member of the Institute of Electrical and Electronics Engineers (IEEE), USA.

Data Quality Evaluation, Outlier Detection and Missing Data Imputation Methods for IoT in Smart Cities

Vera Van Zoest, Xiuming Liu, and Edith Ngai

Abstract Low-cost IoT devices allow data collection in smart cities at a high spatio-temporal resolution. Data quality evaluation is needed to investigate the pre-processing steps required to use these data. Besides data pre-processing, outlier detection techniques are required to detect anomalies in the spatio-temporal IoT dataset. We distinguish between erroneous outliers and events based on spatio-temporal auto-correlation patterns, as well as correlations with other dynamic processes in the environment. We consider missing data imputation to fill gaps caused by sensor failures, maintenance, pre-processing and outlier detection. In this study, we use the temporal covariance structure within the data to impute missing data. We apply the methods for outlier detection and missing data imputation to an IoT testbed for air quality monitoring in the city of Eindhoven, the Netherlands. The methods can be applied in a more general sense to other continuous environmental variables which show a similarly strong spatio-temporal autocorrelation structure.

Keywords Data quality · Outlier detection · Missing data imputation · Air quality · Smart city

V. Van Zoest (✉) · X. Liu
Department of Information Technology, Uppsala University, Uppsala, Sweden
e-mail: vera.van.zoest@it.uu.se

X. Liu
e-mail: xiuming.liu@it.uu.se

E. Ngai
Department of Electrical and Electronic Engineering, The University of Hong Kong, Hong Kong, China
e-mail: chngai@eee.hku.hk

© The Author(s), under exclusive license to Springer Nature Switzerland AG 2021
U. Ghosh et al. (eds.), *Machine Intelligence and Data Analytics for Sustainable Future Smart Cities*, Studies in Computational Intelligence 971,
https://doi.org/10.1007/978-3-030-72065-0_1

1 Introduction

In smart cities, IoT devices are becoming increasingly popular for environmental monitoring, traffic control, and security surveillance. IoT typically consists of low-cost sensors which are used to collect large amounts of data with a high spatio-temporal resolution [1]. This resolution is needed to represent the strong spatio-temporal variability in the data [23]. Although these low-cost sensors appear to provide a cost-effective solution, the data quality of low-cost sensors requires further investigation [33]. Systematic errors in the sensors are relatively easy to correct for, using simple linear regression calibration functions. These calibration functions can be constructed in the lab before deployment by comparison to a reference instrument or reference value [10], or in the field by co-locating the sensor node with a reference instrument [6]. To build correct calibration functions in the field, however, a long calibration period is needed to account for different situations and the full range of expected values. Random errors and interference effects require multivariate regression models for sensor calibration. Interference effects are distortions in the data caused by external factors that cannot be influenced. In environmental monitoring, for example, a sensor may react to other substances or show different results depending on meteorological conditions. By including these in a multivariate regression model, these can partly be accounted for. In the lab, this can be done in a controlled way. The real situation in the field may, however, be much more complex when different variables interact. To account for time-dependent drift effects, i.e. a loss of sensitivity of the sensor over time, non-linear regression models or artificial neural networks may be required [34]. In all cases, calibration functions should be constructed for each sensor individually and updated regularly [39].

Most calibration methods are sensitive to outliers. Outliers, sometimes referred to as anomalies [8], are those observations that differ from the expected observations [3, 37]. These can strongly affect the slope of the calibration function. Therefore, outliers should be removed before the calibration of the sensor data. We present an outlier detection method based on spatio-temporal classification of the observations, to sustain the spatio-temporal pattern of the dynamic process of interest. Next, we look at the challenge of detection of errors versus events. Erroneous outliers are those observations that deviate from the true values. Events are observations that can be detected as outliers, but do not deviate from the true values [37]. Events rather reflect a real change in the measured phenomenon and can therefore be of interest depending on the user perspective. Data cleaning and outlier detection lead to gaps in the data. Missing data imputation methods are needed to fill those gaps for further data analyses such as spatio-temporal predictions, because many of these techniques are sensitive to missing values.

In this chapter, we evaluate the data quality of IoT data, perform outlier detection, and fill missing data gaps. Throughout this chapter, we use the air quality IoT testbed in Eindhoven for illustration of the methods. First, we describe the IoT testbed. Next, we evaluate the data quality of the low-cost sensors and perform pre-processing steps following a standardized approach. Then, we focus on outlier detection methods to improve the thematic accuracy of IoT data, as well as data imputation methods to improve the completeness of the data.

2 Related Work

Outlier detection is widely studied in the literature. However, most studies have focused on either purely temporal outlier detection, or purely spatial outlier detection. Functional outlier detection is an example of temporal outlier detection. In functional outlier detection, time series are compared for a fixed time period. This method has for example been used in an air quality IoT to detect months with unusually high air pollutant concentrations [11, 12, 24, 31]. Functional outlier detection finds outlying vectors of measurements but cannot be used to detect individual outliers. Pure temporal outlier detection may also lead to neglection of a systematic biases and sensor drift [40].

In spatial outlier detection, an individual observation can be compared to other observations within a spatial neighborhood, assuming some similarity of the expected values within a spatial region. An example of spatial outlier detection is kriging, a distance-weighted technique which has been applied to air quality IoT as well [5]. To combine the spatial neighborhood with a temporal neighborhood, spatio-temporal outlier detection may be used [21, 32]. Both spatial and spatio-temporal outlier detection have only been applied in IoTs measuring environmental variables with a low spatial variability. For variables with a higher spatial variability, the spatial covariance structure is not enough to represent the variability within the variable of interest. For air pollutants like NO_2, for example, with a high spatial and temporal variability caused by traffic, more sophisticated methods are needed to distinguish natural differences in locations and time periods from outliers in the IoT data.

Outlier detection leads to gaps in the data. Several methods exist to impute missing data in time series (e.g. [3, 16, 25]), which can deal with longer periods of missing data. In Little and Rubin's fundamental book, different missing data mechanisms and various imputation methods are discussed [22]. One of the widely applied methods is to impute missing data with conditional means [30]. However, estimation of the conditional means of the missing values is non-trivial, given that the stochastic processes monitored by IoT devices are often nonlinear and high-dimensional. Gaussian processes (GPs) are nonlinear and nonparametric models, which are highly flexible to fit various dynamic patterns [36]. Using GPs we are able to estimate the conditional mean for the missing data based on the available observations. Applications of using GPs for imputation are, for example, GP imputation of multiple financial series [41] and GP for crowdsourced traffic data imputation [29]. A recent study [14] combines the GP model with the variational auto-encoder (VAE) model for time-series imputation. The idea is to first use a VAE to embed the data into a latent space of reduced dimensionality, and then model the embedded data with a

GP model. The missing data are inferred by sampling from a conditional distribution in the latent space and then decoded into the original time-series dimensions.

3 IoT Testbed

The IoT testbed is located in the city of Eindhoven, a medium-sized city in the southern part of the Netherlands. The city has a high population density and traffic intensity, elevating levels of traffic-related pollutants such as NO_2 [4]. With media interest in air pollution issues, the inhabitants of Eindhoven have become more aware of the health effects of air pollution. There are, however, only two conventional air pollution monitoring stations in the city. Therefore, the AiREAS association has been set up to monitor the air quality at a fine spatio-temporal resolution. It was founded by a group of engaged citizens, but also involves industrial and governmental partners as well as universities [9].

AiREAS set up an air quality IoT testbed in Eindhoven, referred to as Innovative Air Monitoring Network (Innovatief Luchtmeetnet, ILM). It is the first urban IoT in the Netherlands measuring air quality at a fine spatial and temporal resolution (Fig. 1). After its installation in November 2013, it has been operated continuously since. The network consists of 35 weatherproof 'airboxes'. We refer to the individual airboxes as ILM_x, where x is replaced by the airbox number. Since the total area of the municipality of Eindhoven is approximately 90 km^2, the sensor network is relatively dense. The airboxes are of size 43 × 33 × 20 cm (Fig. 2) and contain an array of sensors. They are manufactured by the former Energy research Centre of the Netherlands (ECN), now part of the Netherlands Organisation for applied scientific research (TNO).

Each airbox measures particulate matter in different sizes (PM_{10}, $PM_{2.5}$, PM_1), ozone (O_3), temperature, and humidity as the air flows through. A large part, 25 airboxes, also measure NO_2 since 2015. Because of the high sensor costs, ultrafine particles (UFPs) are measured at only six nodes in the IoT testbed. The UFP sensors are installed in separate boxes that are attached to the airboxes for power supply and GPRS connection. From November 2016 to February 2017 the UFP sensors were attached to different airboxes every three weeks to cover multiple locations. All AiREAS data is publicly available [2].

The spatial locations of the airboxes were chosen based on several criteria [9]. Most importantly, sampling sites represent locations where humans are exposed. The airboxes are located in the build-up area of the city, near residential areas and schools. The set of locations covers urban background locations in quiet neighbourhoods as well as urban traffic locations near busy roads. One airbox is located outside of the city for regional background monitoring. All airboxes are installed in fixed positions at lamp posts to supply electricity. They are located at 2.5–3 m height, representing human exposure as closely as possible, while minimizing the risk of accessibility by third persons.

Fig. 1 Locations of the airboxes in the city of Eindhoven, the Netherlands

Fig. 2 Airbox attached to
the light pole

At two locations in the city, an airbox is collocated with a reference monitor. The reference monitors belong to the national air quality monitoring network (Landelijk Meetnet Luchtkwaliteit, LML), which is maintained and operated by the National Institute for Public Health and the Environment [28]. The LML sensor network consists of around 60 measurement stations throughout the Netherlands, of which two are situated in Eindhoven. Although the LML has a lower spatial and temporal resolution, the uncertainty of the measurements is expected to be lower than the uncertainty of the ILM measurements. Both reference monitors are located in a traffic location.

After initial lab and field calibration of the sensors, data have been collected since November 2013. There are some gaps in the data for moments in time in which the instruments were removed for testing, adjusting or calibration. The sensors were recalibrated at the end of 2015, together with the installation and calibration of the NO_2 sensors. In this chapter, we focus on NO_2 data acquired in 2016. NO_2 is measured using the electrochemical cell Citytech Sensoric NO_2 3E50 in a differential measurement setup. A switching valve and reagent cartridges are used in front of the electrochemical cell to dry the air. Observations are discarded when temperature and humidity fall outside acceptable ranges.

4   Data Quality Evaluation and Pre-processing

For any spatio-temporal dataset acquired in an IoT testbed, data quality evaluation and pre-processing are important first steps in data analysis. In literature, different elements are included in spatial data quality assessment [15, 35]. In order to assess spatial data quality in a transparent way, we refer to the standard on Data Quality for Geographic Information, provided by the International Organization for Standardization (ISO) in ISO 19157 [35]. Six elements of spatial data quality are defined here: logical consistency, positional accuracy, temporal quality, thematic accuracy, completeness, and usability element. Since IoT data contain attribute values observed in space and time, many aspects in this standard on Geographic Information are relevant for IoT data quality assessment.

Logical consistency: First, we evaluate the logical consistency in all variables of the dataset. Logical consistency deals with the validity of attribute values and the adherence of relationships and compositions between objects to logical rules of structure and compatibility [20]. Data should carefully be checked for their logical consistency, e.g. relative humidity should always have a value between 0 and 100%, whereas wind direction can have valid values between 0 and 360°. While negative temperature values can make perfect sense, negative air pollution values and GPS coordinates should be removed, as they are impossible to occur. For the Eindhoven IoT testbed, we observe negative NO_2 values when the true values were below the limit of detection [40]. Since negative NO_2 values are impossible to occur, several choices can be made in this case: (1) change the negative values to zero, (2) change the negative values to the value which equals the limit of detection, or (3) remove the values from the dataset. The true values are expected to be between zero and the limit of detection, thus between options 1 and 2. We chose the third option in this case, as we plan to impute missing values later on. Besides that, zeroes have another meaning in this specific dataset: they represent sensor failures and should be removed as well. In total, 2.5% of the hourly NO_2 observations in 2016 was removed for this purpose. Deciding on the upper limits of logical consistency is more challenging, especially when dealing with highly skewed variables such as air pollutant concentrations [17, 26]. In general, more extreme peaks can be expected in datasets with a high temporal resolution (e.g. 1 s to 10 min). In averaged data (e.g. hourly, daily, annually), extreme peaks are less likely to occur. When dealing with low-cost sensors in an IoT, additional reference monitors could be useful to determine local upper limits of logical consistency. Since we will perform outlier detection later on, we only remove extreme values $\geq 3 * max(NO_{2ref})$, where $max(NO_{2ref})$ is the maximum hourly NO_2 concentration measured at the two reference monitors in Eindhoven in 2016. This led to a removal of 0.02% of the hourly observations, which were very likely caused by sensor failures. The evaluation of logical consistency does not stop at evaluating whether single observations are within natural boundaries. A more easily overlooked aspect is the evaluation of variability within a time series. Sensor failures could sometimes lead to frozen concentration values for several hours or days, which is inconsistent with the natural variability in air pollutant concentrations. We removed 1.5% of the NO_2 observations for this reason, aiming to recover the natural variability in the data imputation step.

Positional accuracy: Positional accuracy defines the accuracy of positions of features and is always related to some kind of spatial reference system. It deals with the nearness of the true values in comparison to the observed values in this reference system. When working with mobile sensors, positional accuracy is very important. In personal monitoring, for example, the reported location of a sensor can determine personal exposure. Positional uncertainty is then propagated into exposure estimates and related exposure–response functions. Although the airboxes are installed at fixed locations in our IoT testbed, they are all deployed with a GPS sensor. Since the true location of the airboxes is known, the actual measurements of the GPS sensor are not of much interest under normal operating conditions. However, the positional accuracy of the GPS sensor has to be known to distinguish between GPS uncertainty and real movements of the sensor. The latter could be related to displacement for calibration or maintenance. To get an estimate of the positional accuracy of the GPS sensor, we can collect GPS coordinates at a fine temporal resolution for an extended period of time, in which we know the sensor has not been moved. Usually, the uncertainty in the GPS coordinates is within a few meters, while the distance travelled for calibration or maintenance is several orders of magnitude larger. We removed all NO_2 observations of an airbox on days it was moved from its original position for maintenance or calibration.

Temporal quality: Temporal quality explains the quality of temporal attributes and relationships. It consists of the accuracy of a time measurement, temporal consistency, and temporal validity. Van Oort [35] adds three elements which are not in ISO 19157 but which were present in the European pre-standard ENV 12656 [7]: last update, rate of change, and temporal lapse. The temporal lapse represents the average time between change in the real world and change representations in the data, and is thus related to the temporal resolution used when averaging the air pollutant concentrations over a period of time (e.g. ten minutes, hourly, daily). Continuous variables like air pollution tend to change fast over time, with changes in traffic patterns and meteorological variables. The complex mixture of air pollutants makes it hard to model its spatio-temporal variability, while sampling rates of sensors are not able to keep up with the quick chemical reactions in the atmosphere. The sensors in the Eindhoven air quality IoT testbed have a sampling rate of 10 min, which is a relatively high temporal resolution compared to the reference monitors which measure NO_2 at hourly resolution. However, a trade-off needs to be made between temporal quality and quantitative attribute accuracy. Measurements with a high temporal resolution are typically noisy. Averaging the observations to a lower temporal resolution helps to reduce noise in the data. We chose to average the NO_2 data to hourly resolution, as this matches the resolution of the reference data.

Thematic accuracy: Thematic accuracy refers to classification correctness, non-quantitative attribute correctness, and quantitative attribute accuracy. Quantitative attribute accuracy refers to the closeness of the value of a quantitative attribute to the true value [18]. This 'true value' often refers to a value of a reference dataset, measured at the same spatio-temporal location, which is accepted to be true. Quantitative attribute correctness is one of the most debated issues of low-cost sensor data. Low-cost air quality sensors tend to be more sensitive to the interference of other pollutants and meteorological factors, compared to more expensive reference monitors. Besides that, electrochemical gas sensors typically lose sensitivity over time, referred to as drift. These issues need to be considered in automatic calibration methods which correct for drift and interference effects [39]. Outlier detection techniques are needed to flag values that differ from the expected value. The applied methods are explained in further detail in the next section.

Completeness: In terms of completeness, an important issue for air quality sensor networks is missing data. Data could be missing due to sensor failures, displacement, maintenance or manually removed during data quality evaluation. Besides sensor failures producing no data, observations are removed due to logical inconsistency, positional inaccuracy or outlier detection. The low budget for maintenance of the IoT testbed, in combination with a long distance between the testbed and manufacturing location, led to long periods of inactivity (e.g. a few months up to a full year) for some sensors to ensure transport of multiple sensors at the same time. We perform missing data imputation for the smaller gaps (e.g. max. 24 h), as described in this chapter.

Usability element: Usability element refers to the suitability of the data for a specific application. Careful evaluation of the data quality is required to make well-informed decisions on the applications and goals for which the sensors can be used. The data quality of the collected sensor data influences the quality of the output models, maps and exposure estimates. All previously mentioned elements can be used to describe and assess the usability of the data. The overall data quality and usability can be improved by improving the data quality of any element.

5 Outlier Detection

Consider an IoT with sensor nodes at unique and fixed locations $s \in 1...S$, measuring an environmental variable x at time points $t \in 1...T$ where S is the number of sensor nodes and T is the number of time points. Each spatio-temporal observation can thus be denoted by $x_{s,t}$. We use a square root transformation to transform x, such that x follows an approximately normal distribution with mean μ and standard deviation σ. This allows for a symmetric evaluation of outliers both on lower and higher ends of the distribution, where

$$\left(x_{s,t} < L\right) \vee \left(x_{s,t} > U\right). \tag{1}$$

Here, $L = \mu - z \times \sigma$ and $U = \mu + z \times \sigma$, where z is a constant value determining the size of the confidence interval, which can be chosen depending on the required strictness of the outlier detection. For a 95% confidence interval, $z = 1.96$ is used [21], while $z = 3$ gives a more strict confidence interval [24, 32]. Here, we will use $z = 2.97$, corresponding to a 99.7% confidence interval.

Equation 1 assumes that $x_{s,t}$ can be evaluated against a globally stationary μ and σ of x, independent of the spatial and temporal location s, t. This assumption is however not valid in the dynamic nature of environmental variables such as air pollution. Some studies therefore only use the observations in a spatio-temporal neighborhood around s, t to compute μ and σ. Therefore, we adopt the outlier detection method based on spatio-temporal classification of s, t as described in van Zoest et al. [40].

The observations are divided into spatio-temporal classes. Each spatio-temporal class consists of observations taken at similar spatial locations and in similar time periods. Which time periods should be grouped together, and which spatial locations be grouped together, depends on the dynamic process of x and is spatio-temporal variability. In terms of urban NO_2 concentrations, the spatio-temporal variability is mostly governed by traffic patterns. Therefore, spatial locations can be divided into urban traffic and urban background locations. Temporal classes depict traffic patterns throughout the day. To detect outliers, each observation is assigned to a spatio-temporal class c, now denoted as $x_{s,t,c}$, and then evaluated against the lower and upper thresholds of that class:

$$\left(x_{s,t,c} < L_c\right) \vee \left(x_{s,t,c} > U_c\right) \tag{2}$$

where $L_c = \mu_c - z \times \sigma_c$ and $U_c = \mu_c + z \times \sigma_c$, are based on the class-specific mean μ_c and standard deviation σ_c of the normal distribution underlying the truncated normal distribution of x. Van Zoest et al. [40] describe how to obtain these parameters using maximum likelihood estimation.

As a result, we can detect outliers depending on the spatio-temporal class of the observation, as shown in Fig. 3. Here, $L_c \leq 0$ so L_c is fixed to zero. However, U_c shows a strong spatio-temporal variability, of which the temporal pattern is visible in the figure. The rush hours cause peaks twice a day, in which higher concentrations

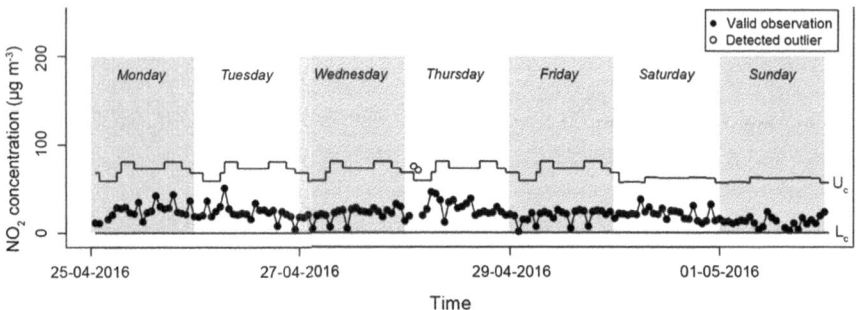

Fig. 3 Temporal variability of L_c and U_c and detected outliers at an urban background location in the network. Adopted from van Zoest et al. [40]

are expected, and the five weekdays clearly stand out compared to the two weekend days at the end of the time series plot. Two outliers were detected in this week at this urban background location, where $x_{s,t,c} > U_c$ based on the spatio-temporal location of the observations. If these observations would have occurred during rush hour or at a traffic location, thus in a different class c with higher expected values, U_c would have been higher and these observations would be within the limits of expected values.

Determining whether the detected outliers are events or errors, requires some knowledge on the environmental variable of interest and its correlation with other factors. In this case, we consider the air pollutant NO_2, of which the temporal variability is largely governed by meteorological processes and chemical reactions with other gases in the atmosphere. Spearman's rank correlation coefficient between the variable of interest x and possible covariates $v \in V$, where V is the set of covariates, may help to detect erroneous sensor observations. Here, $V = \{ozone, temperature, relative\ humidity\}$. Ozone is measured at a central RIVM monitoring location in the city [19, 27]. Temperature and relative humidity are measured by a central monitoring station from the Royal Netherlands Meteorological Institute [19, 27]. NO_2 is typically negatively correlated to temperature and ozone (O_3), as seen in previous analyses [38] and naturally explained by the chemical reaction between NO_2, O_3 and ultraviolet radiation [13]. The correlation coefficient between NO_2 and relative humidity is typically positive, indirectly caused by the negative correlation between temperature and relative humidity. This behaviour is reflected in the first two rows on Table 1. Sensors which are prone to sensor failures causing errors, will show abnormal correlation coefficients for outliers in x. The last two rows in Table 1 show an example for two nodes with outliers. While the correlation coefficients in the non-outlying observations are in the expected directions, the outlying NO_2 observations have correlation coefficients in the opposite directions. We observe highly positive correlations with O_3 and temperature, whereas the correlation with relative humidity is strongly negative. These sensors may therefore be particularly prone to errors. Note that we can only draw conclusions at the node level here, not on individual observations. However, the location of the sensor node and

Table 1 Spearman's rank correlation coefficients between NO_2 and covariates in V, for non-outlying and outlying measurements in nodes ILM_22 and ILM_30. All correlation coefficients are significant at significance level $\alpha = 0.05$

	Ozone	Temperature	Relative humidity
NO_2 non-outlying measurements in node ILM_22	−0.04	−0.46	+ 0.09
NO_2 non-outlying measurements in node ILM_30	−0.14	−0.58	+ 0.14
NO_2 outlying measurements in node ILM_22	+ 0.65	+ 0.73	−0.62
NO_2 outlying measurements in node ILM_30	+ 0.42	+ 0.49	−0.44

its individual characteristics could influence whether it is prone to errors or likely to detect events. More accurate results could be obtained when retrieving covariate data at the node level. Although we have these measurements available at the node level, we consider the central reference monitoring stations more reliable, as errors may occur simultaneously for different sensors in one node.

We can also explore the spatial and temporal covariance structure to distinguish between errors and events. Environmental variables like air pollutants have a strong spatial and temporal autocorrelation. Errors usually occur in one node at a time, and do not influence neighbouring nodes in the network. While sensor drift and events usually appear gradually over time, sensor failures usually occur suddenly, causing strong peaks in the observed values. Anomalies in the spatio-temporal variance–covariance matrix may therefore help to detect erroneous outliers [40].

Figure 4 shows a time series of a week of NO_2 observations at node ILM_26, an airbox at an urban traffic location. Figure 5 shows the corresponding temporal autocorrelation function (ACF) plots, (a) before and (b) after removal of the outliers. In (a), the ACF is clearly disrupted by outliers, showing no significant temporal

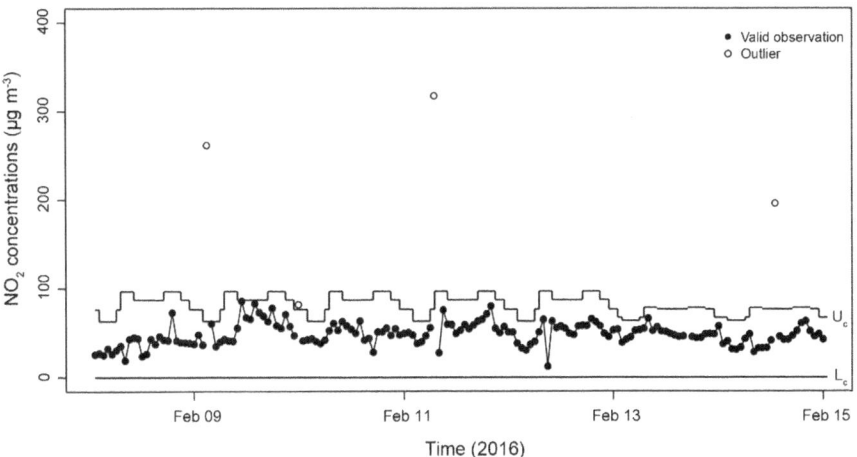

Fig. 4 Time series of a week of NO_2 observations at node ILM_26 (urban traffic location). Adopted from van Zoest et al. [40]

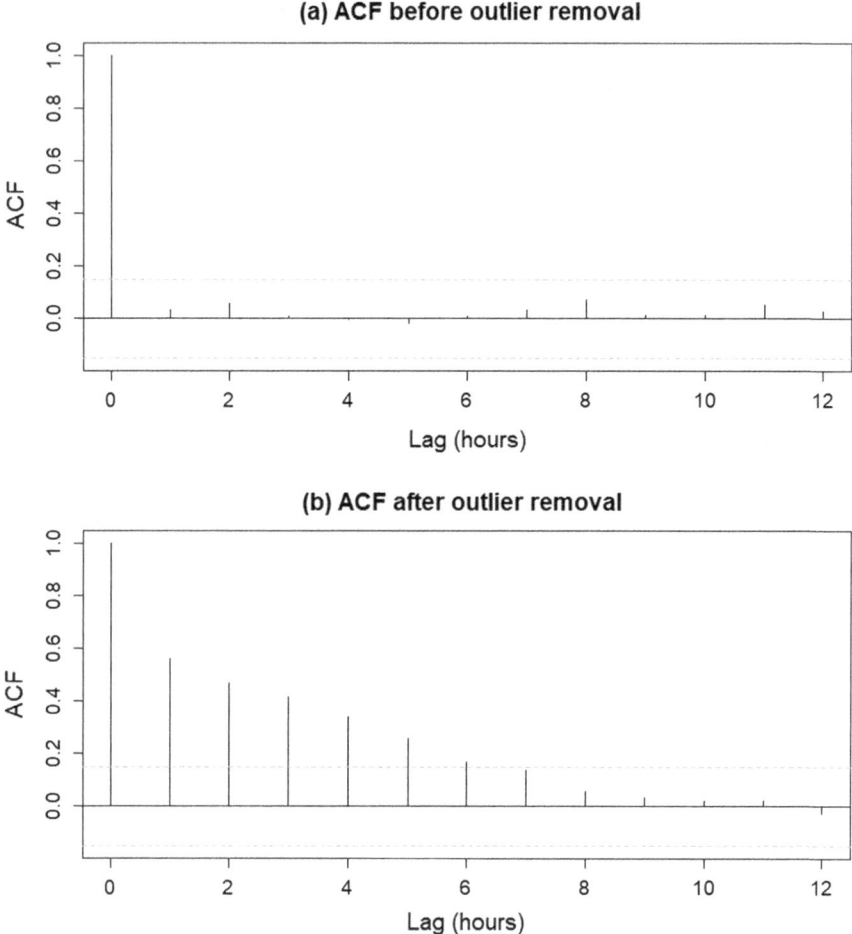

Fig. 5 Temporal ACF **a** before outlier removal and **b** after outlier removal. The same dataset is used as illustrated in Fig. 4. The grey dashed lines represent the 95% confidence interval; significant lags show ACF values crossing this line

autocorrelation. After removal of outliers, (b) shows a temporal ACF as expected, with significant autocorrelations within the first six hours. Based on this exploration, it is likely that the outliers in this particular node and time period are caused by sensor failures rather than more gradually appearing events which would not influence the ACF this strongly. Similarly, the spatial and spatio-temporal covariance structure can be used to detect erroneous outliers, as described in more detail elsewhere [38].

6 Missing Data Imputation

In this section we address the completeness indicator of IoT data via non-linear statistical imputation. Let $x_{s,t}$ be the available measurement of sensor s at time $t \in \mathcal{T}$. Assume that a measurement at time t^*,

$$\left\{ x_{s,t^*} | t^* \in \mathcal{T}^* \right\},$$

is missing due to sensor or communication failure. The goal is to impute the missing data x_{s,t^*}, based on the information from the available measurements. Using Bayesian inference, the posterior of x_{s,t^*} given $x_{s,\mathcal{T}} = \left\{ x_{s,1}, ..., x_{s,T} \right\}$ is,

$$p(x_{s,t^*} | x_{s,1}, ..., x_{s,T}) \propto p(x_{s,t^*}) p(x_{s,1}, ..., x_{s,T} | x_{s,t^*}),$$

where $p(x_{s,t^*})$ is the prior, and $p(x_{s,1}, ..., x_{s,T} | x_{s,t^*})$ is the likelihood.

Therefore an important element for imputing the missing data is to define a probabilistic model which specifies the underlying data generation process for both $x_{s,t}$ and x_{s,t^*}. In this study, we use the Gaussian process (GP) to model the dynamic of NO_2 in an urban environment. A GP is a stochastic process, and an arbitrary set of finite samples from the GP follows a joint Gaussian distribution. A GP is uniquely defined by its mean and kernel functions, $\mu(t)$ and $k(t, t')$:

$$x_{s,t} \sim GP\left(\mu(t), k\left(t, t'\right) \right).$$

The mean function describes the mean value at different time t. For instance, the average NO_2 concentration varies at different hours of the day. The kernel function describes the covariance between the NO_2 concentrations at two time points, t and t'. To design a GP model that characterizes the dynamic patterns of the underlying data process, the mean and kernel functions need to be learned from the available observations. For example, we select a periodic kernel function.

$$k\left(t, t'\right) = \sigma^2 \exp\left(-\frac{2\sin^2(\pi |t - t'|/P)}{l_1^2} \right) \exp\left(-\frac{(t - t')^2}{2l_2^2} \right),$$

where P is the period (in this case, a 24-h period), and $\{\sigma, l_1, l_2\}$ are the unknown hyper-parameters. The values of the hyper-parameters can be estimated from the available observations using the maximum likelihood estimation (MLE) method.

Based on the GP model learned from the available observations, we can evaluate the posterior distribution $p(x_{s,t^*} | x_{s,1}, ..., x_{s,T})$, which is a conditional Gaussian distribution. The posterior mean and covariance for the missing data are given by

$$m_{s,t^* | \mathcal{T}} = m_{s,t^*} + \Sigma_{t^*,\mathcal{T}} \Sigma_{\mathcal{T},\mathcal{T}}^{-1}(x_{s,\mathcal{T}} - m_{s,\mathcal{T}}),$$

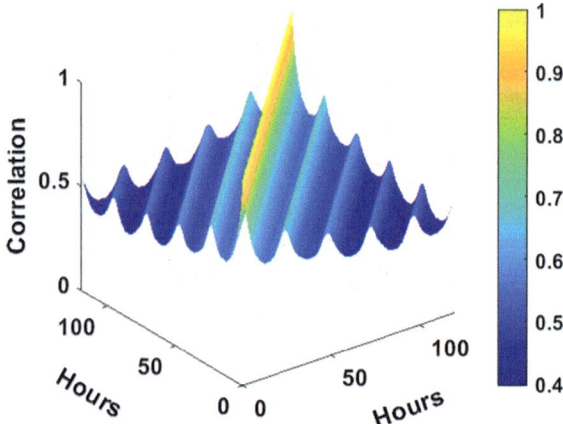

Fig. 6 An example correlation matrix visualized in 3D plot

$$\Sigma_{s,t^*|\mathcal{T}} = \Sigma_{t^*,t^*} - \Sigma_{t^*,\mathcal{T}} \Sigma_{\mathcal{T},\mathcal{T}}^{-1} \Sigma_{\mathcal{T},t^*}.$$

The elements in the covariance matrices are given by the kernel function. For example, a normalized covariance matrix of 5×24 data points is visualized in Fig. 6. In the correlation matrix we observe peaks of the daily periodic pattern.

An example of imputing missing data of 24 h is visualized in Fig. 7. In this example, we assume that the measurements of September 1, 2016 are lost, and use the available measurements from the closest four weeks to impute the missing data (two weeks before and two weeks after the missing date). The imputed data follows the dynamic pattern of the missing data and has satisfying accuracy during most of the 24 h. Furthermore, we provide an uncertainty measure of the imputed data, using the 10 and 90% percentiles of the samples drew from the posterior distribution.

Finally, we repeat the imputation process in the example for all days in the dataset. The root mean square errors (RMSEs) for each hour of the day are visualized using the boxplot in Fig. 8. The overall imputation error remains at a fairly low level,

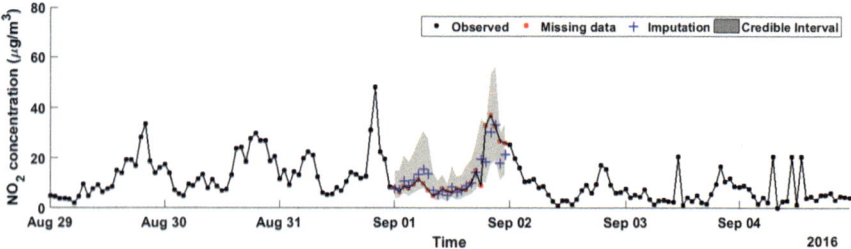

Fig. 7 Imputed 24-h data for September 1, 2016, using data from 2 weeks before and 2 weeks after the missing date. The credible interval shows the area between 10 and 90% percentiles

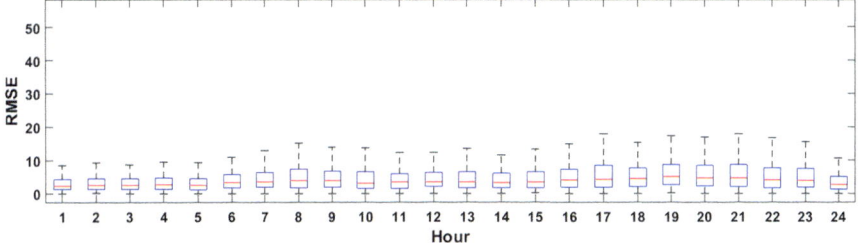

Fig. 8 RMSE of 24-h imputation for the entire year of 2016

considering the high variances of the dataset. However, for some time (rush hours) of the day, the imputation errors are a bit higher, due to the greater variability in the NO_2 concentrations of those hours.

7   Discussion and Conclusions

Various types of environmental data can be acquired in smart city IoT. The use of low-cost sensing systems allows the collection of data at a high spatio-temporal resolution. The data quality of the obtained data, however, needs further investigation. This helps to avoid dissemination of incorrect data, which may result in underestimation of health effects or unnecessary unrest.

After data pre-processing, spatio-temporal outlier detection can be used to detect anomalies with respect to the expected values in space and time. Within the set of detected outliers, we can distinguish between errors and events based on spatial and temporal autocorrelation patterns, as well as correlations with other dynamic processes in the environment. Validating the results remains a challenge to be considered in future work. Local information, for example using video cameras and machine learning algorithms to detect possible causes for outliers, may help to validate the results and make final decisions.

Missing data gaps, caused by sensor failures, maintenance, replacement, pre-processing or outlier detection, can be filled using missing data imputation methods. Missing data imputation is based on the temporal covariance structure within the data. In this book chapter, we presented how to impute IoT missing data with GP models. The GP models can capture non-linear dynamics and provide uncertainty-aware imputation results. For the future work, it is interesting and important to investigate how to adapt the GP models for non-stationary IoT data. Possible solutions are, for example, designing non-stationary mean and covariance functions, and applying a meta-process to select different models from a set of GPs (i.e., mixture of GPs).

In this chapter, the methods are applied to an IoT testbed for air quality monitoring in the city of Eindhoven, the Netherlands. The methods are more widely applicable to continuous environmental variables which show a strong spatial and/or temporal autocorrelation structure.

References

1. Ahlgren, B., Hidell, M., & Ngai, E. C. (2016). Internet of things for smart cities: Interoperability and open data. *IEEE Internet Computing, 20*(6), 52–56. https://doi.org/10.1109/MIC.2016.124.
2. AiREAS. (2016). AiREAS API v2 Documentation. Retrieved from https://data.aireas.com/docs/.
3. Basu, S., & Meckesheimer, M. (2007). Automatic outlier detection for time series: An application to sensor data. *Knowledge and Information Systems, 11*(2), 137–154.
4. Beelen, R., Raaschou-Nielsen, O., Stafoggia, M., Andersen, Z. J., Weinmayr, G., Hoffmann, B., et al. (2014). Effects of long-term exposure to air pollution on natural-cause mortality: An analysis of 22 European cohorts within the multicentre ESCAPE project. *The Lancet, 383*(9919), 785–795.
5. Bobbia, M., Misiti, M., Misiti, Y., Poggi, J.-M., & Portier, B. (2015). Spatial outlier detection in the PM10 monitoring network of Normandy (France). *Atmospheric Pollution Research, 6*(3), 476–483. https://doi.org/10.5094/apr.2015.053.
6. Borrego, C., Costa, A. M., Ginja, J., Amorim, M., Coutinho, M., Karatzas, K., et al. (2016). Assessment of air quality microsensors versus reference methods: The EuNetAir joint exercise. *Atmospheric Environment, 147,* 246–263. https://doi.org/10.1016/j.atmosenv.2016.09.050.
7. CEN/TC 287. (1998). ENV 12656:1998 Geographic Information-Data description-Quality. In: Technical Comittee 287 of the Comité Européen de Normalisation (CEN).
8. Chandola, V., Banerjee, A., & Kumar, V. (2009). Anomaly detection: A survey. *ACM computing surveys (CSUR), 41*(3), Article 15.
9. Close, J. P. (Ed.) (2016). *AiREAS: Sustainocracy for a healthy city. The invisible made visible phase 1*. Springer.
10. Dong, M., Iervolino, E., Santagata, F., Zhang, G., & Zhang, G. (2016). Silicon microfabrication based particulate matter sensor. *Sensors and Actuators a-Physical, 247,* 115–124. https://doi.org/10.1016/j.sna.2016.05.036.
11. Febrero, M., Galeano, P., & Gonzalez-Manteiga, W. (2007). A functional analysis of NOx levels: Location and scale estimation and outlier detection. *Computational Statistics, 22*(3), 411–427. https://doi.org/10.1007/s00180-007-0048-x.
12. Febrero, M., Galeano, P., & Gonzalez-Manteiga, W. (2008). Outlier detection in functional data by depth measures, with application to identify abnormal NOx levels. *Environmetrics, 19*(4), 331–345. https://doi.org/10.1002/env.878.
13. Fenger, J. (2009). Urban air pollution. In C. N. Hewitt & A. V. Jackson (Eds.), *Atmospheric science for environmental scientists* (pp. 243–267). Chichester: Wiley & Sons Ltd.
14. Fortuin, V., Baranchuk, D., Rätsch, G., & Mandt, S. (2020). *Gp-vae: Deep probabilistic time series imputation*. Paper presented at the International Conference on Artificial Intelligence and Statistics.
15. Guptill, S. C., & Morrison, J. L. (Eds.). (1995). *Elements of spatial data quality*. Oxford: Elsevier Science Ltd.
16. Harrell. (2018). *Function aregImpute; package Hmisc 4.1-1*. Vanderbilt University School of Medicine, Nashville, Tennessee.
17. Huang, Y., Organ, B., Zhou, J. L., Surawski, N. C., Hong, G., Chan, E. F. C., & Yam, Y. S. (2018). Remote sensing of on-road vehicle emissions: Mechanism, applications and a case study from Hong Kong. *Atmospheric Environment, 182,* 58–74. https://doi.org/10.1016/j.atmosenv.2018.03.035.
18. ISO/TC 211 Secretariat. (2013). Text for ISO/FDIS 19157 Geographic information-Data quality, as sent to the ISO Central Secretariat for issuing as FDIS. In *ISO/TC 211 Geographic information/Geomatics* (Vol. ISO 19157).
19. KNMI. (2016). Uurgegevens van het weer in Nederland-Download. Retrieved August 31, 2016, from https://projects.knmi.nl/klimatologie/uurgegevens/selectie.cgi.
20. Kainz, W. (1995). Logical consistency. In S. C. Guptill & J. L. Morrison (Eds.), *Elements of spatial data quality* (pp. 109–137). Oxford: Elsevier Science Ltd.

21. Kracht, O., Gerboles, M., & Reuter, H. I. (2014). First evaluation of a novel screening tool for outlier detection in large scale ambient air quality datasets. *International Journal of Environment and Pollution, 55*(1–4), 120–128. https://doi.org/10.1504/ijep.2014.065912.
22. Little, R. J. A., & Rubin, D. B. (2019). *Statistical analysis with missing data* (Vol. 793). John Wiley & Sons.
23. Liu, X., Xi, T., & Ngai, E. (2016, 19–21 Sept. 2016). *Data modelling with gaussian process in sensor networks for Urban environmental monitoring.* Paper presented at the 2016 IEEE 24th International Symposium on Modeling, Analysis and Simulation of Computer and Telecommunication Systems (MASCOTS).
24. Martínez Torres, J., Garcia Nieto, P. J., Alejano, L., & Reyes, A. N. (2011). Detection of outliers in gas emissions from urban areas using functional data analysis. *Journal of Hazardous Materials, 186*(1), 144–149. https://doi.org/10.1016/j.jhazmat.2010.10.091.
25. Nguyen, P. L., & Hoogerbrugge, R. (2014). Methods used to compensate for the effect of missing data in air quality measurements. National Institute for Public Health and the Environment (Bilthoven).
26. Ott, W. R. (1990). A physical explanation of the lognormality of pollutant concentrations. *Journal of the Air & Waste Management Association, 40*(10), 1378–1383. https://doi.org/10.1080/10473289.1990.10466789.
27. RIVM. (2016). Landelijk Meetnet Luchtkwaliteit: Gevalideerde data. Retrieved from https://www.lml.rivm.nl/gevalideerd/index.php.
28. RIVM. (2019). Air Quality Network. Retrieved from https://www.luchtmeetnet.nl/uitleg.
29. Rodrigues, F., Henrickson, K., & Pereira, F. C. (2018). Multi-output Gaussian processes for crowdsourced traffic data imputation. *IEEE Transactions on Intelligent Transportation Systems, 20*(2), 594–603.
30. Schafer, J. L., & Schenker, N. (2000). Inference with imputed conditional means. *Journal of the American Statistical Association, 95*(449), 144–154.
31. Sguera, C., Galeano, P., & Lillo, R. E. (2016). Functional outlier detection by a local depth with application to NO(x) levels. *Stochastic Environmental Research and Risk Assessment, 30*(4), 1115–1130. https://doi.org/10.1007/s00477-015-1096-3.
32. Shamsipour, M., Farzadfar, F., Gohari, K., Parsaeian, M., Amini, H., Rabiei, K., et al. (2014). A framework for exploration and cleaning of environmental data-tehran air quality data experience. *Archives of Iranian Medicine, 17*(12), 821–829. Retrieved from <Go to ISI>://WOS:000347754100006.
33. Snyder, E. G., Watkins, T. H., Solomon, P. A., Thoma, E. D., Williams, R. W., Hagler, G. S., et al. (2013). The changing paradigm of air pollution monitoring. *Environmental Science & Technology, 47*(20), 11369–11377. https://doi.org/10.1021/es4022602.
34. Spinelle, L., Gerboles, M., Villani, M. G., Aleixandre, M., & Bonavitacola, F. (2015). Field calibration of a cluster of low-cost available sensors for air quality monitoring. Part A: Ozone and nitrogen dioxide. *Sensors and Actuators B: Chemical, 215*, 249–257. https://doi.org/10.1016/j.snb.2015.03.031.
35. Van Oort, P. (2006). *Spatial data quality: from description to application.* (Ph.D thesis). Wageningen University, Wageningen, the Netherlands, Wageningen, the Netherlands. Retrieved from https://edepot.wur.nl/38987.
36. Williams, C. K. I., & Rasmussen, C. E. (2006). *Gaussian processes for machine learning* (Vols. 2–3). Cambridge, MA: MIT press.
37. Zhang, Y., Hamm, N. A. S., Meratnia, N., Stein, A., van de Voort, M., & Havinga, P. J. M. (2012). Statistics-based outlier detection for wireless sensor networks. *International Journal of Geographical Information Science, 26*(8), 1373–1392. https://doi.org/10.1080/13658816.2012.654493.
38. van Zoest, V. M., Osei, F. B., Hoek, G., & Stein, A. (2019). Spatio-temporal regression kriging for modelling urban NO_2 concentrations. *International Journal of Geographical Information Science.* https://doi.org/10.1080/13658816.2019.1667501.
39. van Zoest, V., Osei, F. B., Stein, A., & Hoek, G. (2019). Calibration of low-cost NO_2 sensors in an urban air quality network. *Atmospheric Environment, 210*, 66–75. https://doi.org/10.1016/j.atmosenv.2019.04.048.

40. van Zoest, V. M., Stein, A., & Hoek, G. (2018). Outlier detection in Urban air quality sensor networks. *Water, Air, & Soil Pollution, 229*(4), 111. https://doi.org/10.1007/s11270-018-3756-7.

41. de Wolff, T., Cuevas, A., & Tobar, F. (2020). *Gaussian process imputation of multiple financial series.* Paper presented at the ICASSP 2020–2020 IEEE International Conference on Acoustics, Speech and Signal Processing (ICASSP).

Comparison of the Bias and Weighting of Variables in Neural Networks (ANN) for the Selection of the Type of Housing in Spain and Mexico

Julio Arreola, Damián Gibaja, J. Agustín Franco, and Marcelo Sánchez-Oro

Abstract This chapter compares users' housing characteristics in Spain and Mexico through a multilayer neural network trained for selecting the right type of housing by new users. This research aims to analyze the biases and synaptic weights of the variables that are analyzed. Our results show that data's bias and variables' weighting do not influence the neural network's precision for housing classification. Thus, the housing classification is independent of the biases and captures the housing users' preferences in each country. The results' robustness is done by comparing different neural network feedback architectures to improve accuracy through different training.

Keywords Adequate Housing · Cultural Adaptation · Decision Making · ANN

1 Introduction

The Office of the High Commissioner for Human Rights [1] establishes that "adequate housing" must meet the criteria of (1) legal security of tenure irrespective of the type of tenure; (2) availability of services, (3) affordability; (4) habitability; (5) accessibility (including access to land); (6) location; and (7) cultural adequacy. Thus, as we can see, "adequate housing" is different from "affordable housing." However, the housing sector faces the challenge of generating adequate housing

J. Arreola (✉)
Universidad Autónoma de Ciudad Juárez, Ciudad Juárez, Chih. 32310, México
e-mail: julio.arreola@uacj.mx

D. Gibaja
Universidad Popular Autónoma del Estado Puebla, Puebla, Pue. 72410, México
e-mail: damianemilio.gibaja@upaep.mx

J. A. Franco · M. Sánchez-Oro
Universidad de Extremadura, Cáceres, Ext. 10003, España
e-mail: franco@unex.es

M. Sánchez-Oro
e-mail: msanoro@unex.es

© The Author(s), under exclusive license to Springer Nature Switzerland AG 2021   19
U. Ghosh et al. (eds.), *Machine Intelligence and Data Analytics for Sustainable Future Smart Cities*, Studies in Computational Intelligence 971,
https://doi.org/10.1007/978-3-030-72065-0_2

since public policies focus on developing affordable housing. Although it is clear that housing plays a crucial role in the sustainable development of urban areas, there is an unbalance between environmental, economic, and social factors [2]. Specifically, the previous problem arises, given that the real estate sector ignores individuals' preferences and does not cope with users' satisfaction [3]. Typical public policies on housing construction focus on guaranteeing real estate companies' profitability instead of incorporating social and cultural factors in housing development [2].

This research proposes a mechanism for prioritizing housing characteristics, based on artificial neural networks (ANN), which considers the cultural adequacy (in the terms established by OHCHR-UN-Habitat 2010). We use the types of housing identified in the records of the databases generated by the Continuous Household Survey of the National Institute of Statistics (INE) of Spain [4] and the National Household Survey of the National Institute of Geography, Statistics and Computing (INEGI) of Mexico [5]. We show that the proposed mechanism makes it possible to foresee the type of housing that will meet new purchasers' needs by identifying the attributes that characterize each territory.

The ANN allows us to weigh the relative importance of the dwelling type variables that the users' preferences and determine the last hidden layer's biases for Spain's output neurons and Mexico's networks. For this purpose, we consider a hyperbolic tangent function for hidden layers' training since it is a classification problem. At the same time, the activation function of the output layer is of the softmax type, which is used for multiclass logistic regression [6], or multinomial logistic regression, adequate to the seven criteria of "adequate housing" established by UN-Habitat. Comparing the ANNs with data from Mexico and Spain, predictions about housing suitable for new test records were obtained. As can be seen, the results show bias in the Spanish database, probably attributable to the fact that data collection is carried out by telephone. However, statistically, this bias does not represent an obstacle for the ANN to identify the type of housing suitable for new registrations.

Aside from public policies, most studies on "adequate housing" focus on housing affordability. In this regard, the literature searches for determining the factors that influence the creation of housing prices [7], leaving users' satisfaction in the background [8]. Along these lines, Rahadi et al. [9] use the ANOVA model and linear regression to identify the variables with the most significant impact on housing allocation, while Hassanudin [10] uses the Fuzzy Hierarchical Analytical Process (FAHP). Coloma et al. [11] design an ANN to determine the variables most influencing the final price of housing in medium-sized cities. Similarly, Zabada and Shahrour [12] compare the econometric approach to obtain the value of property began with hedonic modeling and ANNs based on a set of property attributes, internal or external, associated with dwelling from Seville. Choy et al. [13] analyze the influence of users' priorities on housing prices, employing a quantum regression. Concerning price prediction, Choy et al. [13] and Yao et al. [14] do so through neural networks; in both cases, they provide evidence that proximity to transport stations or tourist sites increases prices; at the same time, such a situation generates inequality among the population. Pino-Mejías et al. [15] use multiple linear regression (MLR) and artificial neural networks (ANN) to predict the probability that low-income households will fall into energy poverty when assigned to social housing. Similarly, Zabada and Shahrour [12] use an artificial neural network (ANN) to analyze heating consumption in sizeable social housing stock in the North of France. Although neural networks focus on house prices or energy consumption, as far as we know, we are the first to include cultural adequacy characteristics from affordability criteria.

In general, UN-Habitat [8] points out that both researchers and housing policy-makers have left aside cultural adequacy and have been concerned with analyzing this market, mainly as an economic growth engine and job creation, avoiding other implications that directly affect urban sustainability. Consequently, the omission of the user's characteristics[1] is evident in developing housing promotion plans [8]. Even more, a city with little habitability implies urban degradation and spatial segregation. So, a dominant feature of current cities' development is the dispersion of communities and the underutilization of common areas [16].

2   Methodology

2.1   Artificial Neural Networks

ANNs are a branch of artificial intelligence that allows the study of multifactorial problems by executing functions that analyze a fixed number of inputs and produce a fixed number of outputs [17]. In other words, they are programmable mathematical models whose computational integration is done in parallel. They are made up of layers with multiple connections, emulating neurons' link in the human brain [18]. Most types of ANN architecture include a layer of hidden neurons, which are the bridge between input and output. The connections between neurons have an associated weight value (weight), in addition to bias (threshold). The latter is an additional parameter that helps to fit the model to the given data by controlling the activation function [19]. The weights and thresholds then set the output values of the input value set.

The neural network follows a training process to determine the best adjustment weights and thresholds, see Fig. 1. The bias neuron is added to the start/end of the input and each hidden layer. The values of the previous layer do not influence it, so these neurons have no incoming connections. Mathematically, the training process is expressed as follows:

$$Output = \Sigma\,(inputs * weights) + bias \tag{1}$$

Depending on the type of training [20], networks can be classified into two groups:

---

[1]Cultural Adequacy, in this case, implies that the design characteristics allow the expression of cultural identity to be considered and respected, depending on its ethnic, regional, or urban dimension [1].

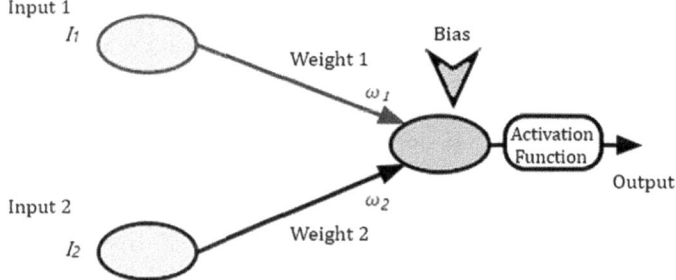

Fig. 1 ANN training process. (*Source* own elaboration with information from Du and Swamy [19], Neural networks and statistical learning, p. 5.)

- ANN with supervised training. They are trained by presenting the outputs they are expected to produce for each set of inputs. The training algorithms then calculate new weights and biases to minimize the error between the desired and the resulting output.
- ANN without supervised training. The training algorithms calculate new weights randomly. These networks are used as classifiers since they are characterized by associating a combination of specific inputs with a single output; in other words, it identifies similarities in the data and reacts based on the presence or absence of each data's common elements record.

It is essential to mention that the selection of the activation function depends on the objective pursued. For example, classification models commonly use the sigmoid activation function, while predictive models assume a linear activation function [21]. The most common activation functions in ANNs are as follows:

- Heaviside function, also known as threshold function or unit step function, is a discontinuous function whose value is positive and zero for negative values.

$$\gamma_{(c)} = \begin{cases} 1 \; if \; c \geq 0 \\ 0 \; if \; c < 0 \end{cases} \qquad (2)$$

- The logistic sigmoid function is an S-shaped curve that is based on an information gain matrix. This matrix suggests a transition from low values to high values through an inflection point.

$$\gamma_{(c)} = \frac{1}{1 + e^{-c}} \qquad (3)$$

- The softmax function (normalized exponential) is generalized by the Bernoulli distribution for multiclass problems. It has a sigmoid type and stops the multiclass classification because it scales the previous inputs in a range between 0 and 1 and normalizes the output layer; that is to say, the sum of all weights is one.

$$\gamma_{(c)_j} = \frac{e^{c_j}}{\sum_{k=1}^{K} e^{c_k}} \qquad (4)$$

- The hyperbolic tangent function is a curve with properties like sigmoid activation, so it is considered an alternative for such activation functions. The outputs are bounded between -1 and 1.

$$\gamma_{(c)} = tanh(c) \frac{e^c - e^{-c}}{e^c + e^{-c}} \qquad (5)$$

2.2 Multilayer Perceptron (MLP)

The multilayer power perceptron (MLP) model is an ANN that consists of a finite number of successive layers, each of which has a finite number of neurons, which connect in the next layer as if they were synapses. In this way, information flows in one direction, from one layer to the next (feedforward). The architecture of an MLP network (Fig. 2) comprises the first layer or input layer. Later there are the intermediate layers or hidden layers, and finally, the output layer is located, whose neurons are also called response nodes [21].

As indicated by expression (1), the neurons in the first hidden layer function as the neurons' weighted sum in the input layer. This function is called the activation function, and the values of the weights are determined by the estimation algorithm, which seeks to minimize errors. If the network contains a second hidden layer, its neurons transform the neurons' weighted sum in the first hidden layer. The previous process is repeated iteratively for each hidden layer until reaching the output layer. MLP models are classified according to the number of hidden layers they have and not by their total number of layers; that is to say, they are classified by the number of

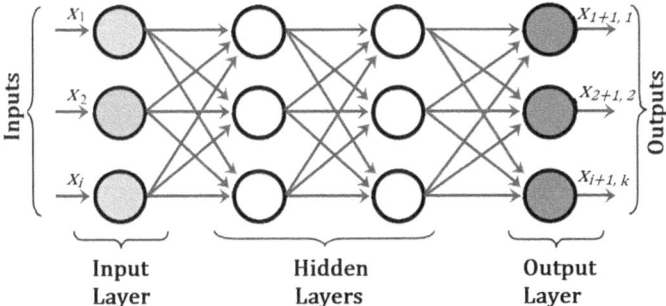

Fig. 2 ANN MLP Architecture. (*Source* own elaboration from ElKessab et al. [22], International Journal of Computer Applications, 107 (21), 25–30.)

layers excluding the input and output layers. The term MLP is applied generically to all models with at least one hidden layer. The rules and regulations that govern the MLP model [21] are the following:

- The input layer has the value x_{0j} as its output from its neuron j.
- Neuron k in layer i receives the output x_{ij} from each neuron j in layer $(i - 1)$. The x_{ij} values are multiplied by the synaptic weights ω_{ijk}, and the products are added together to obtain the next output value. Mathematically, for each neuron k, layer $i + 1$ is constructed as follows:

$$x_{i+1,k} = \gamma \left(\sum_j \omega_{ijk} x_{ij} - \theta_{ik} \right) \qquad (6)$$

where γ is the activation function, θik are the thresholds or biases, and ωijk are the synaptic weights.

Weights and thresholds are determined during the learning or training process; the backpropagation algorithm is the most widely used [21].

Backpropagation refers to how information is disseminated backward, i.e., we start from the output layer and finish in the source layer. The previous procedure allows the neurons' synaptic weights located in the hidden layers to change during training. The variation in synaptic weights is due to the activation function's sensitivity, used in each of the hidden layers. However, in the output layer, the activation function's choice depends on the task's type [23]. The input vectors' results are used for training; their validation relies on the algorithm's multiple iterations.

The model identifies the weights and biases that best fit the network to the behavior of the data. It is important to note that machine learning results from how the algorithm makes intelligent decisions through pattern recognition, learning based on sample data. In other words, a pattern is established between the problem and the answer [24].

3   Model

The construction of the classification and prediction model of the type of dwelling for Mexico and Spain follows these steps:

- Analysis of the information and compilation of the databases.
- Preprocessing of databases and selection of variables to compare.
- Design of the classification model (by using IBM SPSS Modeler 18).
- Evaluation of the results obtained.

3.1 Analysis of the Information and Compilation of the Databases

In this research, the data from the Continuous Household Survey of the National Statistics Institute [4] of Spain and the National Household Survey of the National Institute of Statistical Geography and Informatics of Mexico [14] are used. Both surveys are carried out to know the main characteristics of the population, households, and dwellings. Data from 2017 were selected since they are the most recent records available for Mexico.

3.2 Database Set Up and Selection of Variables to Compare

The records have been purified to homogenize the selected databases. Later, 100,542 records were obtained from Spain and 31,698 records from Mexico. Finally, the set of variables to be tested has been determined, using the following selection criteria:

- Only the variables related to housing present in both databases in Spain and Mexico are considered.
- Considering the above variables, in the resulting unified database, only those related to the 7 UN-Habitat criteria for adequate housing are included (OHCHR-UN-Habitat, 2010).
- Variables with identical data in all records are discarded, and those with capture error or lack of information.

The description of the variables studied, their type, registry values, and their selection justification are presented in Table 1.

3.3 Artificial Neural Network Design

The ANN MLP architecture was designed in the IBM SPSS Modeler software tool that works through the Watson Studio virtual interface. Figure 3 shows the flow of data processing for model construction. For Spain and Mexico, each neural network's modeling uses different tools to establish the ANN architectures. Each device establishes a particular activity for the neural network and is represented by icons in the software.

The architecture flow for each of the networks consists of the following steps:

- **Step 1.** Input records that make up the databases are imported through the Data Allocation icon, allowing working with Excel.CSV files.
- **Step 2.** The partition of the databases' records is determined considering three sets: one for training, another for testing, and a third for validation; The latter is

Table 1 Variable description. *Source* own elaboration from Continuous Household Survey [4] and the National Household Survey [5]

Variable descriptions			
Variable	Type	Value	Substantiation
Housing type (HT)	Nominal	1,2,3,4,5,6 Spain 1,2,3,4,5 Mexico	Dependent variable
Kitchen (K)	Flag	1,2	Fundamental for feeding home users
Number of bedroom rooms (NB)	Continuous	1,2,3,4,…,n	It allows for reducing overcrowding
Number of rooms in the house (N.R.)	Continuous	1,2,3,4,…,n	Consider cultural adequacy by personalizing the home
Number of full baths (FB)	Continuous	1,2,3,4,…,n	Fundamental for the hygiene of home users
Total number of residents (TR)	Continuous	1,2,3,4,…,n	Facilitates determining the type of housing
Type of holding payment (HP)	Nominal	1,2,3,4	Provides legal security
The geographical location (G.L.)	Nominal	Id area	Connectivity and proximity to work centers
Town size (T.S.)	Nominal	1–11 Spain 1,2,3,4 Mexico	It allows us to identify the coverage of services

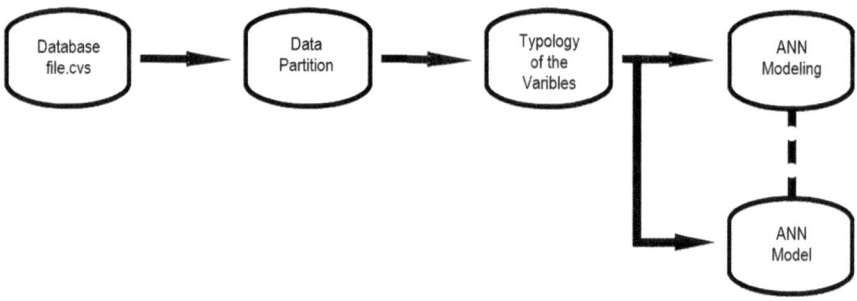

Fig. 3 The data processing flow for model development. (*Source* own elaboration with information from IBM SPSS Modeler [25])

used to determine the end of network training to avoid overtraining. Partition is performed using the Partition icon, with a ratio of 65% training, 25% testing, and 10% validation.

- **Step 3.** The Type icon is used to assign the variables' typology based on their internal components (continuous, nominal, categorical), their role in the hypothesis system (dependent, independent, record identification). The measurement value of each of the variables of interest. The algorithm normalizes the input variables' registers to values between zero and one to minimize the effect of values' size while increasing the learning effectiveness. The software allows adjusting the type of normalization to avoid conflicts caused by various magnitudes' joint treatment.

- **Step 4.** We integrate the ANN modeling icon, which allows us to select the algorithm's features (choice of input variables, the confidence level of the predictions, or the number of cycles to be executed). In this case, we selected 1000 cycles, predetermined by the software. We established a 10% overfit prevention set to track errors during training to prevent the method from modeling the data's probability variance. Finally, the number of hidden layers and the number of neurons in each hidden layer are specified.

3.4   Features ANNs

In this research, MLP RNAs with multiclass input and output neurons are designed; 1 and 2 hidden layers are also used for Spain and Mexico. A hidden layer is recommended for problems seeking low error and two hidden layers for classification problems [11]. However, it has been shown that a single hidden layer is sufficient for most situations (Funahashi [26] cited by Picón [27]).

The input layer of Spain and Mexico's RNAs has eight neurons (one for each variable). In contrast, the algorithm automatically calculates the number of neurons for a hidden layer, so there are 18 neurons for Spain and 16 neurons for Mexico.

Later, we set the number of neurons in two or more hidden layers; in this case, the literature basis ANN construction concerning the number of input and output neurons. Since this research considers a multiclass input neuron, we use the same number of neurons that determine the first hidden layer's algorithm in the second hidden layer. The RNA architecture was replicated, reducing by one unit the number of neurons in the second hidden layer until the best precision was found, thus obtaining the best accuracy with 15 neurons in Spain and Mexico's networks. For the output layer of networks with one and two hidden layers, there are six neurons for Spain and 4 for Mexico, which correspond to the number of housing types in each country. In the case of Mexico, there are five types of housing. However, housing 4 (housing in a rooftop room) is rare; thus, the algorithm omits predictions of this type of housing. The architecture information for each network is described in Table 2. The model architecture information for ANNs is described in Table 2.

In the hidden layers, we a hyperbolic tangent activation function. Concerning the output layer, where the classification type neurons are located, the softmax activation function, since the dependent variable is multiclass, with six types of housing for Spain (Fig. 4) and five types of accommodation for Mexico (Fig. 5) according to the databases of each country. Figures 4 and 5 show, in a dark color, the bias of the records regarding the type of dwelling. The result is expected since both represent the most used homes in each country.

Table 2 Model architecture information. (*Source* own elaboration)

Architecture model information	Hidden layer	
	1	2
Dependent variable	tipo_viv	tipo_viv
Model type	Classification	Classification
Independent variables	8	8
Neurons hidden layer 1	18 Spain, 16 Mexico	15 Spain, 15 Mexico
The hidden layer activation function	Hyperbolic tangent	Hyperbolic tangent
Neurons hidden layer 2		11 Spain, 11 Mexico
Output layer activation function	Softmax	Softmax
Neurons output layer	6 Spain, 4 Mexico	6 Spain, 4 Mexico

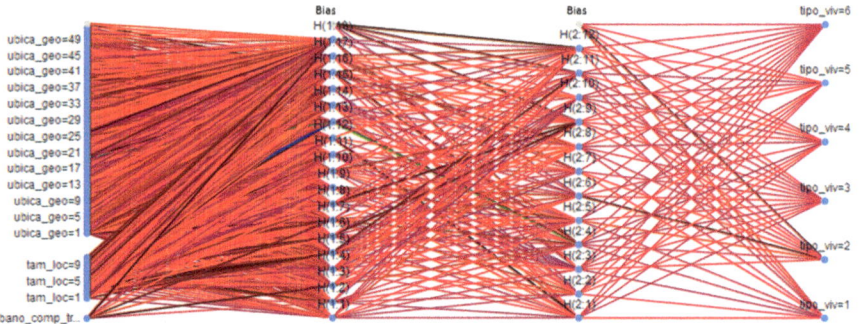

Fig. 4 ANN diagram with four layers (two hidden) from the Spain database. (*Source* own elaboration in IBM SPSS Modeler [25]

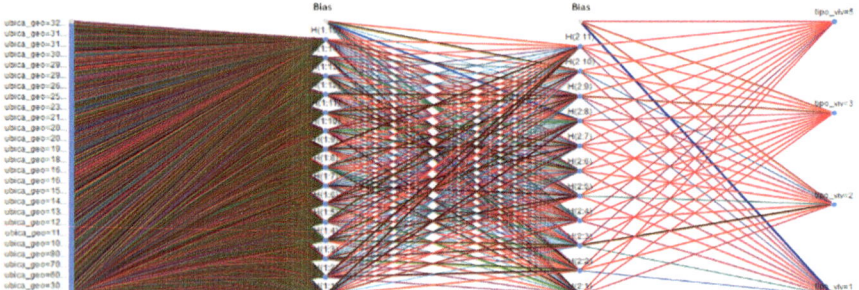

Fig. 5 ANN diagram with four layers (two hidden) from the Mexico database. (*Source* own elaboration in IBM SPSS Modeler [25]

3.5   Evaluation of the Results Obtained

After running the ANNs algorithm, we evaluate training, test, and validation efficiency and precision. In addition to the results, the software provides different evaluation indicators, which are shown in Table 3; This analysis considers three variants of the neural network:

- Actual values of the databases and a single hidden layer.
- Actual values of nominal variables and normalized values of continuous variables from databases with a hidden layer.
- Actual values of databases with two hidden layers.

The indices that we consider for the evaluation of the results are the following:

- **Precision**. Positive predictive value representing the proportion of correct predictions contained in the model. The precision calculation is given by:

$$Precision = \frac{true\ positives + true\ negatives}{(true\ positives + true\ negatives + false\ positives + false\ negatives)} \quad (7)$$

- **True positives rate**. Presents the proportion of correct predictions in positive class predictions. The calculation of the true positives index is given by:

$$True\ positives\ rate = \frac{true\ positives}{(true\ positives + false\ negatives)} \quad (8)$$

- **False-positive rate**. Represents the proportion of incorrect predictions in the positive class. The calculation of the false positive rate is calculated as follows:

$$False\ positive\ rate = \frac{false\ positives}{(true\ negatives + false\ positives)} \quad (9)$$

- **Weighted recall**. It is a sensitivity measure that determines the proportion of correct predictions in a positive class. The exhaustiveness is given by:

$$Weighted\ recall = \frac{true\ positives}{(true\ positives + false\ negatives)} \quad (10)$$

- **Weighted Precision**. Provides weighted mean with weights equal to the class probability. The weighted precision is given by:

$$Weighted\ precision = \frac{true\ positives}{(true\ positives + false\ negatives)} \quad (11)$$

- **F1 weighted measure**. It is a measure of accuracy that considers both accuracy and completeness in calculating the score. The weighted F1 measure is given by:

Table 3 Model evaluation measures. (*Source* own elaboration)

Evaluation measures	Results					
	Spain			Mexico		
Training	Real values 1 hidden layer	Normalized continuous values 1 hidden layer	Real values 2 hidden layers	Real values 1 hidden layer	Normalized continuous values 1 Hidden layer	Real values 2 hidden layers
Precision	0.605	0.599	0.585	0.970	0.970	0.970
True positives rate	0.605	0.599	0.585	0.970	0.970	0.970
False positive rate	0.111	0.112	0.092	0.000	0.000	0.000
Weighted recall	0.766	0.764	0.797	1.000	1.000	1.000
Weighted precision	0.605	0.599	0.585	0.970	0.970	0.970
F1 weighted measure	0.667	0.663	0.664	0.985	0.985	0.985
Test						
Precision	0.598	0.591	0.590	0.970	0.970	0.970
True positives rate	0.598	0.591	0.590	0.970	0.970	0.970
False positive rate	0.120	0.120	0.093	0.000	0.000	0.000
Weighted recall	0.756	0.754	0.798	1.000	1.000	1.000
Weighted precision	0.598	0.591	0.590	0.970	0.970	0.970
F1 weighted measure	0.660	0.655	0.667	1.000	1.000	1.000
Validation						
Precision	0.610	0.599	0.576	0.973	0.973	0.973
True positives rate	0.610	0.599	0.576	0.973	0.973	0.973
False-positive rate	0.111	0.114	0.091	0.000	0.000	0.000
Weighted recall	0.766	0.762	0.790	1.000	1.000	1.000
Weighted Precision	0.610	0.599	0.576	0.973	0.973	0.973
F1 weighted measure	0.671	0.663	0.656	1.000	1.000	1.000

$$F1\ weighted\ measure = 2 * \frac{(precision * weighted\ recall)}{(precisión + weighted\ recall)} \qquad (12)$$

Concerning the results of the output layer's biases, the values presented in Table 4 are obtained for each of the dependent variable classes. Both databases have similar biases since a type of dwelling predominates for each country. Type two predominates in Spain, and type one predominates in Mexico, where there are no registers concerning type four.

Regarding the weights of the independent variables (Table 5), the results of the neural network in Spain show less variation in the type of housing throughout the country; one explanation may be that in urban areas of Spain, housing predominates

Table 4 Output layer bias. (*Source* own elaboration)

Housing type bias					
Spain			Mexico		
Real values 1 hidden layer	Normalized continuous values 1 hidden layer	Real values 2 hidden layers	Real values 1 hidden layer	Normalized continuous values 1 hidden layer	Real values 2 hidden layers
HT2 0.7690	HT1 0.7300	HT1 0.4761	HT1 1.2093	HT1 1.2093	HT1 0.6727
HT1 0.6734	HT2 0.3153	HT2 0.4749	HT2 0.2218	HT2 0.2218	HT2 0.1618
HT4 0.0433	HT4-0.0516	HT5 0.1936	HT3 0.0026	HT3 0.0026	HT3 0.1605
HT5 0.0277	HT5-0.0566	HT6 0.0134	HT5 -0.6392	HT5 -0.6392	HT5 -0.0326
HT3-0.3375	HT3-0.5809	HT3 0.0075			
HT6-0.7563	HT6-1.1073	HT4-0.1650			

Table 5 The weighting of the importance of the variables. (*Source* own elaboration)

Variable weighting					
Spain			Mexico		
Real values 1 hidden layer	Normalized continuous values 1 hidden layer	Real values 2 hidden layers	Real values 1 hidden layer	Normalized continuous values 1 hidden layer	Real values 2 hidden layers
NR 21.51	NR 21.72	NR 23.37	GL 26.42	GL 26.42	GL 25.89
TL 18.16	TL 20.11	TL 20.73	FB 19.29	FB 19.29	NB 19.64
FB 15.99	FB 17.37	FB 16.50	TL 15.41	TL 15.41	TR 14.85
GL 14.03	GL 16.31	GL 13.76	NB 11.13	NB 11.13	FB 12.41
NB 12.52	NB 10.16	NB 12.11	NR 10.52	NR 10.52	NR 10.22
TR 9.17	TR 5.21	K 5.48	TR 9.51	TR 9.51	TL 6.84
K 4.51	K 4.57	TR 4.07	HP 5.97	HP 5.97	K 5.72
HP 4.11	HP 4.55	HP 3.97	K 1.75	K 1.75	HP 4.44

vertical, even in cities with less than 100,000 inhabitants. For this reason, it is considered that Geographical Location (G.L.) is of minor importance when comparing the state of the home in Mexico. So, it is necessary to regionalize the data sample so that G.L. allows the algorithm to identify a weight of more variables consistent with users' preferences.

In the case of Spain, we find data inconsistencies for the variable Number of Rooms (N.R.) that measures the total number of rooms in the dwelling, including Kitchen (K) and Bedroom Rooms (N.B.). For example, in some registers, the number of Bedroom Rooms is greater than the total number of rooms in the dwelling represented by the N.R. variable. However, these records were refined, and the behavior of the models' evaluation measures presented irrelevant changes.

3.6 Discussion

The presented model allows establishing the decision process for selecting "adequate housing" by identifying each user's needs; that is to say, if a new record is included with the characteristics of the dwelling that the user requires, the model determines the type of adequate housing for it. The Mexican database's ANN model was more efficient than the Spanish database's neural network since inconsistency was identified in some records. For example, the number of bedroom rooms exceeded the number of rooms in the entire house. However, when testing with new registers, the network's response assigns a consistent type of housing.

Using analysis tools, we highlight a learning bias from deep networks towards functions such as the softmax activation function that allows an accurate classification of housing types to be made even though the data is not consistent, as in the Spanish database. Intuitively, this property is in line with observing the test and validation partitions that exhibit the same behavior as the training data partition. Furthermore, in the literature, it is recommended to use the softmax activation function with caution since it predisposes that each record is a single class member. However, in this case, its efficiency was compared when comparing this activation function recommended for the layer of output as the identity and hyperbolic tangent functions.

During the ANN model application, 10% of the data was reserved as a prevention setting to overcome the overfitting problem. The best network architecture for Spain's data is obtained using a hidden layer and the variables' real values. The software performs the normalization of each type of variable. For the data from Mexico, the three network architectures used present a good efficiency of the model's precision. In the 3 cases, there is an increase in the evaluation measures for the validation results.

Although ANN's results cannot be interpreted in the same way as in statistical methods, it is possible to affirm that the biases and the layer output's synaptic weights have similar behavior. For similar applications, including this investigation, it is recommended to inquire about the number of dependent variable classes. The Spanish database's inconsistency is not considered the cause of the model's low precision since results showed irrelevant changes when we omit inconsistent records.

4   Conclusions

Neural networks are known to encompass a class of functions capable of adapting even to random input–output assignments with greater precision than other methods. In this work, we present neural network properties that complement this aspect of expressiveness. The perception collected in the databases in the input data has a similar behavior in the output data.

About the results obtained, this research represents the first stage for the design of a housing selection model that considers a more significant number of variables from the perspective of housing selection instead of the housing allocation approach that is the most widely used. However, the model showing in this chapter allows for selecting suitable housing for new users in Spain and Mexico if we considered the same variables. On the other hand, future research is proposed to test the model to determine its efficiency and refine variables' selection.

References

1. ACNUDH-ONU Habitat (2010) El derecho a una vivienda adecuada. Folleto informativo n°21. *Revista de Antropología Social*, *19*, 103–129.
2. Samad, D., Zainon, N., Rahim, F. A. M., & Lou, E. (2017). Malaysian affordability housing policies revisited Open House Interantional. https://doi.org/10.1051/matecconf/20166600010.
3. CONEVAL. (2018). Estudio Diagnóstico del Derecho a la Vivienda Digna y Decorosa 2018. , Ciudad de México.
4. INE. (2017). Encuesta Continua de Hogares. https://www.ine.es/CDINEbase/consultar.do?mes=&operacion=Encuesta+continua+de+hogares&id_oper=Ir.
5. INEGI. (2017). Encuesta Nacional de los Hogares. https://www.inegi.org.mx/programas/enh/2017/.
6. Llinás Solano, H., Arteta Charris, M., & Tilano Hernández, J.: El modelo de regresión logística para el caso en que la variable de respuesta puede asumir uno de tres niveles: Estimaciones, pruebas de hipótesis y selección de modelos. Rev. Matemática Teoría y Apl. https://doi.org/10.15517/rmta.v23i1.22442.
7. Gaspareniene, L., Venclauskiene, D., & Remeikiene, R. (2014). Critical review of selected housing market models concerning the factors that make influence on housing price level formation in the countries with transition economy. Procedia Social and Behavioral Sciences. https://doi.org/10.1016/j.sbspro.2013.12.886.
8. ONU-Habitat: Elementos de una Vivienda Adecuada. https://onuhabitat.org.mx/index.php/elementos-de-una-vivienda-adecuada.
9. Rahadi, R. A., Wiryono, S. K., Koesrindartoto, D. P., & Syawmil, I. B. (2018). External factors influencing housing product price in Jakarta Metropolitan Region. fatores que Influ. o preço. 7, 179–192.
10. Hassanudin, M. T. T., Sakaran K. C. (2016). Prioritisation of key attributes influencing the decision to purchase a residential property in Malaysia: An analytic hierarchy process (AHP) approach. *International Journal of Housing Markets and Analysis*, *9*, 446–467. https://doi.org/10.1108/IJHMA-09-2015-0052.
11. Coloma, J. F., Valverde, L. R., & García, M. (2019) Estimation of rustic housing construction costs through Artificial Neural Networks|Estimación de los costes de construcción de viviendas rústicas mediante Redes Neuronales Artificiales Inf. La Constr. https://doi.org/10.3989/ic.62206.

12. Zabada, S., & Shahrour, I. (2017) Analysis of heating expenses in a large social housing stock using artificial neural networks. *Energies, 10.* https://doi.org/10.3390/en10122086.
13. Choy, L. H. T., Ho, W. K. O., & Mak, S. W. K. (2012). Housing attributes and Hong Kong real estate prices: A quantile regression analysis. *Construction Management and Economics, 30,* 359–366. https://doi.org/10.1080/01446193.2012.677542.
14. Yao, Y., Zhang, J., Hong, Y., Liang, H., & He, J. (2018). Mapping fine-scale urban housing prices by fusing remotely sensed imagery and social media data. *Trans. GIS., 22,* 561–581. https://doi.org/10.1111/tgis.12330.
15. Pino-Mejías, R., Pérez-Fargallo, A., Rubio-Bellido, C., & Pulido-Arcas, J. A. (2018). Artificial neural networks and linear regression prediction models for social housing allocation: Fuel Poverty Potential Risk Index. *Energy, 164.* https://doi.org/10.1016/j.energy.2018.09.056.
16. Ezquiaga Domínguez, J. M., Salat, S., Tojo, J. F., Naredo, J. M., Ainz Ibarrondo, M. J., Bilbao Uribarri, A., et al. (2010). Libro blanco de la sostenibilidad en el planeamiento urbanístico Español. 2nd Natl. Congr. Energy Sp. https://doi.org/10.3989/estgeogr.201126.
17. Eichie, J. O., Oyedum, O. D., Ajewole, M. O., & Aibinu, A. M. (2017). Comparative analysis of basic models and artificial neural network based model for path loss prediction. *Progress In Electromagnetics Research, 61,* 133–146. https://doi.org/10.2528/PIERM17060601.
18. Anand, P.: Bias in machine learning. https://iq.opengenus.org/bias-machine-learning/.
19. Du, K. L., & Swamy, M. N. S. (2014). *Neural networks and statistical learning.* London Ltd: Springer.
20. Ferreira, R. P., Martiniano, A., Napolitano, D., Romero, M., De Oliveira Gatto, D. D., Farias, E. B. P., et al. (2018). Artificial neural network for websites classification with phishing characteristics. *Social Networks, 07,* 97–109. https://doi.org/10.4236/sn.2018.72008.
21. Vanus, J., Fiedorova, K., Kubicek, J., Gorjani, O. M., & Augustynek, M. (2020). Wavelet-based filtration procedure for denoising the predicted CO_2 waveforms in smart home within the internet of things. Sensors (Switzerland). https://doi.org/10.3390/s20030620.
22. ElKessab, B., Daoui, C., Boukhalene, B., Salouan, R. (2014). A comparative study between the K-Nearest neighbors and the multi-layer perceptron for cursive handwritten arabic numerals recognition. International Journal of Computer Applications. 107, 25–30. https://doi.org/10.5120/19140-0117.
23. Marín Diazaraque, J. M. (2007). *Introducción a las redes neuronales aplicadas.* Data Min: Man.
24. Akiyama, R., Kamiyama, K., Kojima, M., Horade, M., Mae, Y., & Arai, T. (2015). Development of multifunctional robot hand for visual inspection equipped with 3D force sensing.
25. IBM Corp. (2020). IBM SPSS Modeler, Version 18.0. Armonk, NY: IBM Corp.
26. Funahashi, K. (1989). On the approximate realization of continuous mappings by neural networks. *Neural Networks, 2,* 183–192.
27. Picón, C. (2011). ¿Son más corruptos los países menos abiertos a los mercados internacionales? Aplicación de un modelo predictivo de clasificación basado en Redes Neuronales. Econ. del Caribe. https://doi.org/10.14482/rec.v0i8.3262.

Artificial Olfaction for Detection and Classification of Gases Using e-Nose and Machine Learning for Industrial Application

R. Manjula, B. Narasamma, G. Shruthi, K. Nagarathna, and Girish Kumar

Abstract An artificial electronic nose (e-nose) is developed, that mimics the human olfactory system, as an alternative to the human nose. In this work, we aim to develop a mini prototype of e-nose and use it for the detection of various types of gases present in the atmosphere. We then use existing machine learning models to carry out the classification task. Our study shows that the proposed e-nose system can find its potential application in various fields such as medical health care to detect chemical gas leakage, industries to detect hazardous gases, a substitute to the human nose when people are suffering from anosmia disorder, etc. We use k-Nearest Neighbours (kNN), Support Vector Classifier (SVC), Linear Regression (LR), Decision Tree (DT) and Random Forest (RF) algorithms to test the classification accuracy. Through the experimentation results we found that random forest model performs better with 97.77% classification accuracy compared to other models such as kNN, SVC, logistic regression and decision tree, whose classification accuracy are 93.33%, 62.22%, 71.11%, and 91.11% respectively. In future, we intend to extend this pilot work to automate the entire task where detected gaseous information by the e-nose is sent directly to the user to its mobile phone via Internet, instantly in real time fashion. We also aim to study using Deep Learning Techniques.

1 Introduction

Machine olfaction begins when human sensing perception ends (or is weak). In this work we attempt to see the feasibility of using machine learning technology [1] and the electronic nose (e-nose)—the smart sniffing technology [2], in detecting and then classify the detected gas odours into categories. Finally, suggest their usage as a potential application in Intelligent Gas-Leakage Detection System (IGLDS) for

R. Manjula (✉)
Department of CSE, SRM University, Guntur, AP, India

B. Narasamma · G. Shruthi · K. Nagarathna · G. Kumar
Ballari Institute of Technology and Management, Ballari, India

© The Author(s), under exclusive license to Springer Nature Switzerland AG 2021
U. Ghosh et al. (eds.), *Machine Intelligence and Data Analytics for Sustainable Future Smart Cities*, Studies in Computational Intelligence 971,
https://doi.org/10.1007/978-3-030-72065-0_3

industrial applications and home applications. For instance, in home application it can be used to detect and raise an alarm to the person if there is smoke due to fire accident, specifically from a burning paper but do nothing if the smoke is due to burning of incense sticks. Similarly, if there is spillage of acetone, may be a kid is playing with it, then there must be system that could intimate this information to the parent. Ignoring or late awareness on the spillage of the acetone might be highly dangerous (flammable chemical if exposed to fire spark).

Electronic noses are developed to mimic the human olfactory system that can detect and characterize odours more effectively than the human sense of smell [3]. It consists of an array of gas sensors of different types that are sensitive to specific types of molecules. These sensors help identify and distinguish between different types of odours. Practically, it is not possible to replace a single human nose with a single gas sensor. To be able to work as well as a human nose or better than that, an electronic nose needs multiple gas sensors.

Gases (both friendly and harmful gases) are present around us and used (friendly gases) in every walk of life, including the air we breathe. We are safe as long as the gas concentration is within the prescribed threshold levels [4]. However, when the concentration level increases beyond these threshold levels, they can be extremely hazardous, with the potential to cause severe harm on the ecosystem or even loss of life. The sources of gases includes nature, industries, drug manufacturers, residential users etc. It is reported that an estimate of about 50% of all pollution is due to and manufacturing and industrial activities [5]. Traditionally, gas detection techniques and technologies have been used by industries to detect the presence of gas leakage [6]. These detectors detect the presence of gases and raise an alarm to the users indicating the sign of danger that will pop up. Sometimes it may just not be enough to identify a particular gas leakage but it becomes more crucial to have additional information such as number of gas types, what is the concentration level of each of these gases, automatic classification and warning etc. In such cases, the traditional gas detection methods such as gas chromatography alone are not sufficient.

We found an alternative method to cater to this requirement—an electronic nose. Electronic noses have been around quite sometime and are still in the research and development phase. Currently, the e-nose has been used in several areas such as agriculture, food processing industries, aroma industries, health sectors etc. [7].

In our work, we attempt to study the potential applicability of e-noses in domain of industrial applications and home applications—the area that are yet to be explored. In particular, we develop a prototype of an automatic gas (or gases) leakage detection system. We use commercially available gas sensors, MQ–series (MQ2, MQ5 and MQ135), Arduino Uno, and other supporting hardware, in the design of e-nose system. Using this model we aim to collect data by exposing the sensors to gases such as acetone, smoke from paper ad smoke from incense sticks. We then perform data processing on the collected data and apply existing supervised machine learning algorithms (such as k-Neartest Neighbours (kNN), Support Vector Classifier (SVC), Linear Regression (LR), Decision Tree (DT) and Random Forest (RF)) [24, 25] to classify the type of gases present in the collected data. We then study the performance aspect of the chosen machine learning models and finally suggest the best model to such applications.

All the existing works have experimented on potential usage of e-nose for various odours/gases/aroma using different varieties of sensors that are mostly costly. Also, we observed that very little work is done on in-door air quality monitoring and hazardous gas vapor detection, using e-nose and machine learning models. In this work we aim to develop *low-cost* e-nose module using commercially available gas sensors to detect and predict the gas odour (both friendly and harmful gases/vapors) for *in-door applications*. The important Features of the proposed work are: (i) The proposed e-nose module is cheaper [26] than the other versions [27], and (ii) Most of the existing work on application of e-nose work on large data-set and suggest the best model for a particular application. In this work, we aim to explore the performance of machine learning models using very limited data. To the best of our knowledge there is no work in the literature on the types of gases we are experimenting with, using the MQ series gas sensors.

Rest of the chapter is organized as follows: In Sect. 2 we presents the existing work on applications of e-nose and in Sect. 3 we give the description of the developed model and the pattern recognition models used in our work. Section 5 presents rigorous analysis of the results obtained through experimentation. Finally, we conclude the work in Sect. 6.

2 Literature Survey

In this section, we present the ongoing work on the applications of e-nose in various fields.

The authors in [8] have designed an e-nose system to detect the quality of wheat. The system detects gas vapors that are released by the wheat while it is in storage and then determines whether the wheat quality has changed by measuring the changes in the gas concentration. Principal Component Analysis (PCA) and improved Back Propagation (BP) algorithms are used on the data set to extract valuable information. The experimental results show that the system can rapidly and accurately detect wheat quality.

E-nose based fruit maturity levels are studied in [9]. The work uses PCA to classify the different stages of the fruit life cycle.

E-noses have also found a place in aroma industries [10]. The work initially describes the performance of a commercial Alpha–MOS 4000 series gas sensor; followed by the analysis, using Discriminant Factorial Analysis (DFA) statistical, of the data on pure-agar and mixture-agar wood oils, by their volatile properties.

A low–cost, non-invasive, and easy-to-use system (based on e-nose and machine learning models) that could help detect and distinguish healthy or diabetic patients is suggested in [11].

The latest research on the usage of e-nose is in the detection of ovarian cancer [12]. Exploration on new ways to "smell" the signs of cancer within the women's body by analyzing patients' breath is going on. An e-nose consisting of an array of flexible sensors is developed that can accurately detect compounds in breath samples which are specific to ovarian cancer. In the future, we can expect to see machine learning approaches for better prediction of the disease.

The work on the application of e-nose for real-time monitoring of thresholds of acetic acid in wine and determining its quality is suggested in [13]. The detection and classification task is carried out using a Multilayer Perception (MLP) neural network. The aim of this work is to speed up the wine spoilage-threshold detection in real-time mode. To this end, the authors propose a raising-window technique that directly works on raw-data (un processed data) to find the necessary patterns (i.e., spoilage-threshold levels) with better performance. The experimental results show that the proposed technique is 63 times faster in classifying the wine quality level (three levels).

A low-cost e-nose and machine learning approach to assess beer aroma profiles is suggested in [14]. The authors use in-house develop sensor-kit—nine sensors housed on a single, small, printed circuit board that is compact in size. The developed kit along with the machine learning model (neural networks NN) and statistical model (ANOVA) is used to evaluate the quality of beers. The statistical method required human experts to qualify the beer rating. Through their findings, the authors conclude that the technology—integration of an e-nose and machine learning modeling— suggested in the work proves to be a reliable and cost-effective tool to assess beer quality in real-time.

A multivariate analysis of volatile organic gases using e-nose and machine learning concepts is presented in [15]. The experimentation is done on the following three types of gas mixtures namely, acetone, ethanol, propane. The analysis was done using three variants of principal components analysis (PCA)—differential, relative, and fractional. The performance of the proposed work is compared with the existing schemes. However, the performance in terms of accuracy of the proposed solution with the existing one is not clearly mentioned. In addition, the authors suggest experimenting with other machine learning techniques such as support vector machine (SVM), k-nearest neighbors (KNN), neural networks (NN), etc., to study the performance parameters.

Sensor-array based sanitation-related malodor detection module called e-nose is developed in [16] that alerts workers who maintain the cleanliness of public toilets. Their technique uses 10 electrochemical gas sensors in an e-nose module and three machine learning techniques namely, Decision Tree (DT), k-Nearest Neighbors (KNN), and Linear Regression (LR) for detection and classification of 180 odor exposures data.

Different to all these applications, we study the presence of gases or vapors in in-door applications. The work in [17] also works on the application of e-nose and neural network models for mixed hazardous in-door gases. However, their work is on prediction using machine learning techniques in presence of interfering gases, which is a different research domain all together.

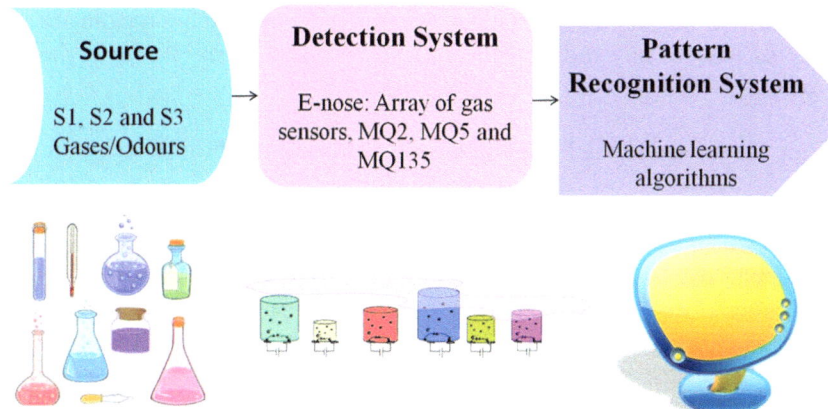

Fig. 1 The proposed e-Nose system

3 Methodology

In this section we present our proposed method. Initially, we describe the developed electronic nose, the components used in its construction, brief specification and the overall module. We then present the background on Machine Learning (ML) technology and describe the ML modules used in the present study.

- **The Proposed Monitoring System**: The proposed electronic nose is as shown in Fig. 1. It consists of three main blocks: (i) The Source System, (ii) Sensor Array (the e-nose) and (iii) Pattern Recognition system. The details of each of these blocks are explained as follows:
- **Source System**: The source to our module could be any kind of odour such as LGP, smoke, aromas, cigar lighter, incense sticks etc. The left most block in Fig. 1 shows the source module.[1]
- **Sensor Array—the e-nose**: The heart of our system is the array of sensors that act as e-nose. The proposed system is built using MQ series gas sensors to detect the source of information (odour), Arduino Uno [18] that reads the analog signals from these gas sensors and converts them into digital signals for further analyses. In a nutshell, the job of this module, the second block in Fig. 1, is to read the samples from the atmosphere and store them in digital value for next stage of processing.
- **Pattern Recognition System**: This is the final block of the proposed system, shown as right most block in Fig. 1. This block act as a human brain whose task is to learn the hidden patterns in the data samples collected by the second stage

---

[1]The images in the second row of the figure were taken from Internet source: http://clipart-library. com/cartoon-computer-cliparts.html.

i.e., the e-nose, find the correlations among the sensor readings and finally train the machine learning model so that it can predict and classify the future detected samples.

Components Used in the Proposed e-Nose Module and Their Description

- **Arduino Uno**: Arduino is an programmable electronic circuit board that can be seamlessly integrated into a wide variety applications [19]. It consists of a microcontroller unit that can be easily programmed to perform tasks such as sensing, detecting, controlling objects, tracking etc. The board has 14 digital input and output pins to read and write data into and out of the board. Each pin operates at 5 V and provides/receives a maximum of 40 mA current. These pins have internal pull-up resistors of 20–50 K Ω.
- **Gas Sensors: MQ2, MQ5 and MQ135**: The MQ2 gas sensor is used for detecting Hydrogen, Propapne, Liquefied Petroleum Gas, Methane, flammable gas, smoke, etc. This sensor indicates that there is smoke present in the surrounding environment by the voltage that it outputs. The higher is the concentration of the smoke particles higher is the output voltage. The heating element within the sensor is powered by VCC and GND from the power supply. The circuit also has a variable resistor whose resistance depends on the smoke quantity in the air. The MQ5 Gas Sensor module is useful for gas leakage detection. It can detect i-butane, LPG, alcohol, methane, smoke, Hydrogen, etc. The sensitivity of this sensor is adjusted using the on-board potentiometer to vary the potential and measure according to the requirement. The MQ-5 gas sensor is make up of tin oxide (SnO_2). When the target combustible gas exist, the sensor's conductivity raises along with the gas concentration indicating the presence of the gaseous material in the surrounding area. This sensor has high sensitivity to Methane, Propane and Butane, etc. The MQ-135 Gas sensors are primarily used in equipments that control the air quality, they are used in the detection of alcohol, smoke, benzene ammonia (NH_3), carbon dioxide (CO_2), acetone etc. The analog pin is used for detecting the gases in parts per million (PPM). This pin is Transistor Transistor Logic (TTL) driven and works on 5 V battery. The end face of the sensor is enclosed by a stainless steel net and the back side holds the connection terminals.
- In addition to the Arudino board and the MQ series sensors, we also need the following components to complete the module: Resistor 330 O, Red LED, Breadboard, connecting wires, cigar lighter, acetone, incense stick.

The layout of the designed module is as shown in Fig. 2.

Machine Learning Technology

Machine learning is a sub-branch of Artificial Intelligence. The name machine learning was coined in 1959 by Arthur Samuel and Tom M. Mitchell provided a formal definition to it: "A computer program is said to learn from experience E concerning some class of tasks T and performance measure P if its performance at tasks in T, as measured by P, improves with experience E" [20]. Machine learning is all about the development of software that can learn for itself. The learning starts with

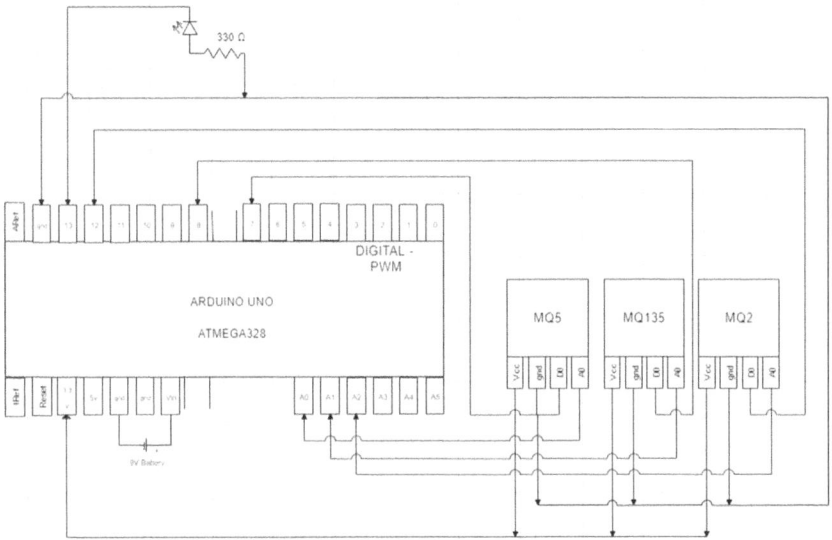

Fig. 2 Layout of the proposed e-Nose system

the observation of the data, the interactions made with the surroundings, and then analyzing the patterns and relations present in the data. The obtained information, after training the system, is then tested and used on future samples that the system encounters. This leads to the automatic learning of computers without any human intervention.

Machine learning algorithms are often categorized into the following types [21, 22]:

- Supervised learning: This type of learning is similar to learning with the help of a teacher. The teacher knows what the learner has to learn by the end of the learning session and accordingly trains the learner. Similarly, the supervised algorithms apply what has been learned in the past to the new upcoming data using the labeled information to predict future events.
- Unsupervised learning: This type of learning is similar to 'learning on self-basis' concept, without the aid of a teacher. This category of algorithms are used when the dataset used for training the model is unlabeled. These algorithms try to explore the hidden pattern in the given data set and draws inferences from the exploration to describe the hidden structures form the unlabelled data.
- Semi-supervised learning: As the name suggests this class of machine learning algorithms has characteristics of both Supervised and unsupervised machine learning algorithms. The 'semi' indicates that the data used for training are both

labeled as well as unlabeled. Typically, the size of the labeled data is less compared to the amount of unlabeled data.

- Reinforcement learning: This type of learning allows machines and software agents to automatically make the decisions based on the learning experience. For every successful learning decision, the algorithm is rewarded and for every unsuccessful decision it is penalized.

Machine Learning Algorithms Being Explored

We brief the details of the machine learning algorithms that we intend to explore and suggest the better model based on the accuracy obtained. In particular we explore the following supervised learning algorithms.

- k Nearest Neighbor (kNN) Algorithm: The algorithm takes in the labelled data and uses the labelled points to determine or learn how to label future input samples. For the classification purpose, it takes the new sample, estimates the shortest distance between the given new sample and the available data, then assigns the new sample to the class to which this distance is minimum. The algorithm uses a parameter k (crucial parameter) as the number of neighbours for comparison.
- Support Vector Classifier (SVC): SVC aims to fit the data that we feed to it and returns a 'best fit' hyperplane that divides i.e., categorizes the data being fed. Thereafter, we can feed some attributes to the classifier to see what the 'predicted' class is.
- Logistic Regression: This is a statistical model that uses a fundamental logistic function such a sigmoid function to model a binary dependent variable, and in advanced cases having more variables. This is basically an improved version of linear regression model, used when linear regression seems a limitation.
- Decision Tree: As the name suggests, decision tree builds classification models in the form of a tree structure. It divides the data into smaller chunks while at the same time a related decision tree is developed incrementally, this is similar to divide-and-conquer concept. Finally, decisions are made accordingly.
- Random Forest: Random forests can be thought of an extension to decision tree. It uses multitude of decision trees at the time of training and giving the outputs. These models try to fine tune the decision trees' habit of over fitting the training set.

4   The Experimental Setup

In this section, we shall initially describe in detail the experimental setup, the procedure followed to collect the data and creation of the data-set. We then describe the machine learning aspects used in our work, followed by realization of the entire analysis.

Table 1 Specification of gas sensors and their threshold values

Gas sensor model	Type of gas odour detected	Threshold value
MQ2	Alcohol, CO (smoke), LPG, Butane (Cigar lighter liquid)	186
MQ5	Alcohol, CO, LPG, Butane, H_2, CH_4 etc.	213
MQ135	NH_3, NOx, alcohol, benzene, smoke, acetone and CO_2	235

4.1 The Electronic Nose and Threshold Criterion for Gas Sensors

Our e-nose system was designed to collect the samples of different gas odours from three odour/gas sources namely smoke from incense stick, smoke from a burning paper and nail polish remover. We used commercially available gas sensors—MQ series—in the development of the e-nose. The gas sensor details are as show in Table 1. In order the detect the gas odours directly from the atmosphere one must ensure that the threshold value for each gas sensor type is chosen correctly i.e., sensors must be calibrated prior to the experimentation. To decide the appropriate threshold values, we exposed each of the three gas sensors (all three together) to air and recorded the samples for 60s. After recording the samples, we obtained around 1500 + samples and we choose exactly 1500 samples of each type. Then we estimate the average of these 1500 values and set them as threshold level for each gas sensor type. The values of threshold levels for each of the three gas sensors are as shown in Table 1.

4.2 The Actual Setup and Data Collection Task

Three different types of odours were chosen: (i) Nail–polish remover as source of Acetone, (ii) Incense stick (with higher molecular concentration measured in parts per million (ppm)) as a source of source of smoke i.e., CO and (iii) Burning paper (with lower molecular concentration) as a source of smoke i.e., CO.

Our setup had three odour detecting sensors (MQ2, MQ5 and MQ135 series) connected to analog inputs of the Arduino Uno board, which in turn was connected to a PC for recording and monitoring purpose. The monitoring process was supported by Arduino IDE that facilitated both the development of the code, that will eventually be dumped into the Arduino board, as well as help us in observing the data that is been captured by the Arduino Uno board. Meanwhile, an additional tool named Tera Term was opened in the back-end to log all the data sampled by the Arduino Uno board and store the logged information in a CSV file format, for future processing. The entire experiment was carried out in a closed chamber (small office cabin) that was devoid of any kind of smoke or other gases prior to the commencement of the experiment. The experiment was carried on 10th May 2019, between 12:00 PM and 5:00 PM. The actual setup in the experimentation room is as shown in Fig. 3 The gas

Fig. 3 The actual setup

odour samples were collected as follows: The experimentation started with sampling of acetone odour. We poured around 5 ml of nail–polish remover (acetone) into a small lid and then exposed the sensors to this odour. We set a stop–watch prior to the sampling process and started the timer just when the sensors were exposed to the input. The gaseous odour was then sampled for 60s. The samples were detected once the threshold value was exceeded and the sampled odour, as sensed by MQ2, MQ5 and MQ135, was converted into respective digital values by the ADC converter, present on the Arduino Uno board, and displayed on the IDE on a PC for human observation. The obtained samples (i.e., the data set) were saved in the form of CSV files.

The process was repeated for other two types of gas odours i.e., for smoke form a burning paper and form an incense stick. This completes the data acquisition phase.

5   Experiment Results and Discussion

Experiments were carried out to evaluate the feasibility of the developed e-nose model in identifying the class of gases. In particular, discriminating between these in-door gas odours: acetone, smoke due to burning paper and smoke due to incense stick. The entire coding was done in Python programming language and standard library tools from open source platform scikit-learn [23] were used in our work.

We employed k-fold (k = 5) cross validation on the datasets and evaluate the following five models: kNN, SVC, LR, DT and RF. To get better results and decide with confidence the best model, the dataset must be randomly shuffled, partitioned into k-folds using k-fold cross validation model and then train the models using each fold one by one. Then the average of all the k folds is computed to get a single representative value. However, prior to training the models the entire data-set of 150 samples is divided into two sets, one for training purpose and the rest of the data-set for testing models predictability on new and unseen data, by the model. We have chosen 70% of the data for training purpose and the remaining 30% is used for testing the models.

The cross-validation step gives the classification accuracy as follows: $KNN = 93.09\%, SVC = 69.54\%, LR = 72.19\%, DT = 84.4\%$ and $RF = 82.31\%$. These values

Table 2 Comparison of accuracy values

Classifier model	Cross validation score	Prediction score
k-NN	93.09	93.33
SVC	69.54	62.22
LR	72.19	71.11
DT	84.40	91.11
RF	87.44	97.77

are for default parameter settings in each of the models. Given these cross-validation scores, our next task is to choose one algorithm that has best score (accuracy) and use it for future predictions on new samples. From the given accuracy values it is seen that kNN performs best compared to all other models. However, this accuracy score alone is sometimes not sufficient to decide on the selection of best model for generalization. Therefore, we further use evaluation metrics such as recall, precision and f1-score to evaluate the developed model. To this end the test data-set, which the trained models have not yet seen, that was set aside will now be used for evaluating the models. This evaluation will give us the final check on the accuracy of the training set. This step will ensure that we are neither overfitting nor underfitting the models (Table 2).

The evaluations of the models are carried out as follows: First develop the machine model using scikit-learn's 'fit' method on the training data-set. Once the model is developed, now predict the target variables using test data-set (i.e., classify the given new samples into appropriate classes using the test data). Once the prediction is done, generate the classification report that has the details of the performance metrics such as precision, recall and F1-score. These are the summary of the confusion matrix. The metric 'precision' measures the number of true positives from the total samples that were predicted positive. It is the ratio of the true-positive predictions to the sum of true-and false-positive predictions. The 'recall' metric measures the number of positive samples captured by the positive predictions. It is defined as the ratio of number of true positives to the sum of true positives and false negatives. Since the decisions cannot be made just using either 'precision' or 'recall' metrics, individually, we must summarize these two metrics using multiclass f_1-score which is given by: $f_1 = 2((precision * recall)/(precision + recall))$. The detailed classification report containing these metrics is shown in Table 3.

5.1 Discussion

Through this experimental analyses, it can be observed that random forest models outperforms the rest of the models when prediction is made on test data with accuracy being 97.77% with f1-scores as shown in Table 3. Random forest model works better than other models. The reason is explained as follows: If the data-set used in the training process for the classification process cannot easily be separable then

Table 3 Classification report

ML model	Gas type (Class)	Precision	Recall	f1-score
kNN	0	1.00	1.00	1.00
	1	0.86	0.92	0.89
	2	0.94	0.89	0.91
SVC	0	1.00	1.00	1.00
	1	0.43	1.00	0.60
	2	1.00	0.06	0.11
LR	0	1.00	1.00	1.00
	1	0.50	1.00	0.67
	2	1.00	0.28	0.43
DT	0	1.00	1.00	1.00
	1	0.91	0.92	0.91
	2	0.92	0.91	0.91
RF	0	1.00	1.00	1.00
	1	0.93	1.00	0.96
	2	1.00	0.94	0.97

any machine learning algorithm will perform poor in classifying the objects being detected. Random forest, built on top of decision trees, is a set of trees each of which is trained on a random subset of training data. The predicted output is stacked together to provide an ultimate value. This aspect of random forest makes it best model compared to others and does not suffer overfitting problem. As a result the classification accuracy will be better and will provide an excellent trade-off for classification performance against data storage space/processing speed.

From the scatter plot as shown in Fig. 4 it is clear that a straight line separating the data sensed by MQ135 and the other two sensors i.e., MQ2 and MQ5 are easily separable, for samples of smoke due to burning paper and smoke due to incense stick. Similarly, a straight line can be easily drawn between the sample detected by MQ2 gas sensor and those samples detected by MQ5 and MQ135, for acetone as input. Whereas it is not possible to draw a clear straight line separating the data samples collected by MQ2 and MQ5 (trials no. 50–150).

The basic version of SVC, used in our work, uses liner hyperplanes to separate classes. A linear classifier will give poor classification accuracy. Where as the kNN uses non-linear kernel to classify the data-set. That is it can generate a highly convoluted decision boundary as it is driven by the raw training data itself. Hence, we see that kNN performs better with 93/33% accuracy than SVC model with only 62.22% of accuracy.

Decision trees tend to overfit on the training data and hence perform poorly when compared to random forest models. In this case the accuracy of decision trees is 91.11%. The prediction accuracy of logistic regression is 71.11%.

Fig. 4 Scatter plot of the data-set

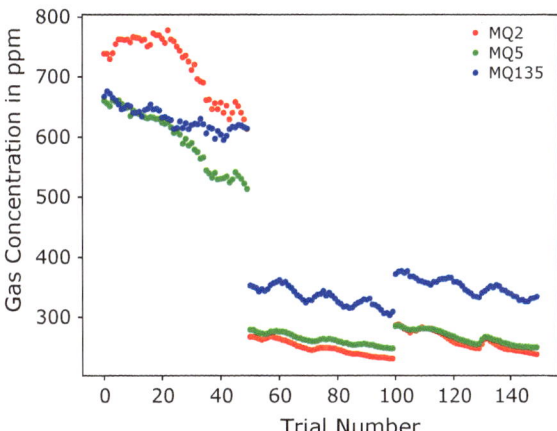

6 Conclusion

The artificial olfaction has witnessed its applicability in various domains such as agriculture, agronomic, fruit quality detection, etc. In this work, we have developed an e-Nose system that could detect various gases from the in-door environments and then classify the gases into different types, using machine learning techniques. In addition, we have also studied the performance of popular machine learning algorithms. Our analyses show that the Random Forest model performs better than the rest of the models considered in this work, the justification being given.

The current form of e-nose could be used in indoor applications such as houses and industries etc., for detecting the leakage of acetone (which is flammable when exposed to fire spark) or raise an alarm only when there is smoke due to a fire accident but not due to incense sticks. As a part of our future work, we would like to extend this work to test with a large number of samples, more number of gas types, influence of the number of gas sensors, temperature, and humidity, concentration levels, interference gases/odors, etc., on the prediction accuracy of the machine learning models and present a comparative study. Later, we also aim to find avenues to make this prototype a bio-implantable device that could help aid persons suffering from *anosmia* (nose blindness) disorder.

References

1. James, G., et al. (2013). *An introduction to statistical learning* (Vol. 112). New York: Springer.
2. Baltes, H., Lange, D., & Koll, A. (1998). The electronic nose in Lilliput. *IEEE Spectrum, 35*(9), 35–38.
3. Davide, F., & Holmberg, M. (2001). LundstrÃűm I. 12 Virtual olfactory interfaces: Electronic noses and olfactory displays. In G. Riva, & F. Davide (Eds.), *Communications through virtual technology: Identity community and technology in the Internet age*. Amsterdam: IOS Press.

4. Vallero, D. A. (2014). *Fundamentals of air pollution*. Netherlands: Elsevier Science.
5. Industrial Gas Pollution. Retrieved May 23, 2019, from https://www.eartheclipse.com/pollut ion/terrible-effects-of-industrial-pollution.
6. Gas Detector. Retrieved May 23, 2019, from http://www.thomasnet.com/articles/instruments-controls/How-Gas-Detectors-Work.
7. Wilson, A. D., & Baietto, M. (2009). Applications and advances in electronic-nose technologies. *Sensors, 9*(7), 5099–5148.
8. Lu, W., Yu, W., Gan, C., Liu, Q., & Li, J. (2015). Application of electronic nose technology in the detection of wheat quality. In *2015 International Conference on Intelligent Transportation, Big Data and Smart City*, 19 December 2015 (pp. 133–136). IEEE.
9. Baietto, M, & Wilson, A. D. (2015, January). Electronic-nose applications for fruit identification, ripeness and quality grading. *Sensors, 15*(1), 899–931.
10. Lias, S., Ali, N. A., Jamil, M., Zainal, M. H., & Ab Ghani, S. H. (2015). Classification of pure and mixture Agarwood oils by Electronic Nose and Discriminant Factorial Analysis (DFA). In *2015 International Conference on Smart Sensors and Application (ICSSA)*, 26 May 2015 (pp. 7–10). IEEE.
11. Sarno, R., & Wijaya, D. R. (2017). Detection of diabetes from gas analysis of human breath using e-Nose. In *IEEE International Conference on Information & Communication Technology and System (ICTS)*, 31 October 2017 (pp. 241–246).
12. Overian Cancer shorturl.at/etKU5. Retrieved May 23, 2019.
13. Gamboa, J. C., da Silva, A. J., de Andrade Lima, L. L., & Ferreira, T. A. (2019). Wine quality rapid detection using a compact electronic nose system: Application focused on spoilage thresholds by acetic acid. *LWT, 1*(108), 377–384.
14. Viejo, C. G., Fuentes, S., Godbole, A., Widdicombe, B., & Unnithan, R. R. (2020, April 1) Development of a low-cost e-nose to assess aroma profiles: An artificial intelligence application to assess beer quality. *Sensors and Actuators B: Chemical, 308*, 127688.
15. Rahman, S., Alwadie, A. S., Irfan, M., Nawaz, R., Raza, M., Javed, E., & Awais, M. (2020, June). Wireless e-nose sensors to detect volatile organic gases through multivariate analysis. *Micromachines 11*(6), 597.
16. Zhou, J., Welling, C. M., Vasquez, M. M., Grego, S., & Chakrabarty, K. (2020, June 12). Sensor-array optimization based on time-series data analytics for sanitation-related malodor detection. *IEEE Transactions on Biomedical Circuits and Systems, 14*(4), 705–714.
17. Zhang, J., Xue, Y., Sun, Q., Zhang, T., Chen, Y., Yu, W., Xiong, Y., Wei, X., Yu, G., Wan, H., & Wang, P. A miniaturized electronic nose with artificial neural network for anti-interference detection of mixed indoor hazardous gases. *Sensors and Actuators B: Chemical, 326*, 128822.
18. Arduino Uno. Retrieved May 5, 2013, from https://components101.com/microcontrollers/ard uino-uno.
19. Arduino For Beginners. Retrieved May 17, 2010 from https://www.makerspaces.com/arduin ounotutorial-beginners/.
20. Michie, D., Spiegelhalter, D. J., & Taylor, C. C. (1994). Machine learning. In *Neural and statistical classification,* 17 February 1994 (Vol. 13, p. 1298).
21. Sivanandam, S. N., & Deepa, S. N. (2008). Genetic algorithms. In *Introduction to genetic algorithms* (pp. 1537). Berlin, Heidelberg: Springer.
22. Reinforcement Learning. Retrieved May 23, 2019, from http://reinforcementlearning.aidepot.com/.
23. scikit-learn. Retrieved November 22, 2020, from https://scikit-learn.org/stable/index.html.
24. Machine Learning Mastery. Retrieved November 22, 2020, from https://machinelearningmas tery.com/start-here/.
25. Müller, A. C., & Guido, S.(2016, September 23) *Introduction to machine learning with Python: a guide for data scientists*. O'Reilly Media, Inc.
26. Gas Sensors MQ Series. Retrieved December 7, 2020, from https://www.electronicscomp.com/ mq2-flammable-gas-smoke-sensor-module-india.
27. Gas Sensors TSG Series. Retrieved December 7, 2020, from https://bit.ly/3noDyjL.

Role of Machine Learning in Weather Related Event Predictions for a Smart City

Muhammad Azmi Umer, Muhammad Taha Jilani, Khurum Nazir Junejo, Sulaman Ahmad Naz, and Conrad Walter D'Silva

Abstract Weather related event prediction is always a fascinating problem for scientists due to its importance in different sectors of life. This chapter has used machine learning algorithms to predict events like rainfall, thunderstorm, and fog in a large metropolitan city. The study proposed here has particularly focused on the long-term event predictions which is currently missing in the state of the artwork. Different machine learning algorithms mainly Random Forest, Gradient Boosting Classifier, Logistic Regression, and others were used to learn the model. Five years of meteorological data was used for this purpose. Different algorithms showed accuracy more than 90%, among which Random Forest outperformed the other algorithms by achieving the highest accuracy.

1 Introduction

Weather has a significant influence on people's daily lives. Nevertheless, it cannot be changed and controlled by humans, still, it can be predicted. In recent years there has been growing interest in weather forecasting since it has an important application in

M. A. Umer (✉)
DHA Suffa University, and KIET Karachi, Karachi, Pakistan
e-mail: azmi.umer@dsu.edu.pk

M. T. Jilani
Karachi Institute of Economics and Technology, Karachi, Pakistan
e-mail: m.taha@pafkiet.edu.pk

K. N. Junejo
Love For Data, Karachi, Pakistan

S. A. Naz · C. W. D'Silva
DHA Suffa University, Karachi, Pakistan
e-mail: sulaman.ahmed@dsu.edu.pk

C. W. D'Silva
e-mail: conrad.dsilva@dsu.edu.pk

© The Author(s), under exclusive license to Springer Nature Switzerland AG 2021
U. Ghosh et al. (eds.), *Machine Intelligence and Data Analytics for Sustainable Future Smart Cities*, Studies in Computational Intelligence 971,
https://doi.org/10.1007/978-3-030-72065-0_4

49

different fields such as agriculture, transportation, utility-services, military and even for general public. Predicting and determining the state of the weather is critical in these fields, since weather information such as short-term forecasts and long term forecasts, can help authorities and companies save humans lives and protect assets from losses. According to the World Bank, losses due to weather-related issues are reached up to $200 billion in the last decade [1]. Similarly, it is more critical for developing countries where due to limited resources and lack of infrastructure such losses can be increased substantially. For instance, in Pakistan, heavy rains and resulting floods cause large casualties every year. It is reported in [2] that in the year 2020, only in the city of Karachi 30 people died in various incidents due to heavy rains.

This chapter emphasizes the use of Machine Learning techniques for weather related event predictions. Machine Learning has very vast applications in self-driving cars [3], health-care [4, 5], cyber-security [6, 7] and industrial decision making [8]. The weather company of IBM is also using machine learning techniques for weather forecasting [9]. The Deep Thunder project initiated by IBM [10], with the aim to optimize the business process with the short-term weather forecasting is another example. The project is not only monitoring the weather but it is also focused on initiating alerts and notifications to the end-user before any event to reduce the impact on business. Weather plays a significant role in our daily life. It has significant impacts on almost every part of our daily life. Due to its importance man has a keen interest in weather forecasting. Planning an event and other important tasks are dependent on weather forecasting. There are several numerical methods that exist for this purpose but are restricted to the very short span of a week, long term forecasting is still a major issue. Numerical weather prediction was used in [11] for short term wind power forecasting. They also proposed an error correction method to deal with the prediction uncertainties but it is limited to short term forecasting. A study reported in [12] used numerical weather data for wind power forecasting. They did feature extraction by utilizing unlabeled numerical data to generate new attributes and they embedded these new attributes in a supervised forecasting model though this model was also intended for short term wind power forecasting. This chapter focuses on the long-term weather forecasting that covers the entire year and could be extended to a greater time span as well.

Machine learning techniques have been explored for weather forecasting. Machine learning techniques uses the past behavior of any system to predict its future behavior. Machine learning techniques have been used in several different applications like in Bioinformatics [13], Pattern Recognition [14], Computer Vision [15], Intrusion Detection [16, 17] etc. This chapter has used several machine learning algorithms like Support Vector Machine, Neural Networks, Decision trees, and others to predict the weather related events on a long term basis. The complete experimental analysis was performed on the Karachi city. It is a metropolitan city of Pakistan and has prime importance in the growth of the country. This model could be used for any city or country by doing slight variations in the model if needed. This model could have important applications in the development of Smart City for developed countries, also it could have important implications in the electronics items like Smart Umbrella.

One of the important characteristics of this model is that it can forecast on a long-term basis which is missing in the current state of the artwork as per the knowledge of authors.

Organization: Section 2 explains the problem statement. Related work is described in Sect. 3. Methodology is described in Sect. 4. Results and discussion are presented in Sect. 5 while Sect. 6 has outlined the conclusions and explains the possible future work.

2 Problem Statement

In the past few decades, scientists and researchers have made various efforts that are fully focused on current weather parameters, which are mainly humidity, temperature, air pressure, and precipitation while providing the weather forecast on short-term weather conditions [18]. However, the long-term or long-range which is more focused on forecasting the trend of weather for the specified duration of a period [19], has not gained much interest by the research community. Due to this, the long term forecast is not extensively studied by the researchers in general, even though it has a great impact on the planning of various activities within the businesses [20]. A reliable and accurate long-term weather forecast can help individuals, organizations and businesses to plan their activities that are highly dependent on weather conditions. This can be even more useful for certain mega-events, such as the cricket world cup. Recently, it is evident that various matches that are scheduled for cricket world cup 2019 were cancelled due to rain and bad weather conditions, thus making it most weather-affected cricket world cup in the history [21]. This event is organized by the International Cricket Council (ICC) once in every four years, and it requires rigorous planning and scheduling activities before the commencing of the event. However, by carefully considering the weather forecasts it is possible to plan such activities in a period that is not much vulnerable to bad weather conditions. In general, the planning of such events starts quite before (like two to three years) before the event. The information on weather during the planning phase will help organizers to schedule the event in a period that will not be affected by bad weather conditions.

3 Related Work

Weather forecasting includes the prediction of humidity, wind, temperature, and other related attributes. All these attributes have prime importance in weather forecasting which in turn have significant applications in different areas. For example, the wind has been a major source for power generation and the prediction of wind would ultimately lead to the prediction of power. Data mining approaches used in [22] to predict the power of wind farms. Different models for the short term and long term were developed in it. They analyzed the prediction performance on two different models. One was a direct prediction model which predicts the power generation directly through weather data. The other was an integrated prediction model which first predicts the wind through weather data and then predicts the power generation through wind predictions. They claimed that the direct prediction model provided better performance as compared to the integrated prediction model.

Regression Analysis including Linear Regression and LASSO (Least Absolute Shrinkage and Selection Operator) were applied in [23] to predict the solar power using weather information. Similarly the study reported in [24] also used the Artificial Neural Network for the similar purpose.

Forests have prime importance in the economy of a country as it is among the major sources of energy. In the year 2020, a natural disaster occurred which not only had a severe impact on the economy but also resulted in the loss of lives of many innocent animals. It was the Australian Bush Fire [25]. Many regions of Australia were severely affected by the fire. The Fires gradually started in 2019 but became more drastic and dangerous in 2020. Huge areas were destroyed and a lot of human lives were lost too. The weather conditions started to deteriorate from late 2019 which got out of control in early 2020. Therefore careful monitoring of forests is an important task. It is normally being done quite smartly by different countries though there are some issues related to forests e.g. forest fires. It causes great economic losses to the country and ultimately to the whole world. Forest fires are highly dependent on attributes like wind, temperature, humidity, etc. Machine learning techniques were used in [26] to predict forest fires. They used the Support Vector Machine along with the attributes of temperature, relative humidity, rain and wind to predict the burned areas of small fires.

Agricultural countries are particularly focused on weather forecasting because they are highly dependent on weather and particularly on rainfall which directly effects the economic growth of the country. Considering the importance of rainfall in agricultural countries, rainfall prediction was done in [27], they utilized the data mining techniques for the prediction of rainfall in Assam, an agricultural state of India. They used the Multiple Linear Regression over the six years [2007 to 2012] meteorological data of Guwahati, a city of Assam. They claim that they obtained the acceptable accuracy using Multiple Linear Regression.

Data mining techniques were used in [28] for the prediction of maximum temperature, wind speed, rainfall, and evaporation. They used Artificial Neural Network and Decision tree algorithms to train the model with the data of nine years [2000 to 2009] of Ibadan, a city of Nigeria. They were able to identify the changes in the climatic patterns due to the availability of sufficient data. They also claim that Artificial Neural Network identified the inherent patterns in the data without the complex equations that are normally required when dealing with the patterns of data.

Data mining algorithms were used in [29] to predict the rainfall using major attributes like wind speed, temperature, humidity, etc. They used Random Forest, Support Vector Machine (SVM), Classification and Regression Tree (CART), Neural Networks and K-nearest neighbor to train the model. They also used the frequent pattern growth algorithm for data cleaning purposes. They claim that their weather accuracy was more than 90%.

4 Methodology

The low-cost sensors with high-density sensing resolution may be used to acquire localized weather information with reasonable accuracy. This approach leads to building a prototype of a weather sensing station which will be connected with the local server to acquire localized weather information. The incoming raw data from sensors are stored in the local weather database, which is then used by the machine learning algorithm to generate the forecast for that particular locality. The overall methodology is described in the following subsections.

4.1 Dataset and Feature Selection

The data available in the system was used to form a dataset. Relevant attributes were selected from this dataset. Relevant attributes have key importance in learning a good model. To select relevant attributes an understanding of the data is required at the initial step. For this purpose, several statistical techniques can be applied to study the insights of the data. It includes finding a correlation among the attributes, finding the noise in the data using box plot, covariance, deviations, central tendency, and several other factors. The Box plots of temperature, dew point, humidity, visibility, and wind speed are presented in Figs. 1, 2, 3, 4, and 5 respectively. It is evident from box plots that attribute related to visibility and wind speed have a higher number of outliers in the dataset. Table 1 describes the features used in the dataset.

Fig. 1 Box plots of high, average, and low temperature

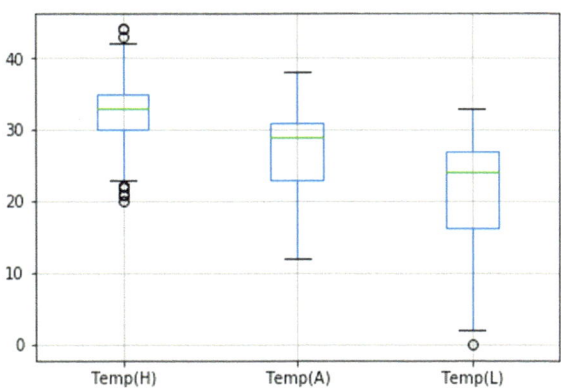

Fig. 2 Box plots of high, average, and low dew point

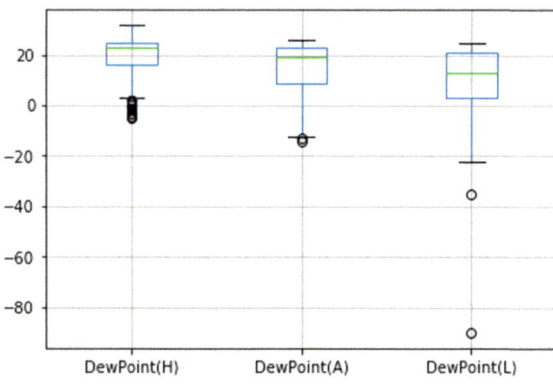

Fig. 3 Box plots of high, average, and low humidity

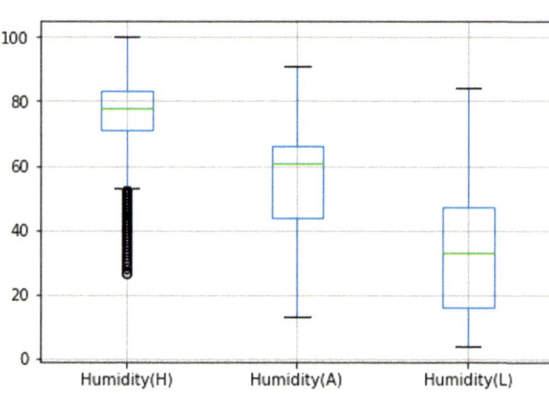

Fig. 4 Box plots of high, average, and low visibility

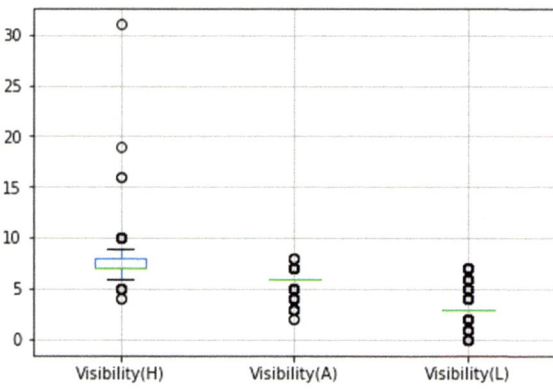

Fig. 5 Box plots of high and average wind speed

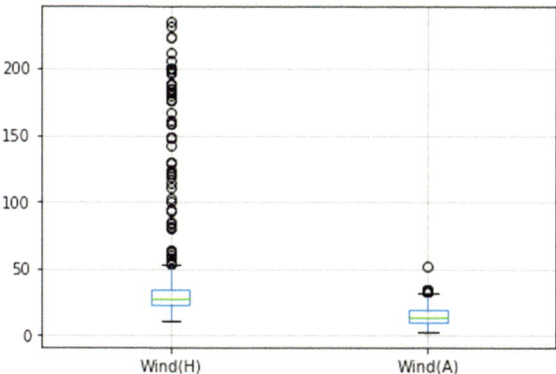

Table 1 Dataset features and their types

Features	Data Type
Temperature (High)	Numeric
Temperature (Average)	Numeric
Temperature (Low)	Numeric
Dew Point (High)	Numeric
Dew Point (Average)	Numeric
Dew Point (Low)	Numeric
Humidity (High)	Numeric
Humidity (Average)	Numeric
Humidity (Low)	Numeric
Visibility (High)	Ordinal
Visibility (Average)	Ordinal
Visibility (Low)	Ordinal
Wind (High)	Numeric
Wind (Low)	Numeric

Class variable or label in this problem is the event which comprises of following four classes:

- Fog
- Rain
- Thunderstorm
- No event

4.2 Selection of Algorithm

The problem being addressed here is related to Supervised Learning (SL). Broadly SL can be divided into regression and classification problems. The selection of the algorithm does not depend only on the type of problem but also on the type of data that is being used to learn the model. In this study, the problem is related to classification. For the classification purpose following classifiers/algorithms were used to learn the model.

4.2.1 Logistic Regression

Logistic regression is used in classification problems and it uses the categorical class variable. There are different variants of logistic regression depending on the class variable; binomial, multinomial, or ordinal. It targets to find the best fit model that can clearly describe the relationship between the dependent and independent variables. Equation 1 describes the mathematical model of logistic regression [30].

$$P(Y_i = 1|X_i) = \frac{e^{\beta^T X_i}}{1 + e^{\beta^T X_i}} \tag{1}$$

4.2.2 Gradient Boosting Classifier

Gradient boosting classifier is used for classification. It normally builds the model in repeated series of rounds. It maintains a set of weights for the training sets. In the beginning, all weights are set equal. Upon each iteration, weights of incorrectly classified examples get increased, so it ultimately gets focused on more and more hard examples available in the training sets. Equation 2 describes the final hypothesis of the AdaBoost algorithm [31].

$$H(x) = sign\left(\sum_{t=1}^{T} \alpha_t h_t(x)\right) \tag{2}$$

4.2.3 Classification and Regression Tree (CART)

CART is the combination of Classification and Regression Trees. Classification trees work on categorical class variables while Regression trees work on the continuous class variables. In this way, CART has the ability to predict the class variable whether it is categorical or continuous [32].

4.2.4 Random Forest

Random Forest has been used in this study for the purpose of classification because it has the ability to build the model based on the combination of tree predictors where each tree depends on the value of random vector sampled independently and it follows the same distribution for all the available trees in the forest. Equation 3 describes the marginal function of random forest [33].

$$mr(X, Y) = E_{\Theta}[I(h(X, \Theta) = Y) - I(h(X, \Theta) = \hat{j}(X, Y)] \qquad (3)$$

4.2.5 K- Nearest Neighbors Classifier

It builds the model by containing the complete training data. Any upcoming example is classified based on the majority class from the k closest examples available in the training data [34].

4.2.6 Support Vector Classifier

In support vector classification, the model is built by nonlinear mapping of input vectors on high dimensional feature space, which in turn is used to construct the linear decision surface. This decision surface has key importance towards the generalization ability of machine learning [35].

4.3 Implementation

To train the model, the available data must be divided into three chunks i.e. training set, validation set, and testing set. Now train the model with the attributes that were selected by previous processes. First train the model using the training data set and then apply that model to the validation set to observe the performance of the model. This is usually a long and iterative process because here it is again required to do feature selection and further data processing as required by the model. After getting the required accuracy from the model the next step is to apply the model to test set, this step gives the overall performance of the learned model, if it is good enough then the model can be applied on the test data.

Python implementation of all the algorithms were used for training the model. Six years [2011 to 2016] Meteorological data of Karachi [36], a city of Pakistan was collected for the experiments. Among which five years data was used to train the model using different classifiers/ algorithms. The remaining one-year data was used to test the performance of the model. Most of the attributes like temperature, dew point, humidity, and wind are numerical attributes. To assess the performance, these attributes were discretized into multinomial attributes but doing so reduced the

Table 2 Algorithms with their accuracy

Classifiers/algorithms	Accuracy (%)
Logistic regression	93.44
Gradient boosting classifier	93.98
Classification and regression Tree	85.51
Random forest	94.26
K-nearest neighbors classifier	90.71
Support vector classifier	89.34

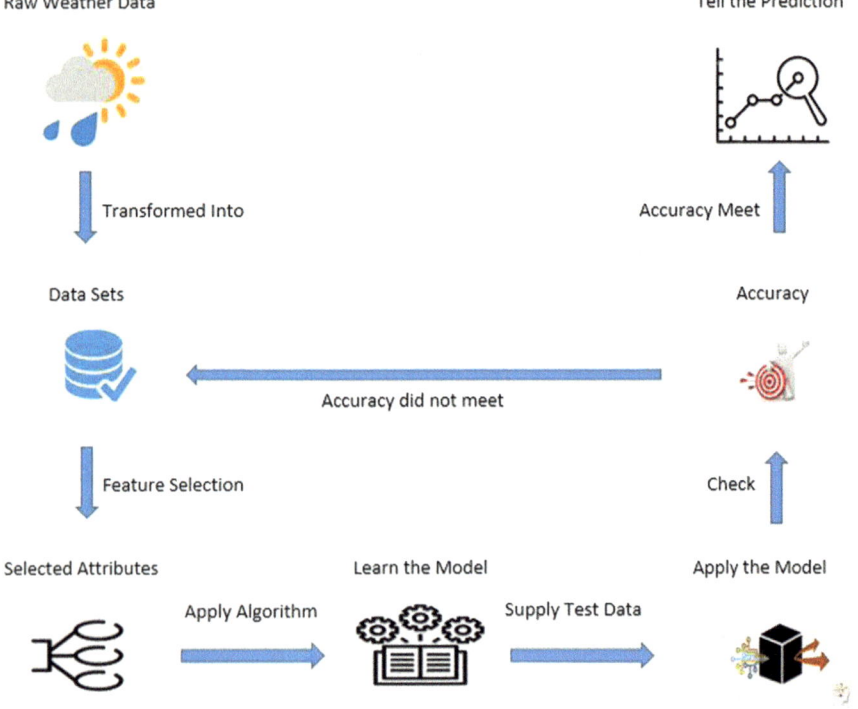

Fig. 6 Workflow

accuracy therefore results presented in Table 2 are calculated when these attributes were in their original form i.e. numerical. Naive feature selection techniques were applied on the dataset to select the appropriate features for the final dataset. Attributes like year, month, and date were present in the original dataset but they are not part of the final dataset. Experiments showed that they did not have any significance in the prediction of the class variable. Figure 6 describes the complete flow of work.

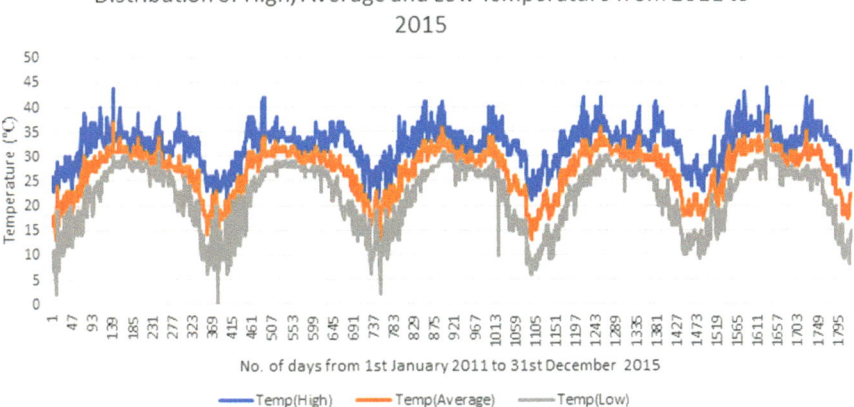

Fig. 7 Temperature distribution from 2011 to 2015

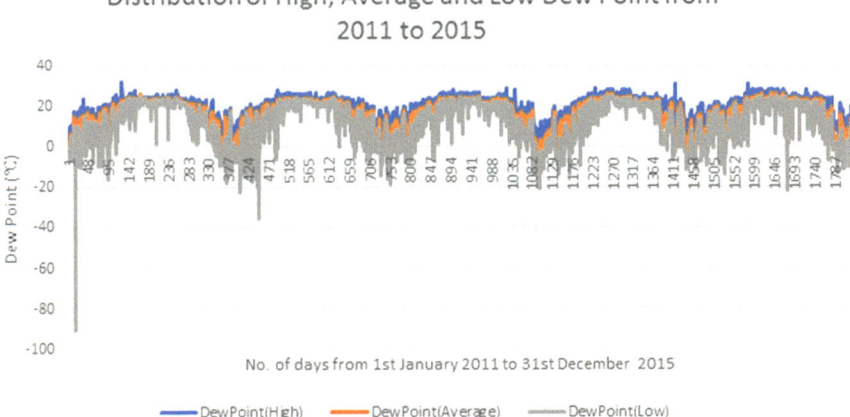

Fig. 8 Dew Point distribution from 2011 to 2015

5 Results and Discussion

Different algorithms mentioned earlier were used to train the model. During the experiments five years of data was used to train the model. The model was tested on 1 year of data. Figure 7 describes the distribution of temperature over a span of five years, it is clearly visible that temperature is normally distributed over the entire year and the same behavior is being observed throughout the five year dataset. Distribution of dew point in Fig. 8 also shows the similar behavior. Figures 9, 10 and 11 describes the distribution of Humidity, Visibility and Wind speed over the span of five years. It can be observed in the distribution of visibility and wind speed that there are some spikes in their distributions which are a little low in case of visibility and high in

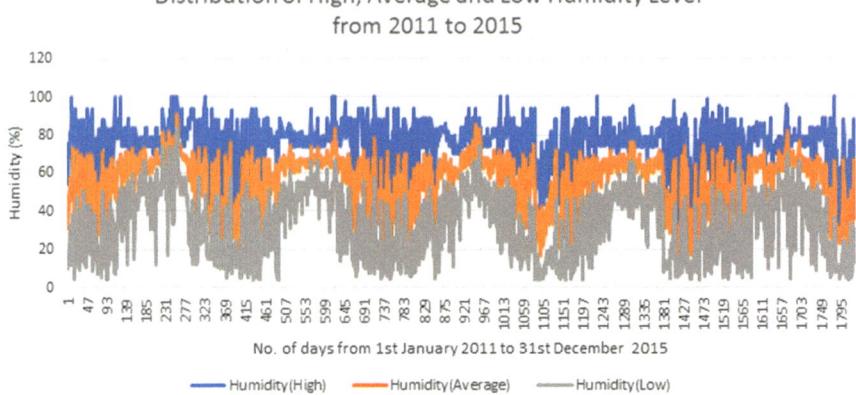

Fig. 9   Humidity level distribution from 2011 to 2015

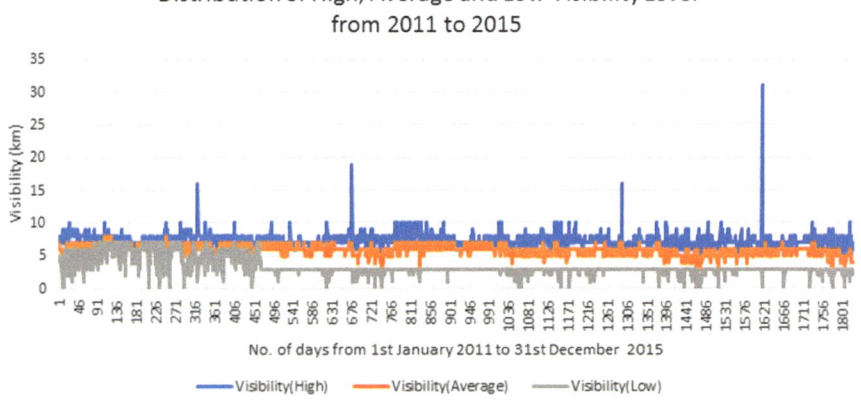

Fig. 10   Visibility level distribution from 2011 to 2015

case of wind speed. These spikes may be due to some noise in the data. Accuracy of different algorithms has been mentioned in Table 2 while Fig. 12 explains the performances of different algorithms.

6   Conclusion and Future Work

The research study was performed on the dataset of Karachi. In future, we may increase the data set by including more cities. We can then compare the results of different cities and validate our claims. We can also increase the time span of the study and as the data set will increase, the chances of error will be minimized. Although

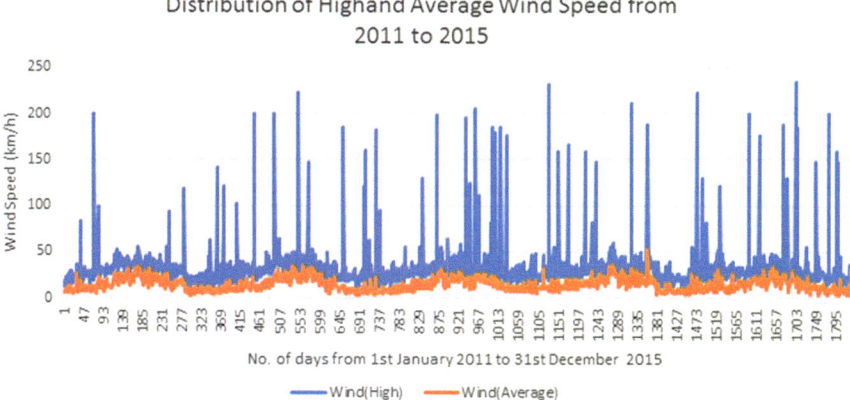

Fig. 11 Wind Speed distribution from 2011 to 2015

Fig. 12 Performance evaluation of algorithms

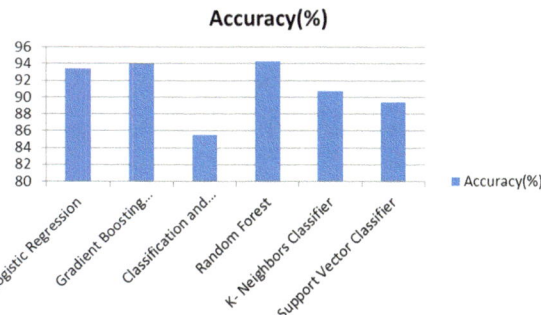

the model showed the maximum accuracy of 94.26% in the case of Random Forest but it should be clear that here the problem we are dealing with is class imbalance. Getting higher accuracy only is not sufficient, here we should focus on the correct predictions of events like fog, rainfall, and thunderstorm as in the current dataset, there are a lot of instances with No Event class. This study may be extended by adding further attributes of weather like Snowfall, Cloudiness, Sunlight, Precipitation, and Aridity etc. This study may have recognition at the national and international level as Weather Conditions is a trending issue in most places all over the world, as severe catastrophic consequences occur if the weather parameters are not laid emphasis upon. This will help the government to take necessary and prompt action for future weather conditions. It can also be used by the government if it intends to deploy weather stations in remote areas of the country. In the future, we intend to solve this problem using Time Series Analysis because it seems more intuitive towards the weather forecasting because it is highly dependent on day to day forecasts.

References

1. Kim, J. Y. (2018). Damages from extreme weather mount as climate warms—world bank report.2018.
2. Dawn News. (2020). https://www.dawn.com/news/1576736.
3. Sze, V., Chen, Y.-H., Emer, J., Suleiman, A., & Zhang, Z. (2017). Hardware for machine learning: Challenges and opportunities. In *IEEE Custom Integrated Circuits Conference (CICC)*. IEEE. pp. 1–8.
4. Ghassemi, M., Naumann, T., Schulam, P., Beam, and A. L., & Ranganath, R. (2018). Opportunities in machine learning for healthcare, arXiv preprint arXiv:1806.00388.
5. Ganguli, I., Gordon, W. J., Lupo, C., Sands-Lincoln, M., George, J., Jackson, G., Rhee, K., & Bates, D. W. (2020). Machine learning and the pursuit of high-value health care. *NEJM Catalyst Innovations in Care Delivery, 1*(6).
6. Ahmed, C. M., Umer, M. A., Liyakkathali, B. S. S. B., Jilani, M. T., & Zhou, J., Machine learning for cps security: Applications, challenges and recommendations. In *Machine Intelligence and Big Data Analytics for Cybersecurity Applications*. Springer, pp. 397–421.
7. Umer, M. A., Mathur, A., Junejo, K. N., & Adepu, S. (2017). Integrating design and data centric approaches to generate invariants for distributed attack detection. In *Proceedings of the 2017 Workshop on Cyber-Physical Systems Security and Privacy*. ACM, pp. 131–136.
8. Lessmann, S., Haupt, J., Coussement, K., & De Bock, K. W. (2019). Targeting customers for profit: An ensemble learning framework to support marketing decision-making. *Information Sciences*.
9. Driving the world's most accurate weather forecasts (2018). https://www.ibm.com/blogs/think/2016/12/accurate-weather-forecasts/.
10. Deep thunder. (2018). http://www-03.ibm.com/ibm/history/ibm100/us/en/icons/deepthunder/.
11. Peng, X., Deng, D., Wen, J., Xiong, L., Feng, S., & Wang, B. (2016). A very short term wind power forecasting approach based on numerical weather prediction and error correction method. In *2016 China International Conference on Electricity Distribution (CICED)*. IEEE, pp. 1–4.
12. Fang, S., & Chiang, H.-D. (2016). Improving supervised wind power forecasting models using extended numerical weather variables and unlabelled data. *IET Renewable Power Generation, 10*(10), 1616–1624.
13. Naresh, E., Kumar, B. V., & Shankar, S. P. et al. (2020). Impact of machine learning in bioinformatics research. In *Statistical Modelling and Machine Learning Principles for Bioinformatics Techniques, Tools, and Applications*. Springer, pp. 41–62.
14. Mshir, S., & Kaya, M. (2020). Signature recognition using machine learning. In *8th International Symposium on Digital Forensics and Security (ISDFS)*. IEEE, pp. 1–4.
15. Athiraja, A., & Vijayakumar, P. (2020). Banana disease diagnosis using computer vision and machine learning methods. *Journal of Ambient Intelligence and Humanized Computing*, 1–20.
16. Umer, M. A., Mathur, A., Junejo, K. N., & Adepu, S., A method of generating invariants for distributed attack detection, and apparatus thereof, Oct. 1 2020, US Patent App. 16/754,732.
17. Umer, M. A., Mathur, A., Junejo, K. N., & Adepu, S. (2020). Generating invariants using design and data-centric approaches for distributed attack detection. *International Journal of Critical Infrastructure Protection, 28*, 100341.
18. Rasouli, K., Hsieh, W. W., & Cannon, A. J. (2012). Daily streamflow forecasting by machine learning methods with weather and climate inputs. *Journal of Hydrology, 414*, 284–293.
19. Weisheimer, A. (2013). If you cannot predict the weather next month, how can you predict climate for the coming decade? 2013, Univ. Oxford.
20. Wagner, A. L., Keusch, F., Yan, T., & Clarke, P. J. (2016). The impact of weather on summer and winter exercise behaviors. *Journal of Sport and Health Science*.
21. Kosky, B. (2019). Rain at the cricket world cup: How the tournament has fallen foul to the weather. 2019, sky News, p. Cricket News.

22. Kusiak, A., Zheng, H., & Song, Z. (2009). Wind farm power prediction: A data-mining approach. *Wind Energy: An International Journal for Progress and Applications in Wind Power Conversion Technology, 12*(3), 275–293.
23. Zafarani, R., Eftekharnejad, S., & Patel, U. (2018). Assessing the utility of weather data for photovoltaic power prediction, arXiv preprint arXiv:1802.03913.
24. Dolara, A., Grimaccia, F., Leva, S., Mussetta, M., & Ogliari, E. (2018). Comparison of training approaches for photovoltaic forecasts by means of machine learning. *Applied Sciences, 8*(2), 228.
25. 2019–2020 Australian Bushfires. (2019). https://disasterphilanthropy.org/disaster/2019-australian-wildfires/
26. Cortez, P., & Morais, A. D. J. R. (2007). A data mining approach to predict forest fires using meteorological data.
27. Dutta, P. S., Tahbilder, H., et al. (2014). Prediction of rainfall using data mining technique over Assam. *IJCSE, 5*(2), 85–90.
28. Olaiya, F., & Adeyemo, A. B. (2012). Application of data mining techniques in weather prediction and climate change studies. *International Journal of Information Engineering and Electronic Business, 4*(1), 51.
29. Taksande, A. A., & Mohod, P. (2015). Applications of data mining in weather forecasting using frequent pattern growth algorithm. *International Journal of Science and Research.*
30. Pohar, M., Blas, M., & Turk, S. (2004). Comparison of logistic regression and linear discriminant analysis: A simulation study. *Metodoloski zvezki, 1*(1), 143.
31. Freund, Y., Schapire, R., & Abe, N. (1999). A short introduction to boosting. *Journal-Japanese Society For Artificial Intelligence, 14*(771–780), 1612.
32. Wilkinson, L. (2004). Classification and regression trees. *Systat, 11*, 35–56.
33. Breiman, L. (2001). Random forests. *Machine Learning, 45*(1), 5–32.
34. Horton, P., & Nakai, K. (1997). Better prediction of protein cellular localization sites with the it k nearest neighbors classifier. In *ISMB*, vol. 5, pp. 147–152.
35. Cortes, C., & Vapnik, V. (1995). Support-vector networks. *Machine Learning, 20*(3), 273–297.
36. Weather underground. weather data. (2017). https://www.wunderground.com/weather/pk/karachi.

Intelligent Vehicle Communications Technology for the Development of Smart Cities

Abdelali Touil, Fattehallah Ghadi, and Khalid El Makkaoui

Abstract Vehicle safety and mobility concerns have become a significant problem for governments and car manufacturers in the last few years. The development of intelligent transportation systems (ITSs) has favorite researchers in industries and academia to put many research efforts into improving road traffic safety and efficiency. The communication systems allowing vehicles to participate in communication networks have advanced significantly with wireless technologies' evolution. Thus, new types of networks, such as intelligent vehicle communications (IVCs), have been designed to establish communication between vehicles themselves and between vehicles and infrastructure. New concepts where IVCs have a crucial role have been proposed, such as smart cities and their subsystems. Smart cities include intelligent road safety and traffic management systems in which road traffic information could be reachable at any time and place. Thus, it was necessary to develop new architectures and standard specifications for telecommunications and information exchange for vehicular networks. For these reasons, this chapter introduces the reader to IVCs and their enabling technologies. We focused on presenting challenges relating to IVC technologies in road traffic environments. We outlined some popular standards, namely, DSRC/WAVE, 802.11p by IEEE, CALM by ISO, ITSC by ETSI. Then, we have explored possible network architectures and applications of IVCs. Finally, we concluded the chapter by pointing to open problems related to smart city applications and future research directions.

A. Touil (✉) · F. Ghadi
Laboratory of Sciences Engineering, Faculty of Sciences, Ibn Zohr University Agadir,
Agadir, Morocco
e-mail: abdelali.touil@edu.uiz.ac.ma

F. Ghadi
e-mail: f.ghadi@uiz.ac.ma

K. El Makkaoui
LaMAO laboratory, MSC team, FP, Mohammed First University, Nador, Morocco
e-mail: kh.elmakkaoui@gmail.com

LAVETE laboratory, FST, Hassan First University, Settat, Morocco

© The Author(s), under exclusive license to Springer Nature Switzerland AG 2021
U. Ghosh et al. (eds.), *Machine Intelligence and Data Analytics for Sustainable Future Smart Cities*, Studies in Computational Intelligence 971,
https://doi.org/10.1007/978-3-030-72065-0_5

1 Introduction

The concept of smart cities refers to cities that make use of information and communication technologies to enhance the quality of urban services or decrease their costs. To become smart, today's cities will have to implement new efficient services in all fields: intelligent transport and mobility, sustainable environment, responsible urbanization, and smart housing. Smart cities provide several services that maximize the social and economic value of digitalization, which makes life better for every community. One of the major challenges is to develop a transportation system that is efficient, easily accessible, safe, and environmentally friendly. This improvement allows a reduced environmental footprint, optimizes the use of urban space, and offers city dwellers a wide range of mobility solutions to meet all their needs.

The term intelligent transport systems (ITSs) was coined several years ago; it is used to describe complex systems that make use of communication technologies to improve and assess the driving experience. ITS services involve vehicles, drivers, passengers, and pedestrians all interacting with each other and the surrounding environment. Decreasing the number of deaths and injuries in the event of traffic accidents is an important goal of many ITS services, as is improving road safety and transport efficiency. Besides, these systems are also being used to reduce the negative effects of transport systems on the environment. ITS is a very dynamic network due to the high mobility of vehicles; traffic information must be exchanged between vehicles and their nearby roadside units accurately and promptly. Such systems can only interoperate successfully if they are based on international standards and have been adopted around the world in multiple implementations. However, they are often seen in the context of the road traffic environment. They can also be used on different types of transportation modes (e.g., rail, air transportation systems, and water).

ITS provide vehicles with different levels of intelligence and prior information about road traffic conditions and other valuable information, which minimize the travel time of vehicles as well as ensure their safety and comfort [34]. ITS applications are widely accepted and initiated in many countries. The applicability is not just limited to the collection of road traffic information, but also for safety concerns and road efficiency usage. Because of its endless possibilities in automotive and other sectors, ITS have recently become a high-profile research field. Thus many research societies and organizations around the world (e.g. IEEE, FCC, ETSI, SAE, ITU) have developed different solutions to fulfil the transportation needs. The collaboration between these actors has led to many research projects worldwide, namely, IntelliDrive safety systems program supported by the U.S. Department of transportation [31], cooperative vehicle-infrastructure systems (CVIS) funded by the European Commission (EC) [29], advanced cruise-assist highway systems (AHS) developed in Japan [9], and has also initiated many consortia (e.g. C2C-CC and CALM) that aim to design, develop and test the feasibility of intelligent vehicles applications. One such well-known example of an ITS is the electronic toll collection (ETC) for the use of toll roads (e.g. Highway). As can be seen from Fig. 1, once the toll system is installed, the system will collect tolls without cash and without requiring vehicles

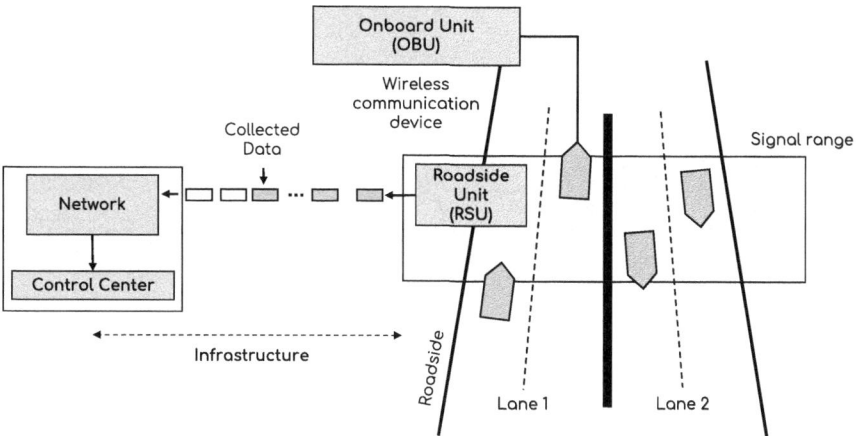

Fig. 1 Electronic toll collection (ETC)

to stop at tollgates. A roadside antenna affixed to each tollgate communicates with vehicles equipped with an onboard unit that automatically deducts the fees from the driver's prepaid account [19]. According to information provided by statistics, five times as many vehicles can go through ETC-equipped toll booths.

The performance of ITS can be improved with recent advancements in information and communication technologies (ICTs), which improve their effectiveness in solving the most complicated transportation challenges. To collect road traffic information that is required to enable driving comfort and road safety, intelligent vehicles (IVs) are one of the main technology for providing the best improvements. In conjunction with the growing need for imparting traffic information, there has been a rapid increase in wireless communication networks to enhance the capabilities of data dissemination while improving their efficiency and facilitating their deployments. IVs may provide considerable benefits by combining ICT with transportation systems. This combination opens the way toward sharing information between vehicles, known as vehicle-to-vehicle communication (V2V), and infrastructure, also known as vehicle-to-infrastructure communication (V2I). Using V2V and V2I communications, vehicles can collaborate with each other or interact with the infrastructure to exchange different types of information, including traffic safety information and environment-related conditions [12], traffic state and mobility [37] or assist navigation [7]. The traffic information obtained from vehicles depends on different attributes and their interrelation, which is pertinence, accuracy, and the level of novelty of the data [6]. This information allows in assisting road network decision-making, monitoring the evolution of traffic, improving the performance of existing infrastructure, and controlling the results of applied plans to enhance road safety [17].

Because of the importance of intelligent vehicle communication (IVC) in the growth of smart cities, this chapter outlines the emerging IV applications field of intelligent transportation systems (ITSs). The chapter is structured as follows. Section 2 provides a brief history of the technology and the potential advantages of the implementation. Section 3 describes the challenges associated with the implementation. Section 4 provides associated standardization activities. Section 5 describes basic architectures and possible deployment scenarios. Section 6 describes the enabling application for IVC to improve traffic safety, traffic management, and assisting drivers. Section 7 concludes the chapter and gives the future research directions.

2 A Brief History

Around 1990, road transportation experts recognized the development of wireless communications, and sensor technologies and started to implement them in traffic and road management systems. Thus was born ITS. Beginning in the late 1990s, ITS systems were developed and used, providing traffic authorities with a wide range of information on real-time road traffic network conditions, which they in turn provided to the public through other platforms. In developed countries, passengers today have access to important amounts of information about driving conditions, whether they are driving their vehicle or riding on public transit systems. Further, ITS have considerably improved the ability of local authorities to react to collisions or other incidents on the road, so that the response delays are minimized. Since one minute of lane blockage typically turns to 10 min of congestion, the advantages of such performances are obvious.

Regarding safety, both researchers, academics, and scientists within automotive industry laboratories have been developing various technologies to assist drivers in avoiding road accidents. In Japan, a tremendous amount of work was done in the 1980s, with the first systems introduced in that business. Still, the costs and capabilities of the technology limited the adoption of these systems for general use. Research and development activities accelerated in the early 1990s by the European project program for European traffic of highest efficiency and unprecedented safety (PROMETHEUS) as part of the EUREKA[1] research actions [42]. The project includes lane monitoring, electronic copilots, and autonomous vehicles. Interestingly, the Japanese started the advanced safety vehicle program to develop advanced collision avoidance systems. Besides, in the United States, both the national automated highway system consortium (NAHSC) and the collision avoidance research programs were launched [39]. Starting in the latter half of that decade, systems introduced to the market in all three continents were, to some degree, the results of these research programs. Named advanced driver assistance systems (ADAS) [26], result introductions continue, and research and development are in full swing for even more developed technologies. The clear result is that we are starting to see technologies

---

[1] EUREKA, is an intergovernmental organization for pan-European research and development funding and coordination.

within vehicles, that can sense critical traffic conditions and react properly in situations where the driver is not. IVs are a reality today, and they will steadily become an essential part of the fundamental fabric of society in the next years.

3 IVCs Characteristics, Application Challenges and Constraints

IVC technology is developed according to the concept of well-known mobile networks. However, the data dissemination in IVC has very different characteristics and implementation challenges compared to mobile networks [16, 22, 43]. This is mainly due to the inherent dynamic behaviour of the vehicular environment, which presents scenarios with unfavourable characteristics for developing communication technologies. Generally, the research works carried out in the context of the mobile networks cannot also be directly applied in IVC. This is because of its distinction (e.g. mobility, density, etc.), which makes the data dissemination approaches less convenient and unsuitable (see Table 1). The quality of the information provided to the driver depends fundamentally on the quality of the communication channel. Therefore, to meet these goals, routing schemes must take into account several effects of IVC behaviour. These characteristics are prompting researchers to develop more effective wireless communication solutions to handle rapid data transfer and provide quality of service over IVC.

The main characteristics that differentiate vehicular networks from other classes of mobile networks can be detailed as follows:

- Dynamic topology variation: the mobility of vehicles is the main factor that differentiates vehicular networks from other types of mobile networks. Because of the significant variations of vehicles speed and acceleration caused by the environments (e.g. urban, rural, and highway areas), the value of speed is on average of 50 km/h in urban areas and up to 120 km/h on the highway or even more, depending on the countries policies. Thus, vehicle movements are relatively unpredictable, and its impact on the connectivity of vehicles remains one of the serious challenges.

Table 1 Comparison between traditional mobile networks and vehicular networks characteristics

Characteristics	Mobile networks	Vehicular networks
Mobility model	Random	Constrained by road network
Node mobility	Low	High
Node density	Sparse	Dense and frequently variable
Topology change	Slow	Frequent and very fast
Energy constraints	Unlimited	Limited
Quality of service	Low	High
Signal range	Up to 100m	Up to 1000 m
Communication type	Unicast	Geocast, broadcast, etc.

- Unbounded network size and high density: the vehicular network constitutes the largest instance of most mobile networks studied in the literature, that usually presume an unlimited network size. It can, in principle, extend over the entire road network and so include many nodes. The number of nodes is not uniform due to a spatiotemporal variation. Urban density is much higher than that in rural areas. For instance, the number of vehicles in an intersection or traffic congestion is greater than that on extra-urban roads where traffic is often smooth.
- Frequent exchange of information: because of the nature of exchanged traffic information, which contains many kinds of safety messages (e.g. prevention and critical alert messages). The information must be directly forwarded from the source to the destination periodically. Thus, many dissemination problems may occur in such a situation, for example, broadcast radiation and communication overhead.
- Sufficient processing capacity, storage, and energy: contrary to the nodes in traditional mobile networks, the batteries in vehicles are self-charging, which provide vehicles with enough energy that can power different onboard devices in the vehicle. Consequently, vehicles will be able to use a large processing capacity and data storage, besides, several network interfaces using different technologies such as Wi-Fi, Bluetooth, DSRC, etc.
- Security and privacy requirements: IVC suffers from different vulnerabilities that could render the threats behaviours targeting the vehicles. To preserve the privacy of transferred sensitive information, it is necessary to prevent any interception, or fake information sent by third-party using malicious devices to meet a specific purpose.

4   Main Standardization Activities

To achieve better interoperability, rapid implementation, and standardization of radio frequencies with which IVs communicate, IVCs become a highly innovative field of research, development, and field trials. Throughout the world, various industry/government projects and activities are designed, to get involved in standards development needed to support the growth of IVs (see Fig. 2). The first developed standards for wireless communication in ITSs typically operate in different frequency bands in different regions around the world [25]: 868/433 MHz in Europe, 915 MHz in the U.S., and 430 MHz frequency band in Asia. These standards are offered as guidelines to develop different services, for example, the electronic tolling system, violation enforcement system, and commercial vehicle management. However, These frequency bands are too low and not suitable for the highly dynamic and fully distributed network environment. Afterward, the IEEE 802.11p standard was established as a new amendment of the IEEE 802.11a LAN standard to add wireless access in vehicular environments (WAVE). IEEE 802.11p is the basis for dedicated short-range communication (DSRC) system, which is mainly designed to ensure timely and low-latency communications. DSRC delivers the required quality

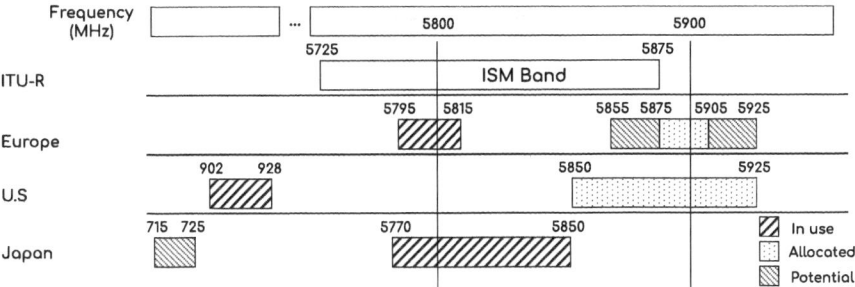

Fig. 2 Activities on ITS radio communications standards in ITU-R ITU Radiocommunication Sector (ITU-R) is one of the three divisions of the International Telecommunication Union (ITU) and is responsible for radio communication. Its role is to manage the international radio-frequency spectrum and satellite orbit resources and to develop standards for radiocommunication systems to ensure the effective use of the spectrum (www.itu.int/ITU-R/). (ISM band Industrial, Scientific, and Medical (ISM) band refer to radio bands reserved internationally for the use of radiofrequency energy for industrial, scientific and medical purposes other than telecommunications.) and in Japan. Japanese Industrial Standards Committee (JISC) is the International Organization for Standardization (ISO) member body for Japan. Its is role is to establish and maintain the Japanese industrial standards (www.jisc.go.jp/eng/). *Source* adapted from [33]

of services for most of IVs applications. There are also other standards proposed by ETSI, called the ETSI TC-ITS architecture, and the CALM architecture proposed by ISO.

In the following section, we present the main standards and specifications, which were specifically designed for IVCs.

4.1 DSRC

The term dedicated short-range communications (DSRC) system refers to one-way or two-way short to medium-range wireless communication channels specifically designed for IVs. To harmonize the use of communications. The U.S. federal communication commission (FCC) decided on October 21, 1999, to assign 75 MHz of spectrum in 5850–5925 MHz frequency band for much dedicated short-range communication applications [38]. DSRC is expected to provide a protocol stack that is suitable for IVs, besides reliable transfer information between vehicles, and vehicles and surroundings infrastructure. As well, on August 5, 2008, the European telecommunications standards institute (ETSI) assigned 30 MHz of spectrum in 5875–5905 MHz frequency band to coordinate the use of radio spectrum and allow the automotive industry to establish radio communication between IVs [13].

As can be seen from Fig. 3, the DSRC spectrum is divided into seven channels of 10 MHz, respectively numbered 178, 172, 174, 176, 180, 182, 184. Channel 178 is known as a control channel (CCH). CCH can be effectively used to exchange control information and support various safety applications. While the other six channels are marked as service channels (SCH) available for general data dissemination with

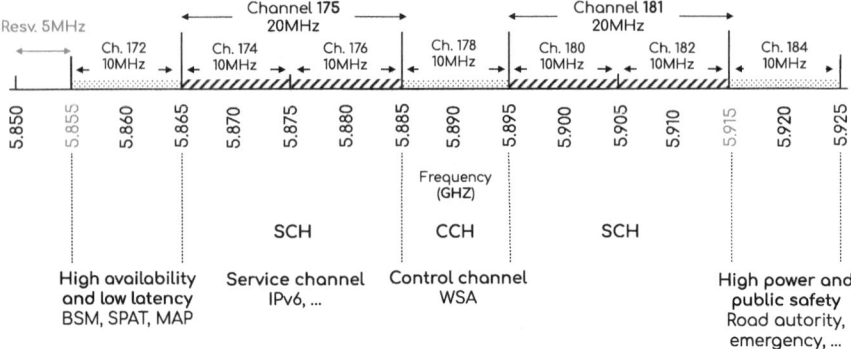

Fig. 3 DSRC spectrum band and channels DSRC transmit different types of messages, including (e.g. BSM Basic safety message is one of the most important messages. It conveys core state information about the sending vehicle, namely its position, dynamics, system status, and size. It also has the flexibility to convey additional information at the discretion of the sender., SPAT Signal phase and timing is a message broadcast by RSUs to provide the current signal status (colour) by lane and when the status is expected to change., MAP Map data message is Broadcasted by RSUs to provide a geometric layout of an intersection., WSA WAVE service advertisements message is used to announce the availability of one or more WAVE services on the service channels. Most DSRC services are expected to be offered by an RSUs, but OBUs can also advertise and offer services.)

lower priority. Channels 172 and 184 are respectively reserved for high availability and low latency, which can be used for high power and public safety. Unlike DSRC in the U.S., the Europe DSRC system uses only the channels 180 as CCH and 172, 174, 176, 178 as SCH.

From a technical standpoint, the European DSRC has many characteristics in common with the U.S. DSRC system. Both are based on the IEEE 802.11p standards, with similar frequency channels and mainly use periodic beacon messages exchanged among vehicles, pedestrians, and roadside sensors located along with transportation systems. Moreover, both DSRC systems support WAVE short message protocol (WSMP) for safety information and IPv6, UDP/TCP for non-safety information [45].

4.2 WAVE IEEE 1609.x Family of Standards

Wireless access in vehicular environment (WAVE) protocol stacks are a set of DSRC based standards which include IEEE 1609.0-4, and SAE J2735[2] message set dictionary [2]. WAVE is specially designed to achieve interoperability and cohesiveness among different vehicle manufacturers. Figure 4 illustrates a schematic view of

---

[2]SAE J2735 Standard includes aspects of defining message sets, data frames and data elements, mainly used by applications to exchange data over DSRC/WAVE based systems.

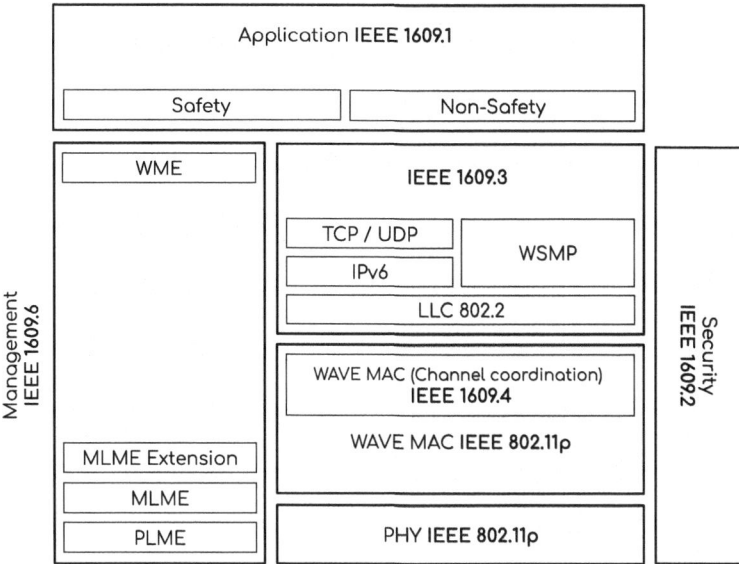

Fig. 4 Illustration of the DSRC/WAVE protocol stack. (WME WAVE Management Entity (WME): provides WAVE networking management services functions, for example, processing the service requests for the higher layers, providing channel assignments, etc., WSMP WAVE Short Message Protocol (WSMP): allows the exchange of data messages among vehicles. it supports rapid varying radio frequency (RF) environment where low latency is an essential requirement, MLME MAC sublayer management entity (MLME): provides channel coordination functions, which are important when there is one or more switching equipment with concurrent alternating operations on the control channel and service channel., PLME Physical sublayer management entity (PLME): performs management of the PHY layer functions in conjunction with the MAC management entity.) *Source* Adapted from [20]

the WAVE architecture. As can be seen, each layer has one or multiple protocols, where each protocol provides specific functionality. This architecture is mainly destined to be used for both safety-critical and non-safety ITS applications. The safety applications (e.g., emergency warning messages) are carried by the WSMP, which is developed to avoid excessive overhead. While the non-safety applications (e.g., traffic management and telematics) use conventional networking protocols, such as IPv6, UDP, and TCP. In general, the WAVE stack is based on the IEEE 802.11p in the lower layers (PHY and MAC) and IEEE 1609.x family of standards in the upper layers (1609.0 for architecture, 1609.1 for application services, 1609.2 for security services, 1609.3 for networking services, and 1609.4 for multichannel and enhanced distributed channel access (EDCA) mechanisms [5]. To ensure interoperability and reusability, it is imperative that DSRC functions properly, exchanges data the same way every time, and uses standardized messages. IEEE 1609.x family of standards was designed specifically to support V2V and V2I wireless interfaces, where each standard facilitates some part of DSRC operations.

4.3   IEEE 802.11p Standard

IEEE 802.11p standard was established as a newly approved amendment of the
IEEE 802.11 traditional standards to support highly mobile vehicular environments
with fast-moving nodes [28]. 802.11p is distinguished by a unique ad hoc mode,
random MAC addresses for privacy preservation, and IPv6 for routing in the network
layer. The unique ad hoc mode allows vehicles to communicate without the need to
establish a basic service set (BSS)[3] in a highly dynamic and fully distributed network
environment where association and authentication are not specified in PHY and
MAC layers but rather handled by the upper layer or the station management entity
(SME).[4] Based on these techniques, the delay incurred by vehicle communications
will be reduced effectively, especially if the vehicles are moving in opposing ways.
Moreover, IEEE 802.11p involves the enhancement of priority categories based on
802.11e and power control based on 802.11h. Prioritization and QoS for safety time-
critical data are addressed with EDCA using many contention windows (CW) size.

4.4   ISO CALM

The international organization designs the communications access for land mobiles
(CALM) for standardization (ISO) TC 204/ working group 16 to define the architectural communications framework of ITS [21]. This set of standards includes many
communication protocols and air interfaces to support a wide range of communication scenarios including different modes of communications (e.g. Vehicle-to-
Vehicle, Vehicle-to-Infrastructure, and Infrastructure-to-Infrastructure) and multiple
techniques of transmissions (e.g. Infrared, GSM, DSRC, Wi-MAX, Bluetooth, etc.).
CALM used an IPv6 convergence layer that separates applications from the communication infrastructure. A standardized set of air interface or access mode protocols
is presented for promoting the use of resources available for short, medium, and
long-range, safety-related communications, utilizing one or more of several media,
with the multipoint transfer.

The CALM model architecture is composed of seven abstraction layers:

– Access layer, which provides transmission media and protocols for communications.
– Network and transport layers, which provide end-to-end transport services for
  applications over several heterogeneous link technologies.
– Facilities layer, which provides application, information, and communication supports.

---

[3]Basic service set is a set of nodes that can communicate with each other within an 802.11 network
with the same medium access characteristics (i.e. Radio Frequency, Modulation Scheme, etc.).

[4]Station management entity refers to the devices controlling the data exchange between nodes of
the network (e.g. Infrastructure).

– Management layer, which includes station management functionality
– Security layer and application layer.

4.5 ETSI ITSC

The intelligent transportation system communication (ITSC) is a communication architecture designed by ETSI to support different transportation scenarios [11]. It is based on ISO/CALM architecture. ITSC is developed to support high mobility of ITS Stations (ITS-S), causing high dynamics topology changes. Besides, ITSC can support many communication technologies, e.g. Internet, public networks, and legacy systems such as 2G/3G/..., Bluetooth and GPS. ITSC enables different classes of prioritization which can be used to provide low latency to road safety applications while providing simple best-effort service to non-safety applications such as technologies for road tolling (DSRC). The main goal of this architecture consists of the development, testing, and deployment of ITSs in Europe based on vehicle to vehicle, vehicle to infrastructure, and vehicle to other road user equipment short-range communication for enhancing road safety and road efficiency. Complementary communication systems like the cellular network are considered where required. The ITSC contains four functional layers, namely, access layer representing OSI layers 1 and 2, network and transport representing OSI layers 3 and 4, facilities representing OSI layers 5, 6, and 7. While the applications layer in the architecture presents the ITS station services used to connect to one or more other ITS station applications.

5 IVCs Network Architectures

IVs act as mobile nodes. They can communicate with each other and stationary infrastructure, hence establishing an autonomous and dynamic mobile network. IVC can be classified into two different categories vehicle-to-vehicle (V2V) and Vehicle-to-Infrastructure (V2I) communications.

5.1 Vehicle-to-Vehicle Communication

V2I, also known as inter-vehicle communication, has been carried out to support vehicle communication. The V2V-based system enables direct or multi-hop communication among vehicles, which allow standalone vehicular communication with no infrastructure support. V2V provides a promising platform for a much broader range of large-scale, highly mobile applications. The prime motivation behind the V2V system is to increase efficacy and safety by avoiding potential accidents and provide the vehicle with different capabilities to exchange its position and speed and other relevant information with surroundings vehicles over ad hoc communication. Vehicles are equipped with different wireless interfaces and software, so-called respectively, onboard unit (OBU) and application unit (AU). The information exchanged with surroundings vehicles allows drivers the capacity to take action to avoid accidents and difficulties. The warning can be either through a seat vibration, tone, or visual display or combinations of these indicators. The V2V communication is based on dedicated short-range communications (DSRC) and has a signal range of more than 300 m and can detect dangers obscured by traffic at the earliest possible state.

The V2V communications can support different applications for road safety [8], for example, electronic brake warning (EBW), on-coming traffic warning (OTW), vehicle stability warning (VSW), and lane change warning (LCW). First, EBW allows alerting the driver when a preceding vehicle performs a severe braking manoeuvre. Then, OTW allows warning the driver of oncoming traffic during overtaking manoeuvres that require the driver to cross into the opposite lane of travel. After that, LCW allows warning the driver who is intending to perform a lane change when it is unsafe to do so. Finally, VSW allows warning of preceding vehicles that have recently required intervention by the vehicle stability control (VSC) system, etc.

5.2 Vehicle-to-Infrastructure Communication

Vehicle-to-Infrastructure, often called V2I, denotes the communication from a vehicle to existing roadside infrastructure or vice-versa. Many different access technologies can be provided to communicate between infrastructure. V2I system enables each vehicle to exchange data with infrastructure services through embedded devices in RSU, to improve road safety, traffic flow, environmental conditions, and social activities in different areas. The RSU is connected with the infrastructure through wired or wireless links, specified by the infrastructure operator. The infrastructure network includes much different networking access (e.g. GSM, LTE, UMTS) and wide ITSs technologies [14]. Generally, V2I communications are not limited to safety and efficiency applications; they can be used for a wide range of applications, including the audio/video downloads, real map navigation, Internet browsing, email access, etc.

V2I-based services can be carried using two different communication systems: Cellular and Wi-Fi-based systems. Because of the development of cellular communication systems for wide-scale deployment and low latency of long-term evolution (LTE), many approaches were published using cellular communication for different V2I-based applications. For example, as indicated in [27], LTE (e.g. evolution from 2G to 4G) can support more than 1000 user that attempts to communicate per second. Therefore, this capacity of processing and decreasing latency will satisfy the requirements of a large variety of safety-critical applications. Wi-Fi-based dedicated short-range communication systems (e.g. 802.11p and 802.11n) incur a delay of only a few milliseconds in most conditions (e.g. density). However, the coverage range of

Fig. 5 Qualitative analysis of cellular network (LTE) and Wi-Fi (802.11) for Vehicle-to-Infrastructure communications

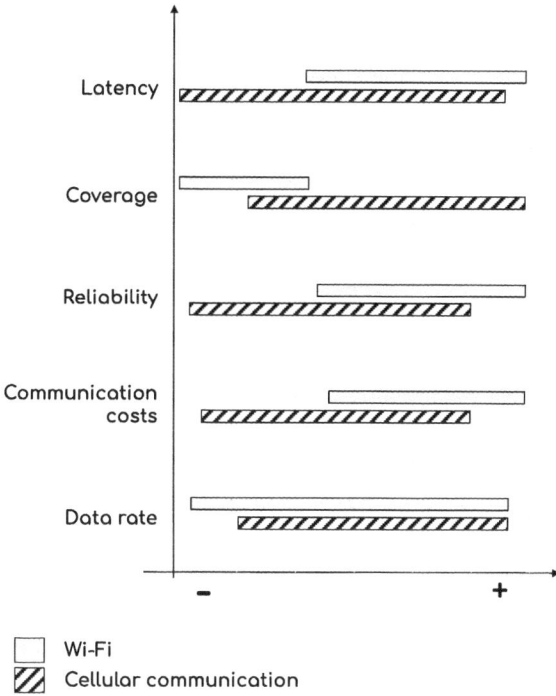

Wi-Fi is up to 200 m, which is smaller compared to cellular communication (up to several kilometres). This is due to the lower transmission power and a higher frequency of 802.11p (5.9 GHz). In contrast, the performance of IEEE 802.11p is higher, because the number of connected devices due to the lower coverage is smaller. In multi-hop scenarios, the coverage range of Wi-Fi communications can be increased, but there are some shortcomings, which affect the delivery delay of packets.

The accuracy and performance of both communication systems depend on the environment and the number of users in the wireless communication range. Figure 5 shows a qualitative analysis of these performance indicators. There are many schemes published in the literature that uses both cellular communication and IEEE 802.11p to establish communication between vehicle and infrastructure [3].

6   VCs Applications and Services

IVC technology tries to achieve efficiency improvements for transportation networks in smart city, by contributing to the solution of several relevant issues. The wireless connectivity between moving vehicles potentially supplies drivers with real-time pre-trip information, which can help to take immediate and corrective measures during the driving process. IV applications are not just limited to traffic management and forecasting systems, but also for road safety and efficient infrastructure usage. Because of its benefits and its endless uses, IV technology becomes indispensable elements and a multidisciplinary field of study. Therefore, many research societies and organizations around the world work together to explore and develop high-quality solutions for applications that involve IVC. These solutions are expected to have great potential to increase the efficiency and safety of ITS. IV applications can be classified into three major categories [23]: Road safety-related applications, efficiency applications, and comfort applications.

Road safety-related applications are mainly used to ensure the safety of the driver and the passengers including the vehicle itself. These applications are supposed to protect the lives of peoples by reducing the number of road accidents. Then, efficiency or traffic management applications play a critical role in ensuring the proper functioning and fluency of road traffic. For this reason, they catch high-interest today as much as in the future since traffic congestion is becoming a more and more serious issue affecting the development of smart cities. Finally, comfort applications aim to provide entertainment facilities and up-to-date information to passengers, which will make the driving experience more convenient. The examples of these applications, will be given in the following subsections.

6.1 Road Safety-Related Applications

The safety applications are highlighted as one of the most important applications of IV technology. They assist in reducing the risk of road accidents and any breach of legal requirements. Safety applications provide drivers with several information services that can help to avoid a collision in dangerous situations [18]. The IV is equipped with sensors to perceive their surrounding environment and with actuators to act in this environment. The perceived information is further disseminated using DSRC (Dedicated Short-Range Communications) to inform other vehicles and potentially available roadside units along its way [1, 23]. Most of these applications are required to broadcast safety messages periodically (e.g. every 100 ms), whereas other categories send event-driven messages when an event occurs. Generally, safety applications have a range of communication between 100 and 1000 m, and a minimum frequency between 1 and 50 Hz.

In response to such requirements, many researchers have been engaged in improving the efficiency and effectiveness of IVCs, for example, the accuracy of system measurements [35], as well as communication in urban environments [15], or concerning the performance and reliability of communications [44].

Road safety-related applications can be classified into two main categories [24]: Cooperative Collision Avoidance (CCA) and Emergency Warning Message (EWM).

The main goal of the CCA system is to avoid traffic accidents on the public road. Vehicles can efficiently warn drivers or even prepare for a dangerous traffic situation, for example, engage pre-crash functionalities like brakes and seat belts to reduce injury. For these reasons, most applications of CCA require stringent communication constraints to avoid packet losses and limit the latency and overhead. Whereas, in EWM applications, the vehicles exchange messages about collisions or hazardous road conditions to other vehicles travelling into a dangerous location. Generally, this type of application needs the warning messages to be kept between vehicles for a long time in the dangerous site.

6.2 Efficiency Applications

Another important class of IV is road efficiency applications. They have a key role in ensuring the proper functioning of the traffic flow and the efficient usage of the road network. Efficiency applications can be used to reduce congestion time at signalized intersections, enable smooth vehicular traffic re-routing to provide better traffic mobility, reduce air pollution in urban areas, and avoids economic losses. In addition to that, the traffic information provided by these applications helps the drivers to plan their time of travel and also in selecting the best path to the destination depending on the received information. These applications differ from safety-related applications in at least two ways. First, they are delay-tolerant applications that not safety-critical, using a carry-and-forward mechanism when partial disconnects from the vehicular network occur [24]. Second, they do not have hard Quality of Service (QoS) bounds. Hence, they are tolerant of packet losses and errors.

They are many potential applications for road traffic efficiency. One prominent implementation is road traffic forecasting. This application can provide relevant traffic information to forecast the traffic situation. The stored data can be used to plan and develop an efficient transportation network. The purpose was to predict where roads should be built or expanded in the future, or where control traffic signals will be needed [36, 40]. Road efficiency applications are typically focused on collecting different types of data [32]: traffic volume, vehicle classification, traffic speed, traffic density, travel time, and so forth.

As can be seen from Fig. 6, IVs are equipped with sensors that record information about environmental data and wireless communication devices that will allow them to cooperate with each other and with infrastructure. Thus, the collected information is sent to the traffic forecast centre (TFC) through roadside units (RSU) [41].

6.3 Comfort Applications

Comfort applications aim to provide the road traveller with information support and entertainment facilities to make the trip more pleasant. They can enable a wide variety of applications ranging from traditional IP-based applications (e.g., voice over IP, media streaming, web browsing) to applications specifically conceived for the vehicular environment (e.g., point of interest advertisements, automatic tolling services, maps download, parking payments) [10]. Most research in this area has focused

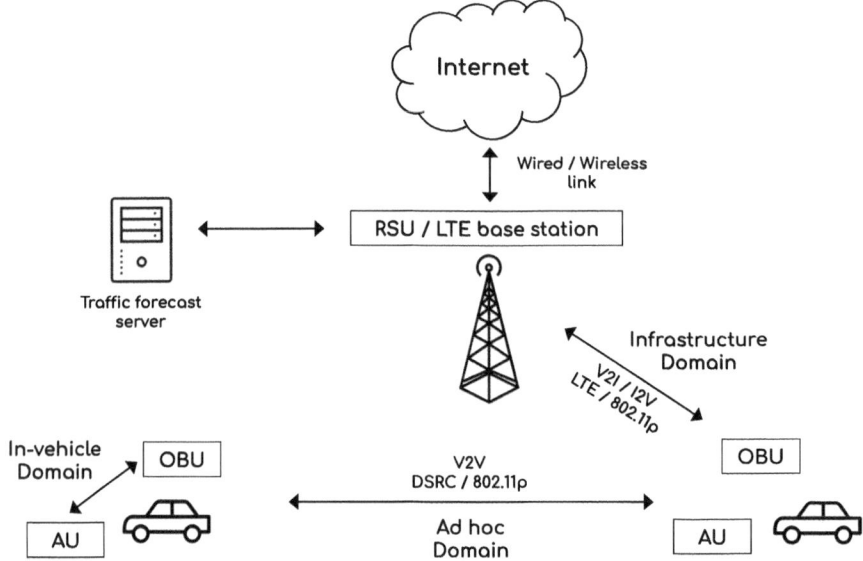

Fig. 6 Vehicular communication architecture: V2V and V2I/I2V communications scenarios based on DSRC, long-term evolution (LTE) or/and Wi-Fi

on Internet technology, which can provide vehicles with Internet access. Recently, several automotive manufacturers integrate Internet access in vehicles through cellular network infrastructures [4]. Generally, comfort applications require high-speed internet connectivity such as WiMAX technology [30].

7 Conclusions and Future Research Directions

7.1 *Conclusions*

One of the most critical difficulties facing the cities development today is private vehicles' societal and environmental impact. Private vehicles are the main reason for road traffic congestion and environmental pollution in big cities. Logistics Industries and Governments are doing everything they can to decrease private vehicle ownership by encouraging them to use other forms of transport systems.

The IV can change all that. A really intelligent, autonomous vehicle can have a drastically different result in its environment. Whether held by private citizens, private transport providers, or the city itself, an IV will have no need for parking, make much more effective use of road infrastructure, and is also likely to be electrically-powered.

IVCs are becoming a reality, driven by road safety needs and by the automotive industry and governments' investments. Their possibilities of new applications are developing very fast and include different types of services for various purposes. Nonetheless, they present many unique and new difficult research challenges ranging from event detection to data dissemination, making research in this field more relevant and attractive. Consequently, IVCs are receiving increased attention from the academic community and car manufacturers, where they are beneficial in providing services such as ITSs and driver and passenger assistant services. In this context, a vehicular network is emerging as a new category of well-known traditional mobile networks, spontaneously formed between moving vehicles and vehicles and infrastructure. They provide numerous useful applications for drivers and passengers, ranging from road safety-related applications to comfort applications. Besides, they are also promising for service providers and many industrial and telecom operators to create new business opportunities.

7.2 Future Research Directions

In such a network, there are still many significant challenges to deal with, which initiate different research topics, such as extend some proposals or consider further research developments. Based on profound studies and considerations, the possible contributions are as follows.

- The future evolution of IVs will indicate to vehicles the accidents, traffic jams, free parking spots in the city, and multiple other services. With this information, the driver assistant devices integrated into vehicles could indicate the drivers where the services are located. These applications could use V2V and V2I communications to get data about the services and communicate it to other vehicles. This would save lives, trip time, petrol, and reduce pollution, which allows the development of sustainable smart cities.
- The future evolution of IVs will indicate to vehicles the accidents, traffic jams, free parking spots in the city, and multiple other services. As an essential part of an intelligent transport system, IVCs produce large amounts of data, making traditional data systems capabilities inefficient, and cannot be used to meet the requirements of various vehicular network applications. This is because these systems do not foresee the fast growth of data and a large amount of redundant information, especially in dense traffic scenarios. Therefore, It is essential to implement big data in such networks, and then one can gain useful insight from a massive amount of operational data to improve traffic safety and transportation efficiency.
- Another interesting research topic is how to use existing long term evolution (LTE) networks to support both periodic and event-driven safety messages. To date, very few researches have studied this topic, and there are still several issues that need an investigation to determine whether or not this technology can be used for road safety and efficiency applications.
- IVs still suffer from many security vulnerabilities. Conventional security techniques are incapable of providing secure IV data sharing (IVDS). The major

issues in IVDS are data accuracy, trust, and data sharing reliability in channel communication. Blockchain (BC) technology is recently utilized to build trust and reliability in peer-to-peer (P2P) networks having similar topologies as IVDS. BC is a distributed system of records for transacting value in a decentralized and P2P network. Technically, BC is a combination of blocks linked by hash algorithms and secure under public-key algorithms. The researchers should improve the security of BC-used cryptographic algorithms, and they should correctly adopt the BC to ensure the security of IVDS.

References

1. Al-Sultan, S., Al-Doori, M. M., Al-Bayatti, A. H., & Zedan, H. (2014). A comprehensive survey on vehicular ad hoc network. *Journal of Network and Computer Applications, 37*, 380–392.
2. Ali, G. M. N., Chong, P. H. J., Samantha, S. K., & Chan, E. (2016). Efficient data dissemination in cooperative multi-rsu vehicular ad hoc networks (vanets). *Journal of Systems and Software, 117*, 508–527.
3. Atat, R., Yaacoub, E., Alouini, M. S., & Filali, F. (2012). Delay efficient cooperation in public safety vehicular networks using lte and ieee 802.11 p. In *Consumer Communications and Networking Conference (CCNC), 2012 IEEE*, pp. 316–320. IEEE.
4. Bayless, S. H.: (2011). Fourth generation wireless-vehicle and highway gateways to the cloud: An evaluation of long term evolution (lte) and other wireless technologies' impact to the transportation sector. Tech. rep.
5. Bettisworth, C., Burt, M., Chachich, A., Harrington, R., Hassol, J., Kim, A., Lamoureux, K., LaFrance-Linden, D., Maloney, C., & Perlman, D., et al. (2015). Status of the dedicated short-range communications technology and applications: report to congress. Tech. rep., United States. Department of Transportation. Intelligent Transportation.
6. Birchall, J., Morris, K., & Beach, M. (2017). M2m communications over LTE—evaluating energy consumption models. In *2017 13th International Wireless Communications and Mobile Computing Conference (IWCMC)*, pp. 536–540. https://doi.org/10.1109/IWCMC.2017.7986342.
7. Caliskan, M., Graupner, D., & Mauve, M. (2006). Decentralized discovery of free parking places. In *Proceedings of the 3rd International Workshop on Vehicular Ad Hoc Networks, VANET '06*, pp. 30–39. ACM. https://doi.org/10.1145/1161064.1161070.
8. Caveney, D. (2010). Cooperative vehicular safety applications. *IEEE Control Systems Magazine, 30*(4), 38–53.
9. Daimon, T., Makino, H., Mizutani, H., & Munehiro, Y. (2008). Study on safety assist information of advanced cruise-assist highway systems (ahs) using vics in blind curve section of urban expressway. *Journal of Mechanical Systems For Transportation and Logistics, 1*(2), 192–202.
10. Dar, K., Bakhouya, M., Gaber, J., Wack, M., & Lorenz, P. (2010). Wireless communication technologies for its applications [topics in automotive networking]. *IEEE Communications Magazine, 48*(5), 156–162.
11. ETSI/EN-302-665: 302 665 v1. 1.1. Intelligent Transport Systems (ITS) pp. 14–15 (2010). Intelligent Transport Systems (ITS) Communications Architecture.
12. Fallah, Y. P., Huang, C. L., Sengupta, R., & Krishnan, H. (2011). *Analysis of information dissemination in vehicular ad-hoc networks with application to cooperative vehicle safety systems, 60*(1), 233–247. https://doi.org/10.1109/TVT.2010.2085022.
13. Festag, A., & Hess, S. (2009). ETSI technical committee ITS: news from european standardization for intelligent transport systems (ITS)-[global communications newsletter] *47*(6), 1–4. DOI https://doi.org/10.1109/MCOM.2009.5116819.

14. Gil-Castiñeira, F. (2015). Opportunistic routing and delay-tolerant networking in vehicular communication systems. In *Vehicular Communications and Networks*, pp. 113–126. Elsevier.
15. Gozálvez, J., Sepulcre, M., & Bauza, R. (2012). Ieee 802.11 p vehicle to infrastructure communications in urban environments. *IEEE Communications Magazine, 50*(5).
16. Hartenstein, H., & Laberteaux, L. (2008). A tutorial survey on vehicular ad hoc networks. *IEEE Communications magazine, 46*(6), 164–171.
17. He, Z., & Zhang, D. (2017). *Cost-efficient traffic-aware data collection protocol in VANET, 55*, 28–39. https://doi.org/10.1016/j.adhoc.2016.09.021. URL http://linkinghub.elsevier.com/retrieve/pii/S1570870516302360.
18. Hedrick, J. K., Tomizuka, M., & Varaiya, P. (1994). Control issues in automated highway systems. *IEEE Control Systems Magazine, 14*(6), 21–32.
19. Hensher, D. A. (1991). Electronic toll collection. *Transportation Research Part A: General, 25*(1), 9–16.
20. Ide, C., Habel, L., Knaup, T., Schreckenberg, M., & Wietfeld, C. (2014). Interaction between machine-type communication and h2h lte traffic in vehicular environments. In *Vehicular Technology Conference (VTC Spring), 2014 IEEE 79th*, pp. 1–5. IEEE.
21. ISO/TC-204: Intelligent transport systems—communications access for land mobiles (calm)—architecture (2010). https://www.iso.org/standard/61570.html.
22. Karagiannis, G., Altintas, O., Ekici, E., Heijenk, G., Jarupan, B., Lin, K., et al. (2011). Vehicular networking: A survey and tutorial on requirements, architectures, challenges, standards and solutions. *IEEE Communications Surveys & Tutorials, 13*(4), 584–616.
23. Kenney, J. B. (2011). Dedicated short-range communications (dsrc) standards in the united states. *Proceedings of the IEEE, 99*(7), 1162–1182. https://doi.org/10.1109/JPROC.2011.2132790.
24. Kihl, M. (2009). *Vehicular network applications and services*. Vehicular Networks: Techniques, Standards, and Applications.
25. Kumbhar, A. (2017). Overview of ism bands and software-defined radio experimentation. *Wireless Personal Communications, 97*(3), 3743–3756.
26. Lu, M., Wevers, K., & Van Der Heijden, R. (2005). Technical feasibility of advanced driver assistance systems (adas) for road traffic safety. *Transportation Planning and Technology, 28*(3), 167–187.
27. Mangel, T., Kosch, T., & Hartenstein, H.: A comparison of umts and lte for vehicular safety communication at intersections. In *Vehicular Networking Conference (VNC), 2010 IEEE*, pp. 293–300. IEEE.
28. Mao, J., Mao, Y., Leng, S., & Bai, X. (2009). A simple adaptive optimization scheme forieee 802.11 with differentiated channel access. *IEEE Communications Letters, 13*(5), 297–299.
29. Mietzner, R. (2007). Cvis-cooperative vehicle-infrastructure systems. *COM Safety: Newsletter for European ITS Related Research Projects, 3*(5).
30. Mojela, L. S., & Booysen, M. J. (2013). On the use of wimax and wi-fi to provide in-vehicle connectivity and media distribution. In *2013 IEEE International Conference on Industrial Technology (ICIT)*, pp. 1353–1358. IEEE.
31. Najm, W. G., Koopmann, J., Smith, J. D., & Brewer, J., et al. (2010). Frequency of target crashes for intellidrive safety systems. Tech. rep., United States. National Highway Traffic Safety Administration.
32. Olariu, S., & Weigle, M. C. (2009). *Vehicular networks: from theory to practice*. Chapman and Hall/CRC.
33. Oyama, S. (2009). Activities on its radio communications standards in itur and in japan. In *1st ETSI TC-ITS Workshop* pp. 1–22.
34. Regan, M. A., Oxley, J., Godley, S.,& Tingvall, C. (2001). Intelligent transport systems: Safety and human factors issues. 01/01 (2001).
35. Shivaldova, V., & Mecklenbrauker, C. F. (2013). Real-world measurements-based evaluation of IEEE 802.11 p system performance. In *Wireless Vehicular Communications (WiVeC), 2013 IEEE 5th International Symposium on*, pp. 1–5. IEEE.

36. Skordylis, A., & Trigoni, N. (2011). Efficient data propagation in traffic-monitoring vehicular networks. *IEEE Transactions on Intelligent Transportation Systems*, *12*(3), 680–694.
37. Sommer, C., Tonguz, O. K., & Dressler, F. (2011). *Traffic information systems: efficient message dissemination via adaptive beaconing*, *49*(5), 173–179. https://doi.org/10.1109/MCOM.2011.5762815.
38. Spivack, A. (1999). Fcc allocates spectrum in 5.9 ghz range for intelligent transportation systems (its) uses p. 1.
39. Stevens, W. B. (1996). The automated highway system program: A progress report. *IFAC Proceedings Volumes*, *29*(1), 8180–8188.
40. Venkata, M. D., Pai, M. M., Pai, R. M., & Mouzna, J. (2011). Traffic monitoring and routing in vanets–a cluster based approach. In *2011 11th International Conference on ITS Telecommunications*, pp. 27–32. IEEE.
41. Wietfeld, C.,& Ide, C. (2015). Vehicle-to-infrastructure communications. In *Vehicular Communications and Networks*, pp. 3–28. Elsevier.
42. Williams, M. (1988). Prometheus-the european research programme for optimising the road transport system in Europe. In *IEE Colloquium on Driver Information*, pp. 1–1. IET.
43. Willke, T. L., Tientrakool, P., & Maxemchuk, N. F. (2009). A survey of inter-vehicle communication protocols and their applications. *IEEE Communications Surveys & Tutorials*, *11*(2), 3–20.
44. Yao, Y., Rao, L., Liu, X. (2013). Performance and reliability analysis of IEEE 802.11 p safety communication in a highway environment. *IEEE Transactions on Vehicular Technology*, *62*(9), 4198–4212.
45. Zheng, Q., Zheng, K., Sun, L., & Leung, V. C. (2015). Dynamic performance analysis of uplink transmission in cluster-based heterogeneous vehicular networks. *IEEE Transactions on Vehicular Technology*, *64*(12), 5584–5595.

Applying Mobile-Based Community Participation Model in Smart Cities

Prafulla Parlewar

Abstract The mobile-based community participation model is a mobile app for citizens to solve everyday problems. The project involved development of mobile app for smartphones integrated with e-Governance for Nagpur Municipal Corporation in Central India as a part of the Smart Cities initiative. The model is an innovative mobile-based system for a multi number of task involving community participation. It provides effective solutions for reporting civic problems to the local authorities. The application of a similar model can solve urban problems and promote community participation in smart cities. Furthermore, this chapter investigates the question on what is a Smart City? How to develop a mobile app for promoting community participation? How to develop a cloud-based model that could be integrated with the existing system of e-Governance? How smart technology is developed by integration of Geographic Information System (GIS) and Global Positioning System (GPS)? The mobile-based community participation model has many benefits. Because of its benefit, it can be easily applied to smart cities.

Keywords Mobile app · Smart cities · e-Governance · Community participation

1 Introduction

Cities around the world are developing smart city technology for improving the urban environment. What is a Smart City? A smart city is an area where smart technology provides solutions with integration of information technology and artificial intelligence. However, urbanization in cities does not relate to only smart technology. To make liveable cities, top to bottom and bottom to top approaches are used by city planners. In this approach, community participation provides bottom to top approach for making citizen-centric urbanization. One such project was developed for Nagpur Municipal Corporation (NMC) in central India know as mobile-based community

P. Parlewar (✉)
City Development Corporation (P) Ltd., Mumbai, India
e-mail: citycorporationindia@gmail.com

© The Author(s), under exclusive license to Springer Nature Switzerland AG 2021
U. Ghosh et al. (eds.), *Machine Intelligence and Data Analytics for Sustainable Future Smart Cities*, Studies in Computational Intelligence 971,
https://doi.org/10.1007/978-3-030-72065-0_6

participation and information system. The application for mobile-based community participation system is developed as a participation and information system for citizens to solve everyday problems. The project involved the development of a mobile app for Android and iPhone integrated with e-Governance. The system is a mobile app for a multi number of tasks with community participation. It provides effective solutions for reporting civic problems to the local authorities and community. The application of similar systems can solve urban problems and promote community participation in smart cities. Furthermore, this chapter investigates the question what is a smart city? How to develop a mobile app for promoting community participation? How to develop a cloud-based model that could be integrated with existing system of e-Governance? What is smart technology? How smart technology is developed by the integration of Geographic Information System (GIS) and Global Positioning System (GPS)?

The system is a mobile app to increase productivity, information sharing and allows municipal corporations to closely participate with the citizens. It allows residents to take pictures of code enforcement violations, attach a description, and submit the information as a request to their city authorities. Furthermore, the mobile app is integrated with the e-Governance of the city to allow citizens to participate in city planning and activities. It facilitates citizens to even apply for registration, pay taxes, get licenses, track applications and various other features of e-Governance. This system is easy to use, scalable and customizable for smart cities. Because of its benefit, this model can be applied to various smart cities.

Many literature give valuable information on development of the mobile apps for citizens services. Badii [1] have presented innovative tool for developing smart city web and mobile apps. Mobile apps literature is available for various topic like urban services, mobility, age-friendly, augmented reality etc. Jog [2] indicated that the key success factors for implementation smart city solutions, will largely depend upon the level of citizen participation and project management capabilities. Also, it is indeed important to investigate the design process for mobile applications in the context of smart cities [3]. Another research indicate, the practical contribution of smart city development by presenting a mobile tool for citizens containing a set of digital services needed in the everyday life of the city [4]. Almao [5] have presented the efforts from the industry to implement age-friendly guidelines in the design of mobile apps for older people. The smart city mobile apps requires user-friendly design, community participation and integrated solution with e-Governance. Moreover, to make the smart city mobile apps successful, it is important to have design built on open standards.

2 Smart Cities

Smart cities are defined as geographic area of human settlement in which society plans, designs and execute citizens urban needs with the technology which takes it's own decision. City is a geographic urban area where citizens undertake social, cultural, and economical progress in the process of urbanization. Generally, cities include complex urban functions. It have complex system of urban housing, public transportation, public utilities, and commercial areas. All of the components of this fabric requires smart solutions. Moreover, these components in smart cities are related to the material infrastructures of cities, with human and intellectual capital [6]. Importantly, a smart solution is the use of technology in which the system undertakes decisions with its own capabilities. Thus, smart cities provide efficiency for the progress of citizens in the process of urbanization.

The smart cities concepts includes: (a) smart society, (b) technology, (c) governance, (d) energy, (e) mobility and (f) sustainability. The smart society in a city provides the citizen's services, social innovations and regeneration. The local government is primarily responsible for the citizen's services related to housing, infrastructure and utilities. Traditionally, to avail these services, citizens will file the request either in physical form or through telephone. In everyday life, there are many types of services that require on-time solutions like electrical, water supply, sanitation, registrations, tax payments, building plan approvals, tree cuttings, licenses etc. Many times these services are delayed and citizens suffer in spite of payment of taxes against the services. To maintain the quality of life, it is important for the local government to deliver service on time. Some other important parts of smart society are innovative solutions for community planning and participation.

The basic approach for city planning are two way: (a) top to bottom approach, and (b) bottom to top approach. Generally, top to bottom approach includes master planning for cities where land is controlled through various statutory obligations on the citizens. One of the significant ways to solve the city problems are bottom to up approach. In this approach, the citizens are involved in the planning, execution, and management of the cities. This process is known as community participation. In a global context, particularly in the USA and Japan, cities apply bottom to top approach for city planning. In Japan, Machizukuri is a process in which community participation provides effective solutions for solving problems in old and newly planned cities. Similarly, in the USA community participation is developed through a variety of innovative models to make the cities efficient. Social innovation is found in the USA with the use of a crowdsensing project developed by MIT Senseable City Lab. One such crowdsensing mechanism is used to understand the dilapidation of urban infrastructure like bridges in Boston, MA [7]. The smarter society also looks into making the city liveable, improving education, empowerment, civic engagements, and community-based regeneration.

Technology for smart cities is developed in fields like geo data, internet of things, big data, network, sensors, security, citizens safety, data science, services integration, e-Governance, mobile apps visualization etc. Geographic Information System (GIS) is particularly important for the local government to understand the complexity of the urban housing, infrastructure and utilities. GIS provides a framework for storing, mapping and analyzing the complex city data. The data related to landuses, water supply, sanitation, electricity etc. can be stored through the GIS technology and analyzed further for providing better services. Globally, many cities are using GIS technology for better solutions to manage land-use, infrastructure and utilities.

Mobile apps have come up in the last two decades. Particularly in the USA, mobile apps provides citizens services easily accessible on android and iPhone. These smart cities applications support the future vision of cities, which aim at exploiting ICTs, namely internet of things technologies (IoT), for value-added service delivery [8]. Significantly, it has provided better services for local authorities. Internet of Things (IoT) indicates objects or devices with embedded electronics that are capable to transfer data over the network without any human interaction. It can be effectively used in cities for security and supply chain complexity. Big Data is to collect, store and process the data. It uses various applications for analyzing, transferring, capturing and searching the data. Some of the smart ways of the use of technology can be with solid waste management, smart water supply systems, smart signages, surveillance etc. The Geomatics Laboratory of the Mediterranea University of Reggio Calabria has developed app to perform a virtual tour with the 3D backdrop model as the main scene and to interact with the surrounding environment [9]. Also, Bostons Street Bump smartphone app, uses a phone's accelerometer to detect potholes without the need for city workers to patrol the streets [10].

Good urban governance is key to the better development of the cities. Some of the smart systems required for the cities are the e-Governance system, economic and fiscal system, efficient urban procurement, transportation management, etc. Various advances are found globally in the e-Governance system for property management, amenities management, disaster management and utilities. Governments around the world are turning cities into smart ecosystems [11]. In the Bandung City of Indonesia, implementing the smart economy model had improved economic governance effectively for market traders [12].

Energy plays an important role in the development of cities. Energy conservation has become now a necessity and global cities are investing to conserve natural resources. The smart ways to conserve energy are in field of solar power, wind power, use of biomass and recycling waste. A considerable number of solar power plants are using the smart grid for efficient use of the energy. Nowadays, many cities have adopted smart metering and electrical supply system. The energy efficiency in cities can be achieved by active and passive approaches which energy efficiency of cities and may allow a sustainable city life [13]. In South Korea, the U-Eco City project use pervasive computing to measure various environmental indices to collect data and improve the urban environment [14].

Smart mobility for cities is also found in various cities. Intelligent Transport System (ITS), public transportation, electric vehicles, parking, urban cycling, self-driving cars etc. can use smart technology. ITS embedded with sensors can identify speed, direction and occupancy information. In the town campus of the University of Bahrain, a smart solution is developed for parking crisis in congested areas [15]. The toll payment collection, flow of toll, utility payment, parking payment etc. can use smart technology. In Dubai, a smart toll system SALIK (Arabic meaning is "clear and moving") is a road toll collection system that is completely automated through the use of ICT [21]. Smart Traffic Management is a system to monitor and controls the traffic signals using sensors to regulate the flow of traffic and to avoid the congestion for smooth flow of traffic. Prioritizing the traffic like ambulance, police etc. is also one application that comes under smart traffic management [16].

3 City Planning at Crossroads

The city planning is the science of making human settlements where citizens achieve social, economic and cultural progress. City planning involves interdisciplinary fields like architecture, geography, engineering, sociology and GIS. It involves planning for the geographic area where land and infrastructure provide a liveable environment for citizens. This liveable environment is developed by controlling land and making infrastructure to provide accurate services for citizens. A city is comprised of housing, commerce, transportation, services and utilities. All these components require accurate planning and management. The housing has a nexus of various economic groups which lives together with social and cultural ethos. How can we make good neighborhood? Does the smart cities concept make a good neighborhood? Recent smart cities solutions for the management of neighborhoods are management of parking and neighborhood safety. Transportation, infrastructure and utilities are important parts of any city. A good transportation system is planned with accurate engineering skills with the latest technology. Some of the smart cities are applying various technologies like highway management and toll management. Infrastructure in cities includes social and physical infrastructure. Generally, social infrastructure include parks and gardens, open spaces and cultural facilities utilized in everyday life by the citizens. The innovative solutions for the management of these facilities with smart technology are essentially required for smart cities. Physical infrastructure includes the water supply, sanitation, solid waste disposal and drainage systems. All these require accurate solutions for planning and management. Utilities are mainly the electrical and telecommunication system.

Community Participation is a subject under city planning in which the cities are planned in the bottom to top approach. Original concepts of community participation were primarily motivated by environmental and societal challenges as well as community participation in governance [17]. However, with changes in technology community participation is possible through the use of cloud services, GIS and many other information technology infrastructure. Successful community participation is the design of a well-integrated work process [18]. Moreover, it is a process of giving some degree of control of a development agenda to a community or a group [19]. Importantly, community participation can generate trust, credibility and commitment regarding the implementation of policies [20].

In general, cities across the globe have a variety of development. The smart city model in countries like the USA and Japan are developed to improve the quality of life. Presently, many cities in these counties are developing smart city models for a better quality of life. On the contrary, cities in developing countries are found to be at crossroads with knowledge of smart technology. Particularly, in South Asia, liveability is more important than smart cities. However, many cities are adopting models of smart cities. Moreover, in developing countries application of the smart technology is more challenging than the developed counties due to the lack of funds and inadequate political and administrative support. In these cases, it is important to have cost effective smart city solutions with good political will to support the implementation of smart city projects.

4 Mobile-Based Community Participation Model

The local government in Nagpur, India planned to design mobile app using a comprehensive community participation infrastructure, constituted to address multiple everyday problem faced by the citizen. The project provided integrated services to citizens in a transparent, effective and efficient manner to bring about complete levels of community satisfaction. The primary goal of this project was to provide services to citizens through an online and easy-to-use mobile service delivery channel. Also, the goal was to ensure accessible, convenient, transparent and timely delivery of services.

The mobile based community participation model for smart cities is based on following components: (a) internet, (b) mobile app, (c) community participation, (d) e-Governance server, (e) GIS server, and (f) cloud servers (Fig. 1). These components interface in real-time with local government and community with mobile and desktop HTTP REST web services. Representational state transfer (REST) services provide interpolation between the internet based system of predefined operations. The main aim of this model is to promote the community participation by giving simple interface on mobile phone with real-time design to interact with local government and community. One of the important advantage of this model is ease of installation for any desired city due to its system architecture.

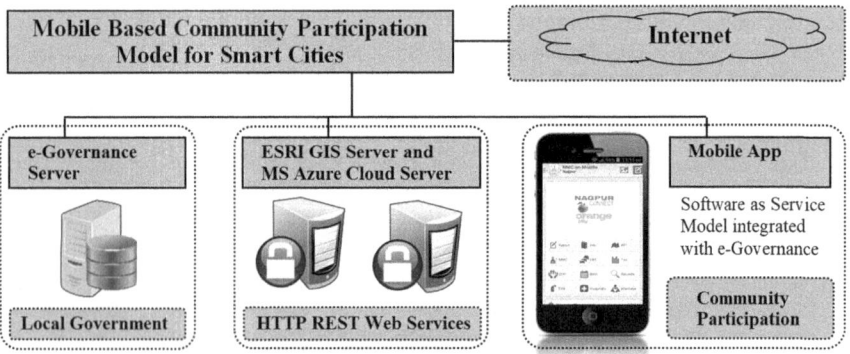

Fig. 1 Mobile based community participation model

4.1 Concept

The project concept involved components like mobile-based service delivery, community participation, information sharing and integration with e-Governance (Fig. 1). The changes in the technology of mobile application provide local government an avenue to offer as many as possible services on the mobile phones to improve services and business on the mobile phone. The proposed model is developed to adapt to emerging technologies and facilitate easy upgrades. Moreover, the model will incorporate more and more services with the emerging technology. This model will be integrated with the mobile service delivery gateway.

The concept looks into citizen participation in reporting and availing various NMC services. The model focuses on resolve citizens issues like road potholes, garbage collection, water supply, sanitation, illegal hoardings, non-working street lights etc. using their mobile devices. The model is proposed to provide information and access to e-Governance services to citizens like procedure and forms, filing applications online, tracking status, registering for birth or death, and paying taxes etc. To meet the requirements of the specified service levels, the backend offices of the NMC was reorganized to become capable to address issues. The following are the salient features of the proposed system of processing of applications: (a) All the reporting activities faced by the citizen will be handled using the administration dashboard. The staff will be assigned to look into the dashboard to see what issues are reported and route the issues accordingly for resolution, and (b) The present system of NMC do not facilitate easy delivery of services to citizens. So, control room with back office staff was established to take care of complaint or issues.

4.2 Service Oriented Approach

The model is developed with a service oriented approach for the NMC. So, the app was designed to deliver the citizens services by enforcing the Service Level Agreement (SLA) with the contractors appointed by the NMC. This integration is developed by the XML and Web service technology. The NMC mobile app work in programmed sequence: (a) massaging, (b) routing of massage, (c) processing of information and (d) transformation of data processed into rich (XML) documents. The present and proposed e-Governance application were integrated with solution of XML with minimum changes. The meta data is represented in XML format. The model will work in online as well as offline mode even without connectivity with internet. It was further supported with Indian language supports like Hindi, Marathi and others with Unicode support in citizen mobile OS.

The e-Governance modules like payment of taxes, birth registration, death registration, payment of water charges, tree cutting permissions, hospital registrations, building plan approval and other services were interfaced with the mobile app model. The model was developed on open standards, for the delivery of many set of services. Moreover, the objective of the mobile app was to develop smarter urban system to provide citizens solution in real-time. The app was an open-ended platform for integration of GIS, urban services, infrastructure development and good governance.

The important part of the system architecture is the scalability of system. As the users increases on the mobile app the system architecture deliver good performance. So, the model works on the concept of scale-up and scale-out in web and data base server. Also, it performs same on the GIS and integration servers. The system architecture scale-up and scale-down is achieved with high number of users and integration of backend system.

4.3 Database Design and Data Transmission

The mobile-based civic participation model data base design is developed on the cloud server. The data transmission and data confidentiality is developed as per the requirement of NMC. Following are the requirements developed in the system architecture: (1) all the data elements maintain standards data definitions in the system, (2) the design intelligently routes the information to required applications and (3) during the transmission of data, and system maintains the security of data. The confidentiality and ownership of the data generated by user or accessed by the user area stored in the NMC server.

4.4 Mobile Interface

Mobile technology is upgrading continuous with 4G to 5G bandwidth. The proposed model is based on latest Wireless Application Protocol (WAP), Hypertext Transfer Protocol (HTTP) and Transmission Control Protocol (TCP). This technology uses development of multiple layer application to facilitate user interface with the mobile. Due to the multiple data layers, this provide easy human and mobile interface. The mobile application is designed to provide interface for authentication, cache data, protect government data, and allow data accessibility to only particular users. The mobile application is integrated with the main e-Governance application. There is a facility to PUSH through and PULL through a mechanism to get and receive information using mobile service. It supports basic devices which run Android, Blackberry, iOS and Windows operating systems. The app have an ability to run on 4th and 5th generation networks.

5 Project Profile

The system is a platform (mobile app) to increase productivity, information sharing and allows municipal corporations to get closely participate with the citizens. It allows residents to take pictures of code enforcement violations, attach a description, and submit the information as a request to their city authorities. The solution is an easy way to notify the authorities about all types of code violations. With a few clicks, citizens can report and monitor problems in their community using their mobile devices. The solution makes it easy and accessible for everyone to communicate concerns to Municipal Corporation in real-time. As the usage of mobile phones is becoming the norm, the future is about mobile reporting and usage of GIS technologies.

The app uses smartphone GPS to identify the location. Citizen can fill information and upload picture to the issue or complaint with detailed information for the city government. This reporting can be used for submitting issues of street pot holes, street lights, garbage, accidents, broken pipelines, water leakages, and many more. They can track, follow, and see details of complaint and issues reported by them to the city. Citizens can create their own geo-fence to receive emails on issue in owns neighborhoods.

With the multi number of features, municipal administrators can have complete control over how they communicate crucial city information to citizens. By adding notifications to map widgets as well as directly emailing registered citizens, they can directly get out information when they need to. From street closings to meeting notices, make sure everyone is informed. This platform empowers everyday citizens to use their smartphones to make their communities a better place through participation.

6 Design of Mobile App

The mobile app is built as a Software as a service (SaaS) model. Software as a service (SaaS) model is cloud hosted which facilitates the on demand software services. The proposed model has two parts: (a) Mobile application and (b) Web portal or Administration Dashboard. The design of the app has following parts: (a) citizens participation, (b) city management, (c) reporting, (d) target, (e) customization, (f) community, (g) local government, (h) media and (i) open API.

(a) Community Participation: The app promotes quality of life in communities by engaging them in the decision-making process. It connects community and government through mobile phones by using Geographic Information System (GIS) in real-time. The basic module of platform provides solutions to problems like garbage, potholes, parking violations, broken sewer or drains, bad roads, encroachments etc. In the advance module, it can be customized to a variety of urban problems like regeneration, traffic and transportation, land use control, building violations, encroachments, project management etc. It provides various solutions to civic engagement and community participation. City official can make vice versa dialog with citizens on mobile phones on city problems, planning and execution. This platform understands the citizen's needs to find a sustainable way to address problems and provide an opportunity to serves as a common ground, raise questions and explore alternative solutions with new social consciousness.

(b) City Management: City administrators can streamline projects easily on their smartphones. The app provides advanced tools of project management linked with the Geographic Information System (GIS) database in cities. City administration can share information to promote campaigns, news, and alerts to citizens. It can even run community program to make real-time community participation. The app can reduce the communication gap between citizens and city administrators by use of real-time technology.

(c) Reporting: The app allows real-time reporting of issues in cities. Citizens can report an issue of everyday problems directly to the city administration. These reports are routed to the city administration for fixing problems in the least possible time through the cloud server. All these reports are integrated with Geographic Information System (GIS) data and e-Governance data (Figs. 2, 3, 4, 5, 6, 7 and 8).

Fig. 2 (1) App flash Screen for iOS OS (2) Home Screen for iOS OS and (2) Issue page for iOS OS

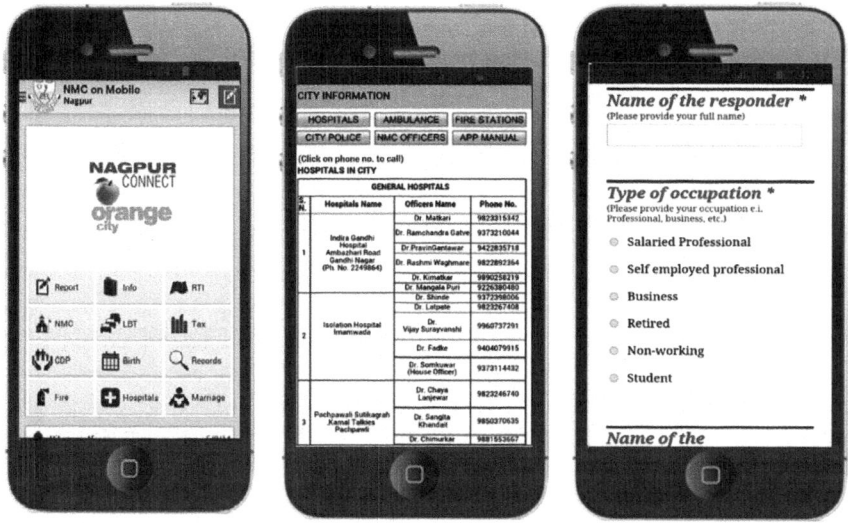

Fig. 3 (1) Home screen on android OS, (2) Contact search screen and (3) Master plan questionnaire

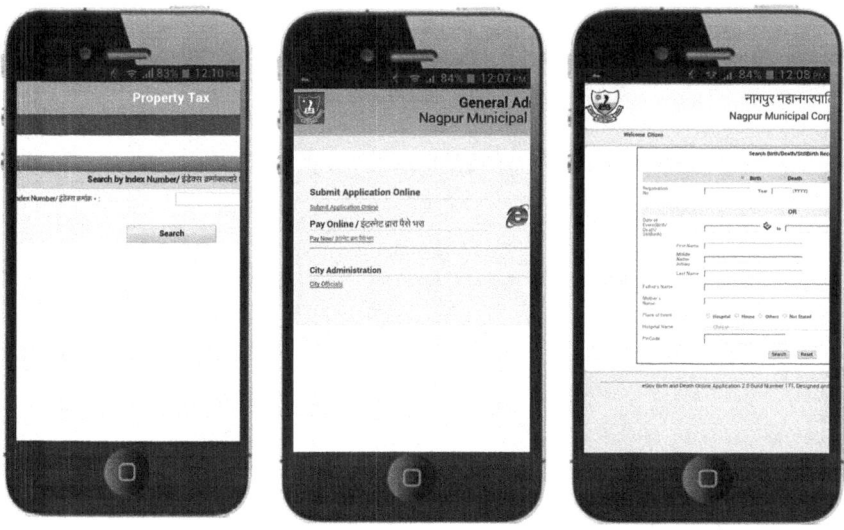

Fig. 4 (1) Property tax payment, (2) Fire NOC and (3) Birth registration

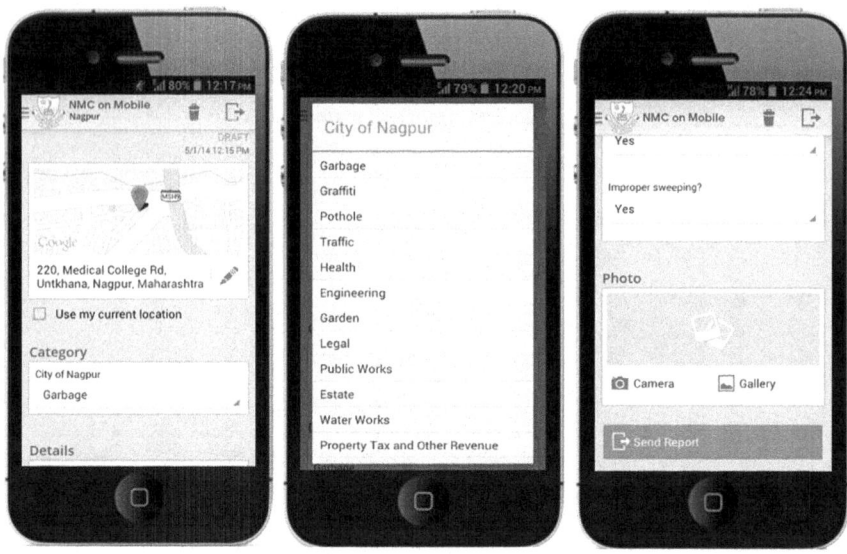

Fig. 5 (1) Issue reporting home page, (2) List of issues and (3) Issue description and photo upload

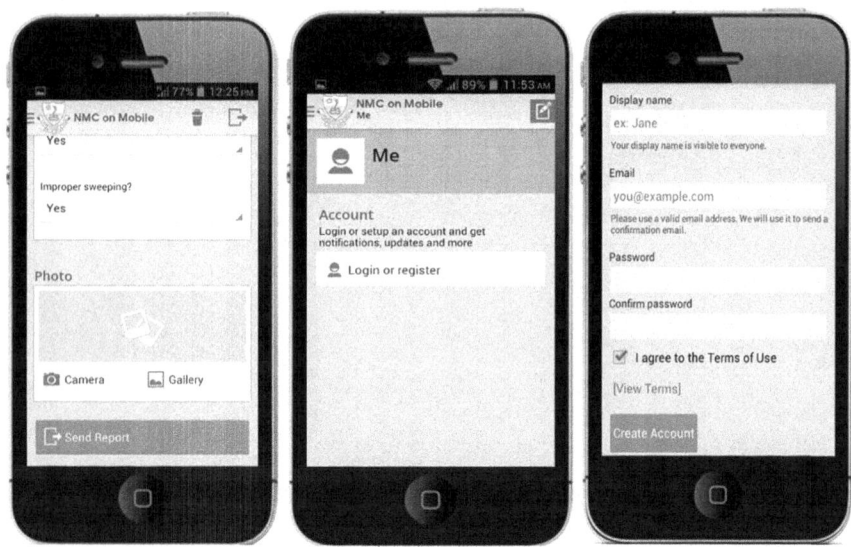

Fig. 6 (1) Send report, (2) Login or registration and (3) Create account

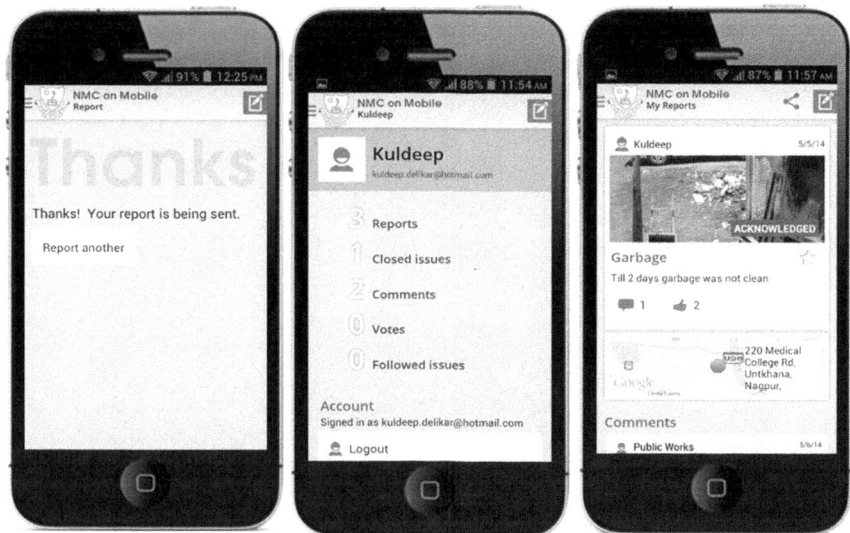

Fig. 7 (1) Report submission, (2) Account information and (3) Acknowledgement of issue

(d) Target: The app is based on latest technology of Geographic Information System (GIS) which locates exact location of an issue. It uses GPS and a camera in a smartphone to identify the exact location. Such accurate data helps city officials to understand the problems easily. Once such real-time data is stored, decision-makers use it for analysis. These targets can be geo-fenced to a vertical perimeter for a real-world geographic area. The app geo-fence can be predefined to city or neighborhood boundary.

(e) Customization: It is an intelligent platform design for integration with internet-scale communities with enterprise quality. It use modern tools of computer learning, user behavior and multi-platform integration. It can be customized to various solutions in cities. Some of the examples are: (i) City Project Management (CPM): City administration can easily control and streamline their projects with powerful project management tools. This improves transparency and accountability of city administration. (ii) Community planning and Participation (CPP): Cities can regenerate communities and encourage community participation in the decision-making process. For example, Central Business Districts (CBD) is geo-fenced to improve business process though information sharing and civic engagement.

(f) Infrastructure Project Management (IPM): Infrastructure companies can improve businesses by organizing projects on the platform.

(g) Community: Citizens and communities are the main focus of the platform. It is available free of cost to them. They can use app as open platform for various issues in cities.

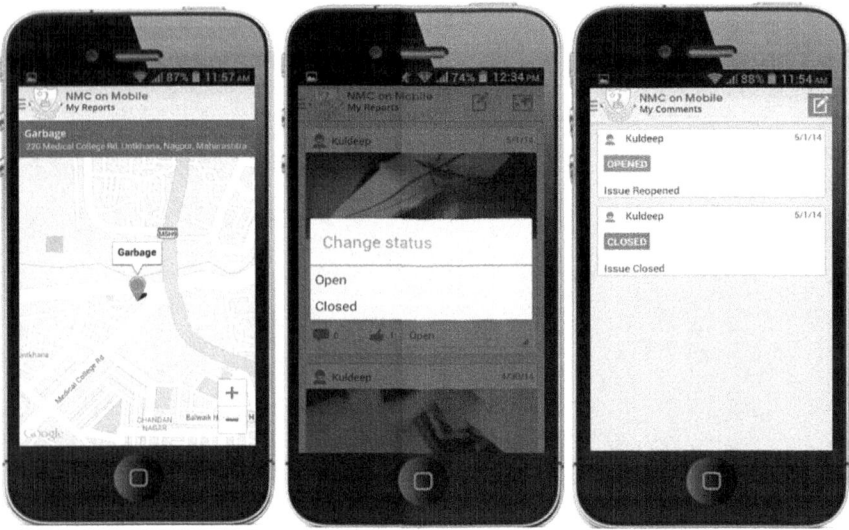

Fig. 8 (1) Issue on map, (2) Open and closed issue and (3) Issue closed screen

(h) Local Government: City governments can use the app to administrate city problems through real-time reporting. This can be branded with city seal. This app
can also have a custom-made web interface through a map widget. Moreover,
it can be integrated easily with Facebook, e-Governance, GIS and various other
applications used by city administration.

(i) Media: Media can leverage this technology to collect everyday data online that
can be used for accurate analysis of city problems. It can plan a significant role
in spreading awareness of local problems that are not solved by the Government.

(j) Open API: Software developers can use platform API as an open platform. They
can extend its core function to various solutions. It is an open-ended platform that
can be used in multi-sectors like infrastructure, construction, urban regeneration,
traffic and transportation, vehicle fleet management, business management etc.

7 Features of Mobile App

The key feature of the mobile app are web portal administration dashboard, zones and
service requests, new service request, issues on map, data exports, canned messages,
email messages, issues by source, issues by service request, issue by employee,
acknowledge issues and generate a work order (Figs. 10 and 11).

(a) Web Portal Administration Dashboard: The web portal administration dashboard
is used by the NMC officer. NMC designated officer can search the issue by the
service request, assignee, status, geography and keyword. It allows the admin-

Fig. 9 (1) Community sharing, (2) Viewing issues and (3) Issue on map

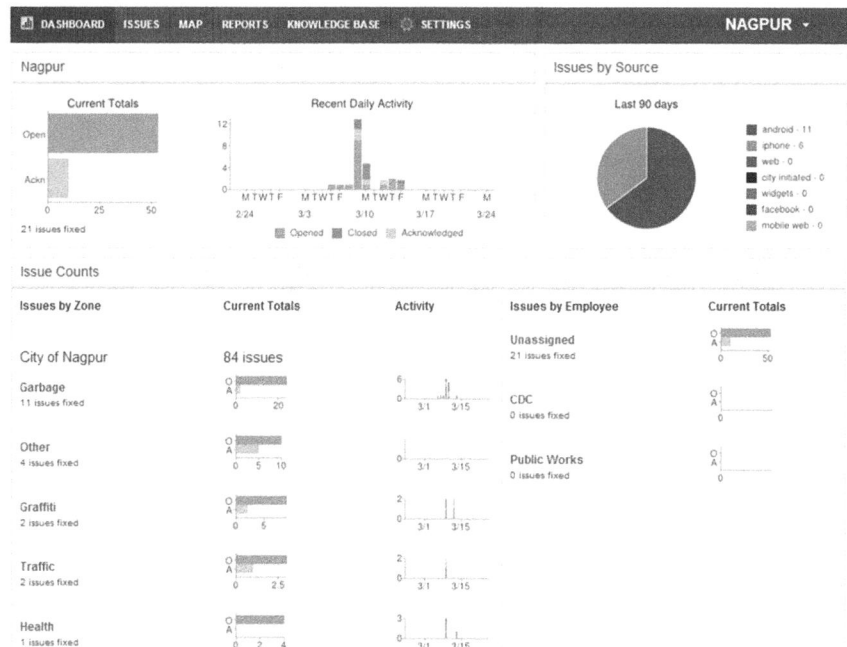

Fig. 10 Dashboard showing reports graphically

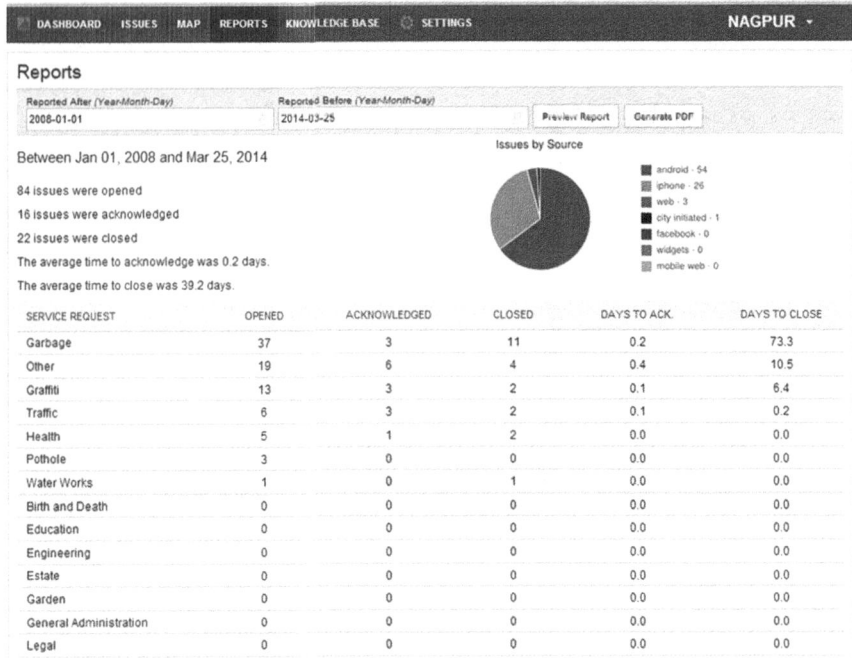

Fig. 11 Dashboard showing reports

istration to view the request details such as pictures or answers to secondary
questions. The administrator can update the status of issues to acknowledged
and closed. Further, a work order can be printed and issues can be assigned
to other users on municipal government accounts. Even it is possible to re-
categorize issues submitted under the wrong request type. Not least but there are
many other features found in the administration dashboard.

(b) Zones and Service Requests: Under the Zones and Service Requests tab citizen
see his city/town/state listed under Zones. When an account is created, the system
associated the exact boundaries of your service area to a citizen's account. In the
pop-down menu citizens can see several service request (SR) types configured by
the city. These service requests give residents the option of choosing a category
when they report issues within city limits via website, widgets, mobile apps and
Facebook app.

(c) New Service Request: The municipal administration Dashboard one can make a
new service request a title such as Health or Sanitation. Now, one can notify the
internal staff about new reports submitted under this request type by entering one
or more email addresses to route notifications to the concerned staff. Multiple
emails can also be forwarded to various staff. Significantly, the system here
used the geo-fencing feature to route the issue to the concerned person. To solve
an issue municipal administration can set service level agreements (SLA) by

request type to track the progress of resolution. SLA times can be set in hours, days and weeks. Moreover, municipal administration can configure automatic assignment to instantly assign all issues reported under your new service request to one person or department associated to your web portal government account.

(d) Issues on Map: The key part of the mobile app is the display of all the issues on the map of the city. A high-level view of all the issues is displayed and integrated with the city's website. A search capability is also provided to the administration of citizens.

(e) Data Exports: Citizen can give the report a title such as City of XX Weekly Report. One can choose his email frequency (daily, weekly, monthly). One can list email addresses, he would like the reports routed to, separated by commas without spaces (publicworks@nmc.com, zonemanager@nmc.com). Moreover, citizen can filter information included in your report by status, service request category, time range etc.

(f) Canned Messages: The canned messages feature allows government users to save preset text to populate the comment field when updating the status of issues. When updating the status of an issue to acknowledged or closed, the citizen is required to input a comment to post publicly on the issue.

(g) Email Messages: Municipal administration can customize emails sent to citizens when issues they have reported are submitted, acknowledged and closed, as well as when new comments are added by the city or by neighbors.

(h) Issues by Source: This keep track of how residents are submitting issues to a city or town with the Issues by Source chart. This chart displays how service requests have been submitted over the last 90 days, tracking points of origin including map widgets for website, iPhone, Android and Facebook apps, as well as city initiated reports submitted by staff on behalf of the residents.

(i) Issues by Service Request: To view current totals and recent daily activity by service request type under Issues by Service Request. After click on any of the service request titles, one can navigate to a page listing all issues reported by citizens.

(j) Issues by Employee: The Issues by Employee list is also displayed for each staff person or department with a portal's government account. When issues are automatically or manually assigned to a user, a citizen can click their account name to navigate to a list of issues.

(k) Acknowledge Issues: Citizens must enter a comment to update the issue by entering text into the comment box. One can also attach a photo or video to comment when updating the status. The comment will also be included in an email sent to the reporter letting them know the updated status of their report (Figs. 7 and 8).

(l) Generate Work Order: The generate work order feature helps the municipal administration on processing the issue to the concerned person (Fig. 9).

8 Smart Technology—GIS and GPS

One of the important parts of this system is the capability of smart technology. Now, what is this smart technology? This smart technology is the ability of the system to take its own decision. This ability is designed by integrating GIS and GPS in the mobile app. To undertake this task in real-time, this ability is developed by using geo-referencing of GIS. The app has accurate capabilities to detect the location of the individual. It takes it's own decision to identify the location and send it to the concerned administrative zone. This capability makes the app smart. Thus, when a person moves from one administrative boundary to another, geo-fencing routes the information to the concerned administrative zone. Hence, the system takes its own decision to route information to the concerned zone administrator. The GIS shape file defines a geo-fence area. Furthermore, in this app, GIS is a major contributor to make the technology smarter.

Geo-fencing is a geographic boundary used to define zones, wards, pre-prefectures, etc. in the city. This geo-fence are developed through use of GIS Shape files. The geo-fencing allows the citizen to detect which geo-graphic ward or zone he/she is located. Nagpur city is divided into ten administrative zones (Fig. 12). These are known as wards. Each ward functions as decentralized administrative units for all the local administration and development of the city. The app automatically detects where you are and then one can report an issue from the incident site. The GPS system locates the issue on the map. This is viewable on smartphones and web portals.

To design the geo-fence of these ten wards, World Geodetic System (WGS) and Universal Transverse Mercator (UTM) coordinate systems are used in the GIS environment. WGS is a geographic coordinate system. It uses a three dimensional spherical surface for providing locations on the earth. UTM is plane coordinate grid system for the map projection. The Nagpur city lies on following projections: (a) Indian 1954/UTM zone 48N: EPSG Projection and (b) Kalianpur1975 UTM-44N.

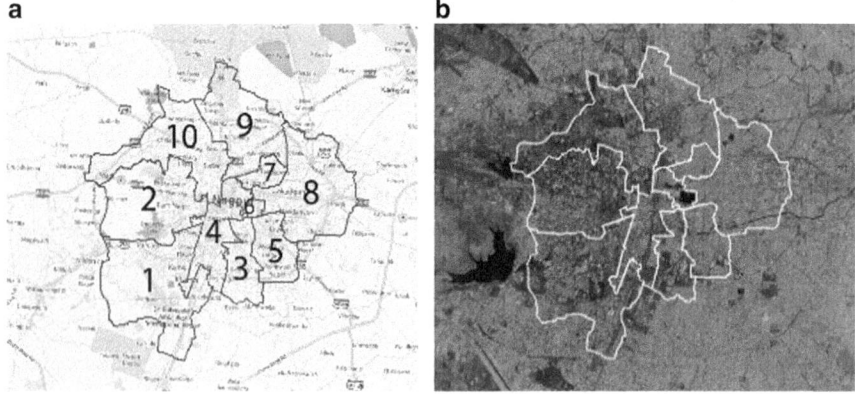

Fig. 12 **a** Geo-fencing ten administrative zone of Nagpur City and **b** Geo-referencing shape file

9 Mobile App—Benefit

The design of the mobile app for NMC is easy for use by citizens with a very simple interface. The mobile platform is a mobile first design strategy to completely maximize customer satisfaction on mobiles. Mobile-apps exist because they provide optimal solutions to interact and deliver required data to users on mobile platforms. This app is custom made for a mobile phone. The Android apps are capable to download based on the latest mobile OS version running on the phone; iOS apps are downloaded based on platform capabilities, windows apps are also downloaded based on device capabilities.

The proposed app provides monitoring and accountability for municipal government. If a pipeline is repaired authority can upload photos for citizens. It can monitor the progress of project through app. Thus, it increases accountability of the projects. The generation of the work order benefit the government. If the work order is responded by the vendor, the portal has the capability of raising the flag and resolves the issue immediately. The on-site reporting allows citizens to report issues from the incident site itself with their mobile phones, rather than going to home and open their desktops/laptops to report issues. Other benefit is accessibility. Apps are natively installed and always accessible to users to store/update information. The backend is based on global cloud and available 24 × 7 irrespective of backend integration site like e-Governance portal. Also, apps are distributed through native appstores and available 24 × 7. The app can be distributed by the government's own website as well. Importantly, ability to upload images/videos with a smartphone camera provides easy information sharing. The mobile platform has access to the smartphone's native functionality like camera, gyroscope integrated into app. For example, this community participation app allows citizens to click and post pictures with annotations with a single button. The platform is geared for simple and efficient use of mobile. Internet platforms do not have access to native components like camera, gyroscope. The app allows easy configuration of issue types. The administrators can add/edit/remove issue types and the questions related to the issue type. The app allows clear actionable reporting to ensure all the issues reported by the citizens are addressed by the officials. The app also allows business intelligence. Out of box user friendly reports can be generated on the app and web portal. Moreover, easy customization is possible on the platform. App can be customizing in future for GIS application. The app can also be customized for using government school monitoring on a weekly and monthly basis.

10 Conclusions

The city planning is the science of making human settlements for the social, economic and cultural developments of citizens. City planning involves interdisciplinary involvement of fields like architecture, geography, engineering, sociology and GIS.

The smart cities concepts includes: (a) smart society, (b) technology, (c) governance, (d) energy, (e) mobility and (f) sustainability.

The NMC project concept involved components like mobile-based service delivery, community participation, information sharing and better quality IT infrastructure. It is also integrated with e-Governance application for online municipal tax payment, registrations, information, master planning etc. The proposed system is designed to establish an extremely smooth citizen interface to promote community participation. The focus in the design has been on maximizing citizen convenience in reporting for various NMC services to resolve issues like potholes, encroachment, illegal hoardings, garbage etc. using their mobile devices.

The application of mobile based community participation model in smart cities has following advantages: (1) scalability, (2) ease of use, (3) GIS and GPS integrated smart technology, (4) community participation. In Nagpur city, the model was successfully implemented and adopted by the community for everyday problems. It is indeed important to develop a user-friendly mobile app for smart cities. Moreover, cities require a community-based approach to shape liveable cities. So, the application of a mobile-based community participation model can be one of the bottoms to top approaches for smart cities.

References

1. Badii, C., Bellini, P., Nesi, P., & Paolucci, M. (2019). *A smart city development kit for designing Web and mobile Apps*, pp. 1–8.
2. Jog, Y., Venkatesh, R., Pandit, A., Bhadauria, R. S., & Padmavati. (2017). Understanding role of mobile apps in smart city services. *International Journal of u- and e- Service. Science and Technology, 10*(4), 27–38.
3. Farias, R. S., de Souza, R. M., McGregor, J. D., et al. (2019). Designing smart city mobile applications. *Empirical Software Engineering, 24*, 3255–3289.
4. Kunttu, I. (2019). *Developing smart city services by mobile application, Conference: ISPIM Connects Ottawa—Innovation for local and global impact*. Ottawa: Canada.
5. Almao E. C., Golpayegani F. (2019). Are mobile apps usable and accessible for senior citizens in smart cities?. In J. Zhou, G. Salvendy (eds.) *Human Aspects of IT for the Aged Population. Design for the Elderly and Technology Acceptance*. HCII 2019. Lecture Notes in Computer Science, Vol. 11592, Springer.
6. Alessandria, F. (2021). Smart city: The citizen protagonist. In C. Bevilacqua, F. Calabro, L. Della Spina (eds.) *New Metropolitan Perspectives. NMP 2020. Smart Innovation, Systems and Technologies*, Vol. 178. Springer.
7. Matarazzo et al. (2018). Crowdsensing framework for monitoring bridge vibrations using moving smartphones. *Proceedings of the IEEE, 106*(4).
8. Dlodlo, N., Gcaba, O., & Smith, A. (2016). Internet of things technologies in smart cities. *2016 IST-Africa Week Conference, Durban*, pp. 1–7.
9. Barrile, V., Fotia, A., Bernardo, E., & Bilotta, G. (2021) Geomatic techniques: A smart app for a smart city. In C. Bevilacqua, F. Calabrò, L. Della Spina (eds.), *New Metropolitan Perspectives. NMP 2020, Smart Innovation, Systems and Technologies*, Vol. 178, Springer.
10. Harford, T. (2014). Big data: A big mistake? *Significance, 11*, 14–19.
11. Guenduez, A. A., Singler, S., Tomczak, T., Schedler, K., & Oberli, M. (2018). Smart government success factors. *Swiss Yearbook of Administrative Sciences, 9*(1), 96–110.

12. Prayogo, I., & Ramadhan, R. (2020). Smart city and its application. *IOP Conference Series: Materials Science and Engineering, 879.*

13. Akcin, M., Kaygusuz, A., Karabiber, A., Alagoz, S., Alagoz, B., & Keles. C. (2016). Opportunities for energy efficiency in smart cities. *4th International Istanbul Smart Grid Congress and Fair (ICSG), Istanbul*, pp. 1–5.

14. Lee, S., Oh, K., & Jung, S. (2009). The carrying capacity assessment framework for ubiquitous ecological cities in Korea. In *2nd International Urban Design Conference, Gold Coast 2009, Australia.*

15. Gazder, U., Shehabi, A., Fardan, K., Alkhabbaz, H., & Ibrahim, Z. (2019). *Smart parking solution for congested areas: Application of vertical rotary systems, 2nd Smart Cities Symposium (SCS 2019), Bahrain*, pp. 1–5.

16. Bhatt, D., & Tiwari, M. (2019). *Smart traffic sign boards (STSB) for smart cities, 2nd Smart Cities Symposium (SCS 2019), Bahrain*, pp. 1–4.

17. Muster, S. et al. (2017). How to involve inhabitants in urban design planning by using digital tools? An overview on a state of the art, key challenges and promising approaches. *21st International Conference on Knowledge Based and Intelligent Information and Engineering Systems, KES2017, Procedia Computer Science*, vol. 112, pp. 2391–2405.

18. Stelzlea, B., & Noenniga, J. R. (2017). A database for participation methods in urban development. *International Conference on Knowledge Based and Intelligent Information and Engineering Systems, KES2017*, 6–8 September 2017, Marseille, France, Procedia Computer Science 112, pp. 2416–2425.

19. Nelson, N., & Wright, S. (1995). *Power and Participatory Development: Theory and Practice* (pp. 1–18). Intermediate Technology Publications, London.

20. van Empel, C. (2008). The effectiveness of community participation in planning and urban development. *WIT Transactions on Ecology and the Environment, 117.*

21. Akre, V., & Yankova, V. (2019). Smart city facilitation framework (SCFF) and the case of Dubai smart city. *2019 International Conference on Computational Intelligence and Knowledge Economy (ICCIKE), Dubai, United Arab Emirates*, pp. 576–580.

Improving KNN Model for Direct Marketing Prediction in Smart Cities

Stéphane Cédric Tékouabou Koumétio[ID] **and Hamza Toulni**[ID]

Abstract Nowadays, data mining is widely used to mine business information and make them very strategic for decision. Many data mining techniques can be according to the large variety of datasets. Our paper aims to optimize the prediction of telemarketing target calls for selling bank long-term deposits in smart cities using improved KNN model. The dataset used come from Portuguese retail bank which addressed from 2008 until 2013, data on its clients, products and social-economic attributes not without ignoring the effects of the financial crisis. From original set of 150 features which has been explored, only 20 realistic features which are known before the call are retained for this work including label. We use optimized KNN algorithm which is one of the simplest data mining techniques based on similarity that are used for prediction. To optimize its performance and to accelerate its process, we propose speeded KNN algorithm based on preprocessing and significant attributes filtering and normalization. It also includes another improvement based on variation of neighbors (k) and similarity model which insures a more accurate classification. Thus, the contributions of this paper are three-fold: (i) introduce high preprocessing method separately each type of features and imputation of missing values of attributes; the automation of features normalization (ii) the selection of realistic and most significant attributes which are known before the call, and (iii) the combination of optimal k and similarity models which gives the best performances. Our approach largely improve the performance of other algorithms used, with average more than 96.91% of f1-measure and vary reduced time processing.

S. C. Tékouabou Koumétio (✉)
Faculty of Sciences, Department of Computer Sciences, Chouaib Doukkaly Univercity, B.P. 20, 2400 El Jadida, Morocco

Center of Urban Systems (CUS), Mohammed VI Polytechnic University (UM6P), Hay Moulay Rachid, 43150 Ben Guerir, Morocco

H. Toulni
GITIL Laboratory, Aïn Chock Hassan II University of Casablanca, Casablanca, Morocco

EIGSI, 282 Route of the Oasis, Mâarif, 20140 Casablanca, Morocco

© The Author(s), under exclusive license to Springer Nature Switzerland AG 2021
U. Ghosh et al. (eds.), *Machine Intelligence and Data Analytics for Sustainable Future Smart Cities*, Studies in Computational Intelligence 971,
https://doi.org/10.1007/978-3-030-72065-0_7

Keywords Bank telemarketing · KNN · Direct marketing targeting · Smart cities · Machine learning · Supervised classification · Similarity

1 Introduction

Advancement of computer technology with significant increase in computing power and the growth of database which collect and store huge amount of information and make them very useful and strategic for competitive companies [1, 2]. Banks reach the customers information a multitude of channels, including mail, e-mail, and phone, in person for sharing information about products or services [3]. for direct costumers targeting, data mining or Knowledge Discovery in Databases (KDD), which is pulling considerable research largely in various fields including Database Design, Pattern Recognition, Statistics, Machine Learning and Information Visualization [4].

Data mining have been used widely in direct marketing to identify prospective customers for new products, by using purchasing data, a predictive model to measure that a customer is going to respond to the promotion or an offer [5]. Thus, data mining has gained popularity for illustrative and predictive applications such as in banking direct telemarketing processes with many techniques.

Banking direct telemarketing is an important and complex task that can affect the efficiency of the business because it concerns sales, it must be extremely precise and effective. Its automation would be very advantageous especially since the costs will be reduced in the long run. It could probably go beyond traditional approaches [6]. However, standard algorithms are still limited and no algorithm has proved perfect for both classification and prediction.

From a technical point of view, the problem of direct telemarketing targeting is part of the binary classifications that are at the crossroads of statistics and artificial intelligence [7]. It aims to classify instances into two groups (yes or no) based on realistic classification criteria i.e. which are known before the call. The binary classifiers include two main types of models: predictive (supervised) such as the KNN algorithm [8], in which the class of each instance is known, and exploratory (unsupervised) as the k-means algorithm whose task is the creation of clusters (classes of instances). Supervised classification models have two main stages: training and testing [9, 10]. Subsequently, this paper proposes a new approach that extends the KNN algorithm by a normalization stage and amputation of missing values. The second improvement is based on the elimination of insignificant and non-realistic attributes such as call duration. Finally, the combination of optimal k and similarity models which gives the best performances.

In experiments, each one of these improvements proved quite interesting and contributed into the overall performance. The rest of this paper is structured as follows: Sect. 2 summarizes main approaches applied to bank telemarketing dataset. Section 3 meticulously details the proposed approach; and in Sect. 4, the obtained results are compared to those of most known classification techniques. Finally, the last section concludes this work.

2 Overview

Quite a few studies have been conducted on bank direct marketing domain. Some researchers have used machine learning algorithms to classify the model according to data mining approaches [3, 11–13]. Data mining approaches aim to build a predictive model that labels data into a predefined class (for example "yes" or "no"). Because of all bank marketing strategies are dependent to analyze huge electronic data of customers. It is impossible for a human analyst to evaluate the meaningful knowledge from the vast amount of data. Many researchers use some of popular data mining techniques such as Naïve Bayes (NB), decision trees, support vector machines, and logistic regression. The purpose is increasing the campaign effectiveness by identifying the main characteristics that affect the success [3, 12, 14, 15]. The success of such campaigns is directly proportional to the effective customer participation. The success rate of such campaigns that especially conducted by banks may be enhanced by data mining methods [16].

2.1 KNN Algorithm for Classification Problem

The KNN algorithm is among the most simple and easy to use machine learning algorithms for prediction problems. For a classification problem for example, when the data sets in which the class domains have many overlaps, the classifier built on the basis of the KNN algorithm is more appropriate compared to the others although it is sometimes inappropriate when these are large data sets while still remaining valid in terms of performance [17, 18]. However, when it comes to making classification decisions, the algorithm is only based on the amount of information in the training samples, but takes no advantage of other distance information, its excessive dependence on the choosing the value k leads to instability of the precision. In addition to this, for unbalanced training data, the k closest neighbors to the test set tend to favor the larger class, which negatively affects the classification performance [17, 19]. In this case, the KNN algorithm proved to be very slow when it reviewed all the classification instances each time while being very vulnerable to both dimensionality and irrelevant and correlated attributes. In addition, making a wrong choice of the similarity distance or the value of k degrades the performance of the algorithm even more [9, 20]. Faced with this algorithmic behavior, many approaches have been proposed to overcome these shortcomings, in order to improve the accuracy of the classification and its classification efficiency, although these improvements are still limited in the face of a certain type of problem having the above-mentioned characteristics.

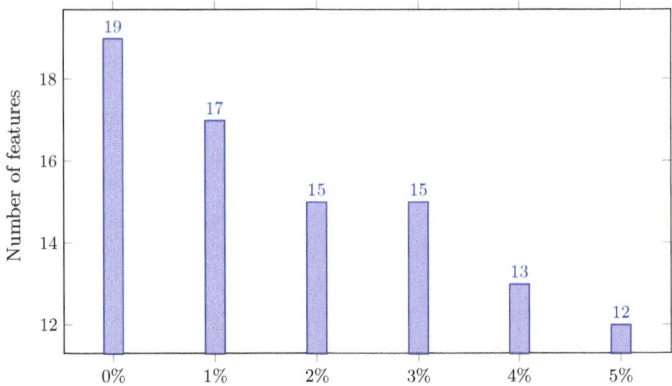

Fig. 1 Variation of significant attributes number according to δ

2.2 Direct Marketing Challenges for Smart Cities

The definitions and characteristics of a Smart City concept are numerous and can often vary from one context to another. According to Jasrotia and Gangotia [21] a smart city is a sustainable and livable city. Harrison [22] defined it as an instrumented, interconnected, and intelligent city while, Lee [23] states that the smart cities utilize the emerging opportunities like growing information and communication technology (ICT) innovations. For Hollands [24] called smart city initiatives as a celebratory label. As shown in Fig. 1. the various dimensions of smart city include Smart Economy (Public expenditure on Research & Development), Expenditure on education, GDP, Unemployment rate), Smart Mobility (Sustainable and safe transportation, Pedestrian lanes, Walkways, Cycle lanes Innovative transportation, Efficient traffic control systems), Smart Living (Availability of recreational and leisure areas, Public libraries, Entertainment centers, Sustainable resource management, Education and Health facilities, Green areas, Availability of ICT), Smart Environment (CO_2 reduction strategies, Efficient use of water and electricity, Availability of green space, Policies to handle urban development, Recycling of waste), Smart People (Education level of citizens, Language skills, Learning programs, Technical skills) and Smart Governance (Number of educational institutes, e-Governance facility, Internet access for citizens, Easy access basic to facilities). All of these dimensions act as pillars which constitute the foundation of smart cities. Hence, it can be said that a smart city is a sustainable, livable, interconnected and intelligent city. Thus, smart cities through the Internet of Things have allowed to enrich, automate and optimize marketing campaigns like targeted telemarketing. Targeted marketing uses machine learning models such as the improved KNN algorithm that we propose in the following section.

3   Proposed Approach

3.1   Bank Telemarketing Dataset

The data is related with direct marketing campaigns (phone calls) of a Portuguese banking institution and provided by The UCI Machine Learning Repository [25]. Data were collected from May 2008 to June 2013 [11]. The total number of training phone contacts database (the sample size) is 41188 and the marketing campaigns were based on phone calls. Often, more than one contact to the same client was required, in order to access if the product (bank term deposit) would be yes ($y = 1$) or no($y = 0$) for subscribe [26, 27]. The data set includes 21 variables (including binary label variable y) where numeric and categorical variables are mixed as we can see in Table 1. For more details of the data, see the information available at The UCI Machine Learning Repository.

3.2   Improving KNN Algorithm for Bank Telemarketing Prediction

This process to optimize KNN algorithm for classification technique consists of 3 big steps which deal meanly with features in order to favore good fitting. The goal of each of these steps is to correct the weaknesses presented by the KNN algorithm for a better predictive classification of clients. The first step in our approach is to introduce appropriate function for coding minimal variables and missing values. Once these digitized values, in the second phase we will standardize all of variables to limit the effects of amplitude and in the third stage we will select the relevant attributes and simply remove the noise variables to optimize performance while reducing execution time. To better explain these three steps, let's consider the data of Table 1, extracted to our data set.

Table 1   Initial data values

Instance	Age	Marital	Housing	Education	Y
I_1	59	Married	No	Professional	No
I_2	39	Single	yes	Basic.9y	Yes
I_3	59	Married	No	university	No
I_4	41	Divorced	No	Unknown	No
I_5	44	Married	No	Basic.4y	yes

3.3 Introducing Functions for Features Transformation

This step consists in transforming the different ones into digital form. Thus the technique separates them into four types: numerical, boolean, nominal and missing values.

▶ **For Nominal features types**: These features are considered as independent features and are directly associated to the decision function for our approach. The particularity of this approach is how it deal with nominal independent features (job/marital). Each value of these two features V_j for the example i can belong to class C_k if its maximum class frequency is reached on the class in the training set; this frequency is given by the formula:

$$V(C_k)_{ij} \leftarrow \begin{cases} 1 \text{ if Max } \dfrac{n_k(V_{ij})}{N_k} \quad k = 0, 1; \\ 0 \text{ else.} \end{cases} \tag{1}$$

where: $n_k(V_{ij})$ is the number of V_{ij} variable j in the class C_k; N_k is the total number of k

So, independent nominal attribute of the testing set takes the value 1 in this class k and 0 in other classes. For example if we consider the Table 5:

- The total number of yes is $N_{yes} = 2$
- The total number of yes is $N_{no} = 3$
- The number of "*married*" in the class "*no*" is: $n_{k(married)} = 2$
- The number of "*married*" in the class "*yes*" is: $n_{k(married)} = 1$

$\frac{n_{no}(married)}{N_{no}} = \frac{2}{3} = 0, 66$; $\frac{n_{yes}(married)}{N_{yes}} = \frac{1}{2} = 0, 5. 0.66 > 0.5$ so, the belonging to the class "*no*" of all marital attributes "*married*" will be 1 and their belonging to class "*yes*" will be 0.

▶ **For Missing Attribute Values.** The original dataset contains several missing values meanly in some categorical attributes, all labeled with the "unknown" label. The particularity of this approach is how these missing values have been treated by using specific imputation techniques [28] which depends to the type of features. It consists to replace all the missing values V_{ij} of the feature V_j by the average if they are scaled or numeric variable or the mode if they are Boolean or nominal variables inside the class k:

$$(V_{ij})_{kmiss} \leftarrow \begin{cases} Mode(V_j)_k \text{ if Boolean or nominal} \\ Mean(V_j)_k \text{ if numeric or scaled} \end{cases} \tag{2}$$

For example if we consider Table 2 bellow: $V_j =$' education' is scaled variable; I_4 takes the value Unknown for education and k= 'no', $Mean(V_j)_{no} = \frac{(3+2)}{2} = 2.5$; The missed value will be replaced by 2.5.

Table 2 Non normalized distances to center

	x_1	x_2	s x_3	x_4	x_5	Y
C_0	39.91	0.163	220.84	984.11	5176.17	No
C_1	40.91	0.152	553.19	792.04	5093.12	Yes
E_{21}	27	0.1	698	999	5228.1	

Table 3 Normalized distances to center

	x_1	x_2	x_3	x_4	x_5	y
C_0	0.28	0.61	0.04	0.99	0.80	No
C_1	0.30	0.56	0.11	0.49	0.79	Yes
E_{21}	0.12	0.32	0.14	1	1	

3.4 Normalization

This step which consists to eliminate the impact of the order of magnitude is very important to optimize machine learning model performance. Considering the two following examples: C_0 which is the center of the first class no, and C_1 which is the center of the second class yes [9].

We can easily estimate the similitude of the instance $E_2 1$ to every one of the classes by calculating the Euclidian distances for each center of the two classes which are: $d(E_{21};C_0) = 480.38$ and $d(E_{21};C_1) = 286.73$. E_{21} is then closer to C_1, we can conclude that E_{21} belong to the class yes regardless of the order of magnitude of the features x_j. In order to overcome such problems, a normalization is adopted for each value v_{ij} of the attribute x_j for the example E_i. It reduces each V_ij to the interval [0,1] by the formula:

$$V_{ij} \leftarrow \frac{V_{ij} - min_j(V_{ij})}{max_j(V_{ij}) - min_j(V_{ij})} \tag{3}$$

With this normalization $d(E_{21};C_0) = 0.96$ and $d(E_{21};C_1) = 1.05$. The example E_{21} belongs, after normalization, to the class no as it is closer to C_0 than C_1. Thus, the normalized data values can be presented as follows Tables 3 and 4.

3.5 Selection of Significant Features

The performance of our approach is sensitive to irrelevant and correlated attributes, so, the threshold δ of the correlation coefficient have been fixed in order to select only significant variables that are not correlated together but with the label variable. For example, the correlation between the class variable y and variables default and loan

Table 4 Transformed and Normalized data values

Inst.	Age	Marital	Housing	Edu.	Y
I1	0.519	1	0	0.875	No
I2	0.272	0	1	0.562	Yes
I3	0.519	1	0	0.231	No
I4	0.296	1	0	0.688	No
I5	0.333	0	0	0.250	Yes

is absolutely less than 0.5%, which makes these variables lessen the performance of KNN. Figure 1 shows the variation of the number of significant attributes according to the threshold.

4   Experimentation and Results Analysis

In the previous section, we detailed the approach to improving the KNN algorithm for the realistic prediction of the results of a telemarketing campaign. In this part we must analyze and discuss the results obtained. We will start by presenting the best metrics for assessing and comparing these performances, then analyze them, compare those of the different algorithms used and finally the performance of previous work.

4.1   Performance Measure

Unlike most research that evaluated the performance of their models in terms of accuracy, which measurement is critical because not related exactly to the target. We therefore preferred f1-measure of correctly classified customers to the Accuracy because this measure directly linked to our target. The f1-measure, is calculated as follows:

$$f1 - measure = \frac{2a}{2a + c + d} \tag{4}$$

a : refers to the set of clients that are correctly predicted "yes", b: refers to the set of clients that are correctly predicted "no", c: is the number of clients wrongly predicted "yes" positive and d : is the number of clients wrongly predicted "no".

4.2 Results Analysis and Discussion

Figure 2 shows that the performances increase according to δ until becoming maximum for a δ of 4% with just 13 significant attributes. This allows to reach maximal f-measure of 97.27% for k = 2 and 97.00% for k = 3.

The proposed preprocessing approach is compared to the four algorithms: LR, NB, ANN and SVM in terms of f1-measure.

Figure 3 shows that the proposed algorithm performed highest f-measure 96.91% for k = 3 with 13 attributes, followed by ANN with 94.32% and then LR with 93.42% followed by SVM (RBF kernel) with 93.35% and NB with 69.95% shows

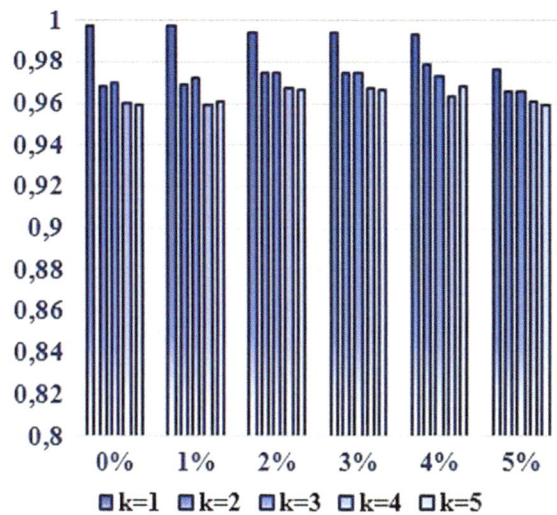

Fig. 2 f-measure for different values of k and δ

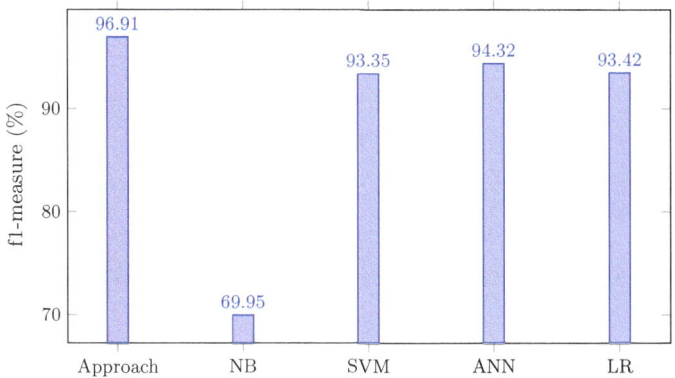

Fig. 3 Comparison of f1-measure of our approach with other algorithms

the weakest performance. Finally, on both accuracy, f1-measure and processing time, the proposed approach has undoubtedly yielded the best performance. Better still, the thresholding made it possible to eliminate the variable "duration" which in reality is not known before the call and therefore taking it into account in the prediction does not allow to have realistic results. The extension of the proposed approach to larger and more complex data sets is the focus of our current research.

4.3 Comparison of the Obtained Results with the State of the Art

Table 5 compares the best performance with that of five cited authors who worked on the same data set. With respectively 21, 17, 21, 8 and 18 significant attributes, [12, 14, 26, 27, 29] obtained their best scores of 93.5%; 93.23%; 92.14% and 89.4% using NN, C4.5, C4.5 and RF respectively. This shows that in many studies the DT C4.5 achieves the best performance and this is not verified in this document where improved KNN model reaches 96.91% correct prediction.

The obtained results show that our pre-processing approach has significantly improved prediction performance compared to the results obtained in previous work on the same DMD database. And even better, the standardization of the variables has improved these performances for SVM, LR, ANN even if it was disadvantageous for DT and much more for NB. However, the best performances (97.3% for accuracy and 100% for f1-measurement) are obtained by combining standardization and weighting of the characteristics and show that DT has a very low processing time.

Table 5 Comparaison des meilleurs performances pour DMD

	Score (%)	AUC	Nb features	Best model
Moro et al. [11, 30]	NA	0,938	22	NN
Elsalamony [12, 14]	90,09	NA	17	DT(C4.5)
Vajiramedhin et al. [26]	92,14	NA	21	DT(C4.5)
Grzonka et al. [2]	89,4	NA	8	Random Forest
Karim et al. [31]	93,96	0,9334	NA	DT(C4.5)
Lahmiri [32]	71	0,59	18	Two-stage system
Tekouabou [15]	69.1	0,5535	18	SVM
Approach	**96,91**	**95,9**	**12**	**Improved KNN**

5 Conclusion

The increasing power of data mining techniques due to the increase in the machines' computing capacity as well as the quantity and the quantity of data collected by the organizations make it essential for any business such as banks and finance that will to use to predict targeted marketing. To do this, optimizing the performance of the models used remains a challenge for both scientific research and companies. In this paper, we have proposed a new optimal model with the KNN algorithm in order to improve the realistic bank direct marketing prediction. It consists to preprocessing of each type of attributes and selects most significant attributes to classify a given instance into one of the two classes (yes or no) order to speed up the KNN process. This process is of major interest especially for large and complex datasets since it reduces both the number of computed distances (k classes instead of n instances), and the number of attributes figuring in these distances since it eliminates insignificant ones. The results of classification show that the proposed algorithm outperforms ANN, LR, NB, SVM with an f1-measure slightly exceeding 96.91% on the considered dataset.

References

1. Farooqi, Md. R., & and Iqbal, N. (2017). Effectiveness of data mining in banking industry: An empirical study. *International Journal of Advanced Research in Computer Science, 8*(5).
2. Grzonka, D., Suchacka, G., & Borowik, B. (2016). Application of selected supervised classification methods to bank marketing campaign. *Information Systems in Management, 5*(1), 36–48.
3. Parlar, T., et al. (2017). Using data mining techniques for detecting the important features of the bank direct marketing data. *International Journal of Economics and Financial Issues, 7*(2), 692.
4. Rahangdale, G., Ahirwar, M., & Motwani, M. (2016). Application of k-nn and naive bayes algorithm in banking and insurance domain. *International Journal of Computer Science Issues (IJCSI), 13*(5), 69.
5. Ayetiran, E. F., & Adeyemo, A. B. (2012). A data mining-based response model for target selection in direct marketing. *IJ Information Technology and Computer Science, 1*, 9–18.
6. Soni, J., Ansari, U., Sharma, D., & Soni, S. (2011). Predictive data mining for medical diagnosis: An overview of heart disease prediction. *International Journal of Computer Applications, 17*(8), 43–48.
7. Tung, H.-H., Cheng, C.-C., Chen, Y.-Y., Chen, Y.-F., Huang, S.-H., & Chen, A.-P. (2016). Binary classification and data analysis for modeling calendar anomalies in financial markets. In *2016 7th International Conference on Cloud Computing and Big Data (CCBD)*, pp. 116–121. IEEE.
8. Qiao, Y.-L., Zhe-Ming, L., Pan, J.-S., & Sun, S.-H. (2010). Fast k-nearest neighbor search algorithm based on pyramid structure of wavelet transform and its application to texture classification. *Digital Signal Processing, 20*(3), 837–845.
9. Cherif, W. (2018). Optimization of k-nn algorithm by clustering and reliability coefficients: Application to breast-cancer diagnosis. *Procedia Computer Science, 127*, 293–299.
10. Cherif, W. (2018). Hybrid reliability-similarity-based approach for supervised machine learning. *Procedia Computer Science, 12*(3), 170–175.

11. Moro, S., Cortez, P., & Rita, P. (2014). A data-driven approach to predict the success of bank telemarketing. *Decision Support Systems, 62,* 22–31.
12. Elsalamony, H. A., & Elsayad, A. M. (2013). Bank direct marketing based on neural network and c5. 0 models. *International Journal of Engineering and Advanced Technology IJEAT, 2*(6).
13. Tekouabou, S. C. K., Cherif, W., & Silkan, H. (2019). A data modeling approach for classification problems: Application to bank telemarketing prediction. In *Proceedings of the 2nd International Conference on Networking, Information Systems & Security,* pp. 1–7.
14. Elsalamony, H. A. (2014). Bank direct marketing analysis of data mining techniques. *International Journal of Computer Applications, 85*(7):12–22.
15. Koumétio, C. T., Cherif, W., & Hassan, S. (2018). Optimizing the prediction of telemarketing target calls by a classification technique. In *2018 6th International Conference on Wireless Networks and Mobile Communications (WINCOM),* pp. 1–6. IEEE.
16. Akbal, E., Dogan, Ş, & Varol, N. (2017). Karar ağaçları ile telefon dolandırıcılığı verilerinin analizi. *Fırat Üniversitesi Mühendislik Bilimleri Dergisi, 29*(1), 171–177.
17. Miao, Z., Yan, T. et al. (2014). An improved knn algorithm for imbalanced data based on local mean. *Journal of Computational Information Systems, 10*(12), 5139–5146.
18. Yong, Z., Youwen, L., & Shixiong, X. (2009). An improved knn text classification algorithm based on clustering. *Journal of Computers, 4*(3), 230–237.
19. Nutanong, S., Zhang, R., Tanin, E., & Kulik, L. (2010). Analysis and evaluation of v*-knn: An efficient algorithm for moving knn queries. *The VLDB Journal, 19*(3), 307–332.
20. Goldberger, J., Hinton, G. E., Roweis, S., & Salakhutdinov, R. R. (2004). Neighbourhood components analysis. *Advances in Neural Information Processing Systems, 17,* 513–520.
21. Jasrotia, A., & Gangotia, A. (2018). Smart cities to smart tourism destinations: A review paper. *Journal of Tourism Intelligence And Smartness, 1*(1), 47–56.
22. Harrison, C., Eckman, B., Hamilton, R., Hartswick, P., Kalagnanam, J., Paraszczak, J., et al. (2010). Foundations for smarter cities. *IBM Journal of Research and Development, 54*(4), 1–16.
23. Lee, J. H., Phaal, R., & Lee, S.-H. (2013). An integrated service-device-technology roadmap for smart city development. *Technological Forecasting and Social Change, 80*(2), 286–306.
24. Hollands, R. G . (2008). Will the real smart city please stand up? intelligent, progressive or entrepreneurial? *City, 12*(3):303–320.
25. Lichman, M. (2013). UCI machine learning repository, p. 40. http://archive.ics.uci.edu/ml.
26. Vajiramedhin, C., & Suebsing, A. (2014). Feature selection with data balancing for prediction of bank telemarketing. *Applied Mathematical Sciences, 8*(114), 5667–5672.
27. Kawasaki, Y., & Ueki, M. (2015). Sparse predictive modeling for bank telemarketing success using smooth-threshold estimating equations. *Journal of the Japanese Society of Computational Statistics, 28*(1), 53–66.
28. Lakshminarayan, K., Harp, and S. A., & Samad, T. (1999). Imputation of missing data in industrial databases. *Applied Intelligence, 11*(3), 259–275.
29. Miguéis, V. L., Camanho, A. S., & Borges, J. (2017). Predicting direct marketing response in banking: comparison of class imbalance methods. *Service Business, 11*(4), 831–849.
30. Moro, S., Laureano, R., & Cortez, P. (2011). Using data mining for bank direct marketing: An application of the crisp-dm methodology.
31. Karim, M., & Rahman, R. M. (2013). Decision tree and naive bayes algorithm for classification and generation of actionable knowledge for direct marketing.
32. Lahmiri, S. (2017). A two-step system for direct bank telemarketing outcome classification. *Intelligent Systems in Accounting, Finance and Management, 24*(1), 49–55.

Prediction of Satisfaction with Life Scale Using Linguistic Features from Facebook Status Updates: Smart Life

Ferda Özdemir Sönmez and Yassine Maleh

Abstract Diener's satisfaction with life scale, SWLS, is broadly used as a measure for estimating global life satisfaction in the literature. Despite the popularity of social media applications and numerous researches linking daily word usage to social sciences, none of the existing studies managed to identify solid negative and/or positive relations or any sign of irrelevance between the Facebook status updates and SWLS results of individuals. The main objective of this chapter is to fill this gap in the literature by investigating the relations between smileys and linguistic features in Facebook status updates and SWLS of individuals and choosing the most appropriate linguistic features among many others. The data analysis procedure presented includes correlation analysis, linear regression and support vector machine (SVM) regression methods. The results indicate that there are significant positive and negative correlations between SWLS and some word groups and the relations differ for male and female users. The procedure exposes that SVM regression is more suitable than the linear regression in identifying the relations between the word groups and SWLS. The results also reveal that the selected word groups can be further used to predict personal attributes using Facebook data. The results of the proposed procedure are discussed and compared with the research findings of the previous studies.

Keywords Facebook profile · SWLS · LIWC · Support vector machine · Smart health

F. Ö. Sönmez (✉)
IEEE, Institute for Security Science and Technology Imperial College, London, UK
e-mail: f.ozdemir-sonmez@imperial.ac.uk

Y. Maleh
Laboratory laSTI, ENSAK, IEEE, USMS University, Beni Mellal, Morocco
e-mail: yassine.maleh@ieee.org

© The Author(s), under exclusive license to Springer Nature Switzerland AG 2021
U. Ghosh et al. (eds.), *Machine Intelligence and Data Analytics for Sustainable Future Smart Cities*, Studies in Computational Intelligence 971,
https://doi.org/10.1007/978-3-030-72065-0_8

119

1 Introduction

Smart city term has a lasting history and has found adaptation in multiple cities. In 2020, due to the worldwide epidemic, people understood the importance of smartness and saw the results of enlarging the smartness at the city and even country level. Satisfaction with life has direct consequences on people's behaviors and decisions. Thus, the prediction of satisfaction of life scale scores (SWLS) using available data would add to the smart city concept, being a smart tool related to a social term.

Facebook was founded in 2004. Since then, there has been an extensive amount of user-generated content on Facebook. However, still there is no solid indicator or function suggested by scientists showing that Facebook user-generated content is related or unrelated to level of individual's overall satisfaction of life. The search for personal life satisfaction from social media data has been an issue for some time, but there is a need to delineate and develop more solid indicators on this subject as previous approaches failed to make a concrete conclusion on such possible relations. Therefore, support vector machine models will be used to gather more information regarding such indicators in this study.

The popularity of social media has grown rapidly in recent years leading to a major change in how business decisions are made. In the first quarter of 2016, Facebook had 1.508 billion active users [1]. This feature of social media has gained significant attention among researchers of various backgrounds. Social media is built up of nodes representing individuals who create a social network and also edges representing types of relations between these individuals. Popular research questions in social media analysis domain are "Who uses social media and why?", "What are their purposes in using social media and how do they achieve this?" A concise taxonomy of social media research considering both social and technical perspectives is provided in Ganis and Kohirkar [2]. In the recent decade, social media analysis has been commonly used in a wide range of disciplines including, prediction of political preferences [3], trend and sentiment analysis [4], and analysis of the nature of communication flows during social conflicts [5].

The recent social media research has focused on personality attributes or mood changes, and attempted to address questions such as; "Do different genders use different words to share same feelings?" or "Do people who have more satisfaction in life have a certain kind of writing style?". As of late, only a few research projects have compiled a sufficient amount of data to support the social media analysis research. One of these projects is the myPersonality project [6], which included data of more than eight million users, including demographics, geolocation, Facebook activity, religion and political views, Facebook status updates, couples data, big5 personality traits, and Satisfaction with Life Scale (SWLS) results. The data have been used by social science researchers in many studies [7] including the investigation of subjective well-being [8].

While the early social science studies mainly focused on the negative part of well-being, "unhappiness", "Happiness' was considered in 1973, and many researchers have focused on the positive part of subjective well-being since then. Few researchers have attempted to measure the well-being and proposed various well-being metrics. These studies brought out some extraordinary subjective prosperity measurements utilized as a component of various, yet unique studies. Some of these are single item scales and some are multi-items scales. Some of the commonly used measurements are: "Bradburn scale of psychological well-being" [9], "Subjective happiness scale" which is also called "General happiness scale" [10], and "Satisfaction with Life Scale" [11]. SWLS was first proposed by Diener et al. as a five-item, seven scale measure to assess one's life satisfaction. The result indicates the degree of agreement of being satisfied. The answers to the query range from strongly disagree (one) to strongly agree (seven).

SWLS originally included 48 items each reflecting a life satisfaction, only 10 of which gave better results in terms of reflecting one's cognitive judgment evaluation process. Later, these items were reduced to five in order to eliminate redundancies of wording and cause minimum cost in terms of alpha reliability [12] . Pavot and Diener [13] revealed the reliability of SWLS for measuring the well-being.

Generally, social scientists use dialect and words to assess individuals' thoughts and emotions. The origins of linguistic research go back to Freud [14]. The recent development in the technology and statistical techniques leveraged language analysis methods to a higher level [15]. Pennebaker et al. proposed the Linguistics Inquiry and Word Count (LIWC) [16] with the aim of categorizing and counting the words in psychological classes. LIWC program has been improved and used in numerous research since then. Many recent studies have attempted to make conclusions related to people's lives using the social media data. One of these attributes is subjective well-being of people. Tov et al. [17] performed a laboratory study in a controlled environment to study the SWLS using LIWC. The results indicated that LIWC positive emotion was correlated with higher levels of self-reported daily positive emotion and lower levels of sadness and stress. Similarly, LIWC anger and anxiety were associated with daily negative emotions, and global life satisfaction was correlated positively with positive emotion and negatively with negative emotion.

In a recent study, Wang et al. [8] examined the validity of Facebook's Gross National Happiness Index (FGNH) as a mood and well-being measurement scale. Wang et al. compared FGNH with SWLS scores and revealed that FGNH is not associated to well-being and does not accurately reflect individuals' moods. Farnadi et al. [18] focused to identify the relation between feelings communicated by Facebook status updates and demographic data of users. The study revealed that women posted status updates more often than men, and expressed more feelings in the status updates. Similarly, the emotions expressed increased with age. Farnadi et al. [18] also mentioned that messages spreading displeasure were mostly posted by men and messages with positive feelings were mostly posted by women.

LIWC was used by Gill et al. [19] to develop an "instinctive feeling detector" software. The detector was used to investigate the content of 50–200 words genuine blog texts which were also examined and grouped by human experts. The results indicated that anger is more frequently used by angry authors, sadness is expressed utilizing a higher rate of misery-inducing words, and positive feeling words are utilized at high rates by joyful creators. Gill et al. [19] claimed that the length of the short blog posts maybe a limitation for the text classification techniques and as the text lengths increase the performance of the text classifiers improve.

The impact of demographics on the web use was mentioned by Dhir et al. [20]. Fichman and Sanfilippo [21] also made the conclusion that in regards to online trolling, men and women have distinctive inspirations for similar behaviours in different communities. Golder and Macy [22] studied a person's daily and seasonal mood changes in different cultures using public Twitter messages. Golder and Macy's mentioned that social media data overcame the limitations of laboratory generated data and offered precise real-time measurements across diverse populations.

Although collecting the social media data is becoming popular, existing studies often use other computerized content analysis to investigate the subjective well-being. Few existing studies use social media data to predict subjective well-being. There are many other studies using social media-originated data to measure various other emotional factors. For example, Coviello et al. [23] measures the contagion of happiness in massive social networks. They took rainfall as an instrument to measure happiness as a negative indicator. They made the calculations in aggregated manner, at city level, to overcome the calculation difficulties of reciprocal influence of people to each other. Bollen et al. [24] investigated assortativity of happiness as a factor similar to the same music tastes and locations using Twitter data. Schwartz et al. [25] examined temporal orientation of users, such as past, present and future using Facebook data. Due to space limitations, these kinds of studies, which use similar data, are not included in this study.

There are also some studies which seem to predict SWLS using similar data (similar type of research), such as Collins et al. [26]. Collins et al. used Facebook data to predict SWLS, however, they took a different approach. They had a two-step method. In the first step, they predicted Big 5 personality features, which are openness, conscientiousness, extraversion, agreeableness, and neuroticism. Later, they used Big 5 personality prediction results to extrapolate life satisfaction of individuals. Maurer [27] examined social media data to find correlations between linguistic statements and Quiet Ego. Khodadady and Dastgahian [28] examined the relationships of subjective well-being to religiosity. Wongkoblap et al. [29] created a multi-level predictive model to explore the relationship of SWLS to depression. Krys et al. [30] investigated the relationship of individualism to the well-being. Since there is no study making a detailed examination of linguistic data aiming prediction of SWLS, the main focus of this study is to fill this gap in the literature by making a detailed examination for various word groups to identify the relations among the subjective well-being and the linguistic features, and the smileys shared by the Facebook users.

The main contributions of this study can be listed as follows:

- Design and use of an experimentally-built unique process consisting of Anova, Filtering, Linear Regression Analysis, Support Vector Machine Regression Analysis, and Graphical Analysis methods.
- Design and use of a new type of analysis based on the occurrence of specific linguistic terms in the two tails of data which has not been used in similar research before.

- Graphical analysis of each selected attribute using SVM results, providing non-linear relationship of independent variables to dependent variable.
- Findings relations of linguistic features to SWLS supported by earlier literature, such as "i", "we", "posemo", "negemo", and "swear" and totally new features such as "family".

The remainder of this chapter is organized as follows. The next chapter is the motivations section. Following the motivations section, there is the research questions section. The fourth section is the methodology section. The fifth section contains the results. The sixth section is devoted to discussions and limitations. Finally, the last section is the conclusions section.

2 Motivations

Golder and Macy's [22] study is related to recognizing a person's daily and seasonal mood changes in different cultures using millions of public Twitter messages. According to Golder and Macy, using social media data overcomes the limitations of earlier generations' habit of using laboratory generated data. Golder and Macy [22] also stated that this approach offers precise real-time measurements across diverse populations. Thus, using social media data for measurement and foreseeing of real-time personal attributes have many benefits for researchers. This led to a new investigation for the prediction of SWLS using social media data.

Humans assemble their future relying on conscious and unconscious decisions. These decisions heavily rely on the predicament they face currently in life. Thus, our current satisfaction level of life is a starting point. For example, if one is not satisfied with his current employment state, his next decision might be searching for new job opportunities or new educational opportunities to obtain new skill sets. Today, it is a known fact that humans reflect their spot in life continuously by way of social media sharings. Although there are some studies of SWLS prediction/investigation using social media data, none of these studies have pointed out the relevancy or irrelevancy of Facebook status updates data to SWLS.

Measuring real-time SWLS using social media data would be crucial for many purposes, including business decisions. The SWLS prediction may also be utilized with additional personal data such as monetary status, hobbies, interests, e-commerce records, and premiums to improve the decision making. SWLS level of an individual may be valuable during business decisions, such as selection of product alternatives, sentimental connection choices, or investment decisions.

Pavot and Diener [12] mentioned benefits of assessing the SWLS in general well-being and health. Blais et al. [31] reported "a strong negative correlation ($r^1/4\_0.72$) between the SWLS and the Beck Depression Inventory [32]. Pavot and Diener also indicated that some future practices, like suicide attempts, can be foreseen by using SWLS. Forecasting SWLS with Facebook data would not only enable crucial input to marketing decisions, but could also provide useful information to different disciplines.

Considering even these limited examples, prediction of SWLS is valuable, and it can be complemented by other knowledge. However, finding possible use cases is another research topic that is not the focus of this chapter.

3 Research Questions

In the literature, few studies have utilized the Facebook status updates for different purposes, however, none of the existing studies managed to find the correct set of attributes used for the prediction of individuals' SWLS using the social networking sharing. Within this context, the main research question of this article is "Is there a correlation between the Facebook status updates and the individuals' fulfillment levels of life and which linguistics features are more suitable to be used in such predictions?". The article also investigates two auxiliary research questions: "Is there a relation between the distinctive linguistic properties of diverse age groups and their SWLS?" and "Is there a difference between the words or smileys of male and female social media users and their representation of SWLS?". Both of these auxiliary questions are related to the primary research question. The potential correlations between the SWLS output and the occurrence of the LIWC word categories are also examined, since the use of certain word groups independent of usage frequency, such as swear words, can also be crucial in forecasting these individuals' satisfaction level.

4 Methodology

The data used in this study is compiled from the myPersonality database application, which is an end product of the myPersonality project [6, 33]. The SWLS table utilized from the myPersonality project has the accompanying attributes: "id" and "swls". The first attribute "id" is generated by the myPersonality application and is unique inside the various databases, indicating a particular participant in the project. The second attribute "swls" is identical to the mean of results of the five inquiries: subjective health, Big Five personality traits, domain satisfaction, eudemonic well-being, and affects, for each participant. The value of a swls test question result is an integer between one and seven. Hence, the average of the five inquiry test results is a floating number between one and seven, where one is the minimum satisfaction from life and seven is the maximum satisfaction from life [34]. The only dependent variable in this study is the "swls" value. The dataset includes swls values with a range of 1.2 (lowest satisfaction) and 6.8 (highest satisfaction).

In myPersonality project, Facebook status update files of each user have been used as an input for the LIWC program. Facebook status update dataset contains a normalized measure of twenty two million status updates of 154,000 users. The words are classified into 4 main categories and 64 subcategories, according to the LIWC program used in the project. The names of the main categories are linguistic processes, psychological processes, personal concerns and spoken categories. For example, the personal concerns category includes elements such as work, achievement, leisure, home and so on. These elements map more than one word, for example, sadness element includes "crying, grief, sad", leisure element includes "cook, chat, movie". LIWC categories are listed in Table 1. The normalized measurements of the 64 attributes are the primary data source of the independent variables of this study. The Smiley data which is also used in correlation analysis is also compiled from the myPersonality project. The smileys dataset includes 144 popular smileys. Smileys were used in 12 million of status updates of 88,000 U.S. Facebook users. A few examples are as follows: ":)", ":(", ":-)", ":-(", ":'-(", ":'-)".

The demographics dataset includes 4.3 million records having a number of variables including age, gender, relationship status, and number of friends. Age and gender attributes are obtained from the Demographics database and number of status updates for the users are compiled from the Facebook Activity database, which is part of the myPersonality database. These attributes do not correspond to linguistics features, but are considered the auxiliary variables. These variables are used as the control variables to examine the research questions.

Table 1 LIWC categories and sample words

Category	Examples	Category	Examples
I—Standard linguistic dimensions		II—Psychological processes (continued)	
Pronouns	I, them, itself	Perceptual processes	See, touch, listen
Articles	A, an, the	Seeing	View, saw, look
Past tense	Walked, were, had	Hearing	Heard, listen, sound
Present tense	Is, does, hear	Feeling	Touch, hold, felt
Future tense	Will, gonna	Biological processes	Eat, blood, pain
Prepositions	With, above	Body	Ache, heart, cough
Negations	No, never, not	Sexuality	Horny, love, incest
Numbers	One, thirty, million	Relativity	Area, bend, exit, stop
Swear words	*****	Motion	Walk, move, go
II—Psychological processes		Space	Down, in, thin
Social processes	Talk, us, friend	Time	Hour, day, o'clock
Friends	Pal, buddy, coworker	III—Personal concerns	
Family	Mom, brother, cousin	Work	Work, class, boss
Humans	Boy, woman, group	Achievement	Try, goal, win
Affective processes	Happy, ugly, bitter	Leisure	House, TV, music
Positive emotions	Happy, pretty, good	Home	House, kitchen, lawn
Negative emotions	Hate, worthless, enemy	Money	Audit, cash, owe
Anxiety	Nervous, afraid, tense	Religion	Altar, church, mosque
Anger	Hate, kill, pissed	Death	Bury, coffin, kill
Sadness	Grief, cry, sad	IV—Spoken categories	

(continued)

Table 1 (continued)

Category	Examples	Category	Examples
Cognitive processes	Cause, know, ought	Assent	Agree, OK, yes
Insight	Think, know, consider	Nonfluencies	Uh, rr*
Causation	Because, effect, hence	Fillers	Blah, you know, I mean
Discrepancy	Should, would, could		
Tentative	Maybe, perhaps, guess		
Certainty	Always, never		
Inhibition	Block, constrain		
Inclusive	With, and, include		
Exclusive	But, except, without		

Table 2 Summary statistical attributes of the dataset

	Male	Female
# of cases	812	1367
# of valid cases	812	1367
Avg. age	24.25	25.05
Total # of status updates	159,944	304,027
Avg. # status updates	196.98	222.4

The descriptive statistics of the data are summarized in Tables 2 and 3. Table 2 summarizes the basic descriptive statistics of the overall data, and Table 3 illustrates the descriptive statistics of selected attributes. These attributes are ordered based on importance plots generated by the Statistica [35, 36] program using F-value analysis results. The statistics of only top 22 attributes are included due to space limitations.

LIWC data include normalized measures for categories, while the Smiley dataset provides frequencies, therefore the Smiley dataset and LIWC were examined separately. Demographics, SWLS data and Facebook activity, were linked with Smiley data and LIWC data independently. 182 tuples with special user identifiers for LIWC and 2291 tuples for smiley were obtained. The LIWC data was joined with the Demographics data to test the effects of gender and age based on the study of Farnadi et al. [18].

A "number of status update" threshold for LIWC linked data was set to dispense the impact of noise on the data. LIWC data was decreased by 1003 tuples (32%), when the users with a minimum number of 50 posts were used. Similar analysis was applied to both datasets to establish a base method. The threshold value was set as 50 posts, which enabled sufficient data points for all the tests, while maximizing the threshold.

Two tails of the data were used to form two separate subsets to examine the potential linguistic usage differences between the individuals with low and high satisfaction levels. The first subset consisted of people having satisfaction level lower

Table 3 Detailed statistical analysis of selected attributes

Male	Minimum	Maximum	Mean	Std. deviation	Female	Minimum	Maximum	Mean	Std. deviation
I	0.06	10.53	3.9133	1.73092	I	0	10.31	4.4837	1.7704
we	0	2.09	0.4034	0.31556	we	0	3.01	0.4183	0.31465
you	0	4.55	1.3505	0.81754	you	0	6.33	1.4468	0.90223
verb	14.88	64.27	34.1101	5.98171	verb	0	80.06	34.1526	5.97708
negate	0	3.78	1.3681	0.52754	negate	0	3.22	1.4302	0.49479
swear	0	3.53	0.4634	0.47483	swear	0	4.17	0.3151	0.35292
family	0	1.46	0.2374	0.20735	family	0	2.33	0.3616	0.30384
posemo	0.03	7.86	3.317	1.13402	posemo	0	10.16	3.8768	1.26514
negemo	0.09	5.07	2.0248	0.81747	negemo	0	6.63	1.9202	0.74848
anger	0	4.1	0.9039	0.56681	anger	0	5.08	0.7282	0.44691
sad	0	1.5	0.3799	0.20643	sad	0	1.79	0.4206	0.21208
discrep	0	2.72	1.2294	0.48624	discrep	0	2.92	1.2747	0.47542
tentat	0	4.35	1.8128	0.72614	tentat	0	4.26	1.7172	0.62831
certain	0	2.58	1.0586	0.45406	certain	0	3.2	1.0463	0.42595

(continued)

Table 3 (continued)

Male	Minimum	Maximum	Mean	Std. deviation	Female	Minimum	Maximum	Mean	Std. deviation
Percept	0.07	5.02	1.8182	0.59213	percept	0	5.28	1.9167	0.62038
hear	0	3.78	0.448	0.27113	hear	0	1.76	0.4437	0.24951
bio	0.07	6.62	2.0759	0.83403	bio	0	6.11	2.2937	0.80428
body	0	2.19	0.7315	0.3759	body	0	2.72	0.7782	0.37075
health	0	2.4	0.5719	0.28738	health	0	2.78	0.635	0.29985
work	0	4.55	1.2894	0.69998	work	0	5.08	1.2077	0.70021
leisure	0.06	4.85	1.3168	0.59702	leisure	0	4.87	1.2784	0.55489
home	0	1.92	0.3977	0.262	home	0	2.6	0.5499	0.34289
c#swl	1.2	6.8	4.238	1.4214	c#swl	1.2	6.8	4.339	1.3799

than two, while the second subset consisted of people having satisfaction level equal or higher than six. These two tails of data were used in some of the statistical tests. The procedure is discussed in detail in the following section.

During the selection of LIWC dictionary attributes, the following decisions are made:

- Standard linguistic dimension attributes are generally excluded due to logically irrelevant expectation to the SWLS with the exception of "i" and "we" attributes. The reason for this exception was the existence of effects of using these attributes in the earlier literature.
- Psychological Processes attributes are included with the logical expectancy that Social Processes, Affective Processes, and Biological Processes attributes would be more likely to be relevant to SWLS, but Cognitive Processes, Perceptual Processes, and Relativity attributes would be less likely to be relevant to SWLS.
- Personal Concern attributes are included with the expectancy to be highly relevant to SWLS.
- Spoken Categories attributes are excluded due to their logically irrelevant expectation to the SWLS.

The data analysis procedure included five parts, Fig. 1. The boxes painted in yellow color is out of scope of this work but part of the myPersonality project. In the first part, LIWC data were merged with Demographic data, and the impact of age and gender on the SWLS scores were examined by Analysis of Variance (ANOVA) tests. The Anova test was used to analyze the variances of groups. This method was based on comparison of means of groups. The formula is given in Eqs. (1)–(3) where:

$$F = \text{ Anova Coefficient},$$

$$\text{Anova } Coefficient \quad F = MST/MSE$$
$$MST = Mean \ treatment \ sum \ of \ squares,$$
$$MSE = Mean \ sum \ of \ squares \ due \ to \ error, \tag{1}$$

$$\text{Mean sum of squares} \quad MST = \frac{SST}{(p-1)}$$
$$SST = \sum (n(x - \bar{x}))^2 \tag{2}$$

Treatment = Specific combination of factor levels whose effect is to be compared with other treatments.

SST = Sum of squares due to treatment, variation attributed to, or between groups

p = Total number of populations,

n = Total number of samples in a population.

SSE = Sum of squared errors

S = Sample variance in a group

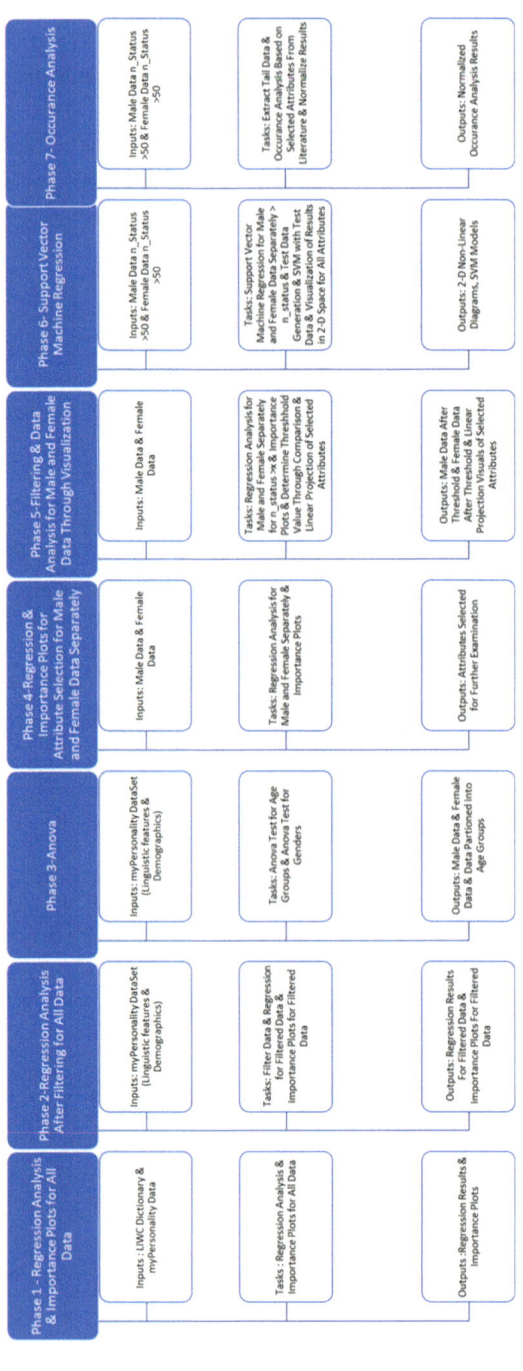

Fig. 1 Summary process methodology

$$\text{Mean square error} \quad MSE = \frac{SSE}{(N - p)}$$

$$SSE = \sum (ni - 1) Si^2 \tag{3}$$

The second part consisted of importance plots, which sorted the attributes according to their importance (F-value) on the dependent variable, SWLS. Lin et al. [37] mentioned that the more people portray their encounters with compromising styles (i.e. more utilization of "we"), the better psychological health they will experience. In the third part of the analysis, multiple linear regression models were constructed to investigate the impact of the frequency of the "we" pronoun compromising style on the well-being, Eq. (4). In this part, to demonstrate how SWLS scores are affected by using pronouns "we" and "I" together with words containing positive and negative feeling, multiple linear regression models were developed.

$$\text{Linear Regression classifier} \quad y = \beta_0 + \beta_1 x_1 + \beta_2 x_2 + \beta_3 x_3 + \beta_4 x_4 + \beta_5 x_5$$
$$+ \beta_6 x_6 + \beta_7 x_7 + \beta_8 x_8 \tag{4}$$

Linear regression analysis method was insufficient to identify the non-linear relations. Non-linear regression requires a priori decision on the class of relations (exponential, logarithmic, power functions etc.), which becomes extremely difficult when dozens of independent variables are considered simultaneously, such as the case of modeling personal attributes using linguistic attributes. Machine learning methods provide several advantages within this context for identifying a mapping function from the independent variables to the dependent variable. In recent years, Support Vector Machines (SVM), with their strong generalization properties, emerged as a powerful machine learning alternative, which is the main focus of this study; to identify a predictive model of personal attributes based on Facebook sharing. Hence, in the fourth part of data analysis, SVM regression models were developed to identify the relations between SWLS and independent attributes. Statistica commercial software was used in data analysis of all parts.

SVM is a non-probabilistic supervised classification and regression method derived from statistic learning theory. Input vectors are represented in a highly dimensional feature space using a previously chosen kernel function to distinguish the classes with decision surface, notably termed as the optimal hyper plane. Special properties of the decision surface empower its high generalization ability. SVM is a machine learning method which is similar to neural networks. Different types of kernel functions can be used in SVM models (linear, sigmoid, polynomial, radial basis function). In SVM, input features are used to predict the depending attribute(s), using the training data. Each individual set of features is called a vector. The goal is to determine a hyper plane, which locates clusters of vectors with the same category of results in the same side of the hyper plane.

A two-dimensional hyper plane, which is for example, a straight line, will be capable of classifying a resulting attribute with two categorical values and two predictor attributes. Additional predictor values will result with additional dimensions in the hyper plane. For instance, a third predictor attribute can be a space similar to a cube which contains a surface similar to a hyper plane. There are several hyper planes (lines for a two-dimensional case) which can make this separation of vectors. The aim is to find the optimal hyper plane. The optimal hyper plane is the line which maximizes the sum of closest margins from both sides of the line, i.e. maximizes the margin between the samples on the two sides of the hyper plane. Kernel functions are necessary to model the real-life situations in which using straight lines or straight surfaces is not adequate to separate the data points. For a two-class problem with n samples and the training set consists of relations such as in Eq. (5).

Sample training set in an SVM problem $(y_1, x_1), ...(y_n, x_n), y_i \{class1, class2\}$ (5)

The solution of hyperplane is the result of the optimization problem shown in Eq. (6) [38] where

$$\xi = \text{slack variables to allow non-separable variables,}$$
$$w = \text{vector and } b = \text{scalar to define optimal hyperplane orientation,}$$
$$\varphi = \text{Kernel function.}$$

Optimum SVM hyperplane error function $\min \frac{1}{2} w^2 + C \sum_{i=1}^{n} \xi_i$

$$\text{subject to: } y_i (w \cdot \varphi(x_i) + b) \geq 1 - \xi_i \text{ and}$$
$$\xi_i \geq 0 (\forall \, data \, points \, xi) \quad (6)$$

Outline of common kernel substitutes are as shown in Table 4.

The recent version of the SVM technique (Cortes and Vapnik) [39] requires the selection of an adequate kernel function. The radial basis kernel usually leads to better results, since it has less parameters, hence the radial basis kernel was used in this study. A ten-fold validation process was used to determine the optimal SVM regression parameters for the kernel functions. Sensitivity analysis was performed to determine the relations between the independent variables and the dependent attributes. In sensitivity analysis, the SVM regression model was used to determine the SWLS scores, while the values of one independent variable was increased between the minimum and maximum values with 0,1 increments, and the values of remaining independent variables were kept constant at their average value. The procedure was repeated for all of the variables to determine the impact of independent variables on the dependent attributes.

Table 4 Kernel function alternatives

Kernel	Formula	Kernel	Formula
Linear	xiTx	Radial basis	$\exp - \gamma \, (\|xi - x\|2)$
Polynomial	$(\gamma xiTx + r)d$	Sigmoidal	$\tanh(\gamma xiT \, x + r)$

The SVM regression models were constructed separately for both genders. There were 426 male and 431 female user records. The initial dataset (n_status >= 50) was grouped as 75% training 25% test cases in the first execution. During the second execution, 100% of initial dataset was used as the training set and sensitivity data as the test set. There are two objectives in constructing the SVM regression models with two datasets. The first one is to confirm the significant relations between the independent attributes and the dependent attributes detected through 2-D plots and the second is to determine the effects of attributes on the SWLS scores through sensitivity analysis.

In the final part of the process, the significance of difference between the occurrence of some linguistics features in user groups with low and high satisfaction were tested by hypothesis tests and the results were compared with the results of the SVM regression. In hypothesis tests, two tails of dataset, users with SWLS score <2 (extremely dissatisfied group), and users with SWLS scores >= 6 (highly satisfied group) were used. The least satisfied group included 173 sample users and the most satisfied group included 426 sample users. Occurrence test was performed by checking the occurrence of 64 LIWC word categories for each sample. If a word appeared once in the Facebook post then the occurrence was considered as one, otherwise it was considered as zero for the sample. The sum of occurrence values for each word category for the least and the most satisfied groups were normalized by dividing the sums to the total number of users in each group. The normalized occurrence values were used in hypothesis tests.

5   Results

In the first part of data analysis, ANOVA tests were performed to identify the impact of age and gender on SWLS. The first test was performed by grouping the data into 7 categories according to age: <20, between 20 and 30, between 40 and 50, between 50 and 60, >70. The impact of age on SWLS scores was not significant ($p = 0.862$, SST $= 4.989$, degree of freedom $= 6$, MS $= 0.832$, F $= 0.426$). However, the ANOVA test results indicated that the impact of gender on the SWLS scores was significant at $p = 0.1$ level ($p = 0.101$, SST $= 5.248$, degree of freedom $= 1$, MS $= 5.248$, F $= 2.695$).

In the second part of the data analysis, correlation analysis and importance plots were used to identify the potential attributes for the SVM model. Spearman Rho was used to measure the correlations between selected attributes and SWLS. Correlations for base dataset (3182 tuples) (n_status >= 0), thresholded dataset (2178 tuples) (n_status >= 50), and gender grouped dataset with threshold (812 male, 1366 female tuples) (n_status >= 50) are presented in Table 5.

Correlations between LIWC attributes and SWLS scores in the base dataset supported some of the initial hypotheses. Use of the pronoun "we" had a positive correlation with the subjective well-being, and also use of pronoun "i" had a negative correlation with SWLS, in parallel with Lin et al.'s study [37]. While positive

Table 5 Spearman correlations among SWLS and selected LIWC attributes

LIWC attribute	SWLS score (n_status >= 0)	SWLS score (n_status >= 50)	SWLS score male (n_status >=50)	SWLS score female (n_status >=50)
i	−0.066**	−0.093**	−0.095**	−0.099**
we	0.065**	0.072**	0.063	0.074**
negate	−0.099**	−0.121**	−0.089*	−0.142**
swear	−0.014**	−0.191**	−0.178**	−0.2**
posemo	0.051**	0.054*	0.0620	0.044
negemo	−0.165**	−0.229**	−0.232**	−0.223**
anger	−0.158**	−0.215**	−0.190**	−0.227**
sad	−0.070**	−0.104**	−0.103**	−0.108**
body	−0.094**	−0.121**	−0.130**	−0.119**
work	0.069**	0.099**	0.096**	0.107**
leisure	0.076**	0.078**	0.0520	0.093**
home	0.067**	0.066**	0.117**	0.027

** Correlation is significant at the 0.01 level
*Correlation is significant at the 0.05 level

emotion was correlated with a greater life satisfaction, use of words that fall into negative emotion, anger and sad categories presented lesser life satisfaction scores. When the data with <50 status updates were removed, the correlations became more significant.

All of the correlations except the correlation of "home" attribute increased, when the data with <50 status updates were removed. Negative emotion and anger were negatively correlated with SWLS at levels >0.2, as shown in Table 5. As the ANOVA test results revealed that gender has an impact on SWLS results, correlations were analyzed for female and male users separately. Noticeable results include, female Facebook users who utilize more negative words, anger words and swear words when contrasted with their male counterparts would probably have less satisfaction with their life.

Tov et al. [17] suggested that the use of positive and negative emotions impacted the SWLS. Similarly, Lin et al. [37] claimed that use of the pronoun "we" resulted with a style which affected the SWLS scores. The impact of these variables on the SWLS were also achieved in this study using myPersonality data, which is illustrated in Fig. 2a.

In Fig. 2a, users with SWLS scores higher than 6.0 (highest life satisfaction) are displayed with black stroked white specks, and users with SWLS scores lower than 2.0 (lowest life satisfaction) are displayed with black specks. The results in Table 5 show that positive emotion loses its significance when data was isolated according to gender. Hence, in the linear projection diagram, both genders were shown in the

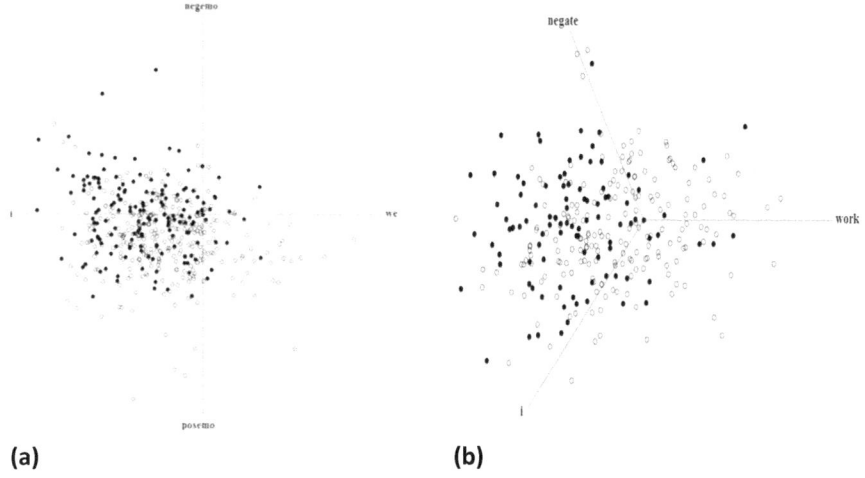

(a)   **(b)**

Fig. 2 Linear projections **a** attributes "we" and "I" on horizontal axis, and "pomotion" on vertical axis. **b** female users: attributes "work", "I", and "negate" (n_status >= 50)

same diagram. Although, there is not a clear-cut for the zones in Fig. 2a, users whose linguistic preferences fall into the 4th region (greater use of positive emotion and the pronoun "we") in general, had the highest life satisfaction.

Figure 2b contains data points identical to that of Fig. 2a however, only female users were included on this projection. Figure 2b indicates that, although negation is negatively correlated with SWLS; regardless of using negation words, female Facebook users using words related to work category had higher SWLS scores. In addition to correlation analysis, a feature significance test based on F-values was performed. The significant attributes are sorted according to their importance and are summarized in Table 6.

In the third part of data analysis procedure multiple linear regression models were developed in which the significant attributes are defined as the independent variables, and SWLS score was defined as the dependent variable. The linear regression models for users with status updates > 0 ($R^2 = 0.07063411$, Adjusted $R^2 = 0.05155185$) and users with status updates >= 50 ($R^2 = 0.10599956$, Adjusted $R^2 = 0.07893426$) had very low R^2. Hence, the models revealed that linear regression analysis did not explain the relation between the SWLS and the attributes adequately. In the fourth part of data analysis, in order to investigate potential non-linear relations and to have models with strong generalization capabilities, SVM regression models were developed for both genders independently.

SVM regression models for male users were developed using radial basis kernel for 19 independent variables (all the attributes determined by feature selection analysis, excluding gender). 542 support vectors were generated for the male users. The observed mean and the predicted mean for the male users were 4.010256 and 4.441462 respectively, and the mean squared error was 2.247173. 2-D plots of independent and dependent variables were generated to illustrate the results of the SVM

Table 6 First 20 LIWC attributes sorted according to their importance (F-values) on SWLS (N = 2178, both genders, n_status >=50)

	Attribute	F-Value	p-value		Attribute	F-Value	p-value
1	anger	7.788	0.000	11	discrep	2.332	0.001
2	swear	7.609	0.000	12	tentat	2.211	0.002
3	negemo	7.094	0.000	13	bio	2.211	0.002
4	negate	2.784	0.000	14	we	2.054	0.005
5	gender	2.695	0.101	15	you	1.858	0.020
6	verb	2.544	0.003	16	relative	1.838	0.015
7	sad	2.499	0.001	17	leisure	1.826	0.018
8	body	2.415	0.001	18	percept	1.787	0.019
9	health	2.414	0.001	19	i	1.781	0.020
10	work	2.367	0.002	20	hear	1.777	0.029

regression. These plots were generated using the test data of original dataset. In the validation phase of the SVM regression models, the predicted data corresponding to each validation data was used to demonstrate the autonomous dependency of independent attributes to the dependent attribute. In each plot, values of one attribute were varied between its minimum and maximum values, while the values of the remaining attributes were kept constant at their mean values.

The plots reveal that the SVM regression models captured significant impacts of the attributes "work", "anger", "leisure", "negemo", "i", "swear", and "sad" on the SWLS, as shown in Fig. 3. The plots were also used to compare the results of the SVM regression models with the correlations determined in the second stage of the data analysis procedure. The plots revealed that the autonomous effects of independent attributes: "work", "swear","we", "leisure", and "negemo" to the dependent attribute are in parallel with the results of correlation analysis (Fig. 3).

SVM regression models for female users were also developed using radial basis kernel for 19 independent variables. 862 support vectors are generated for the female users. The models had an observed mean of 4.232749, predicted mean of 4.435168, and mean squared error of 1.914240. During the validation phase, the autonomous dependency of independent attributes to dependent attribute were analyzed using plots that are based on sensitivity analysis, analogous to the male gender data. The plots of female users were included in Fig. 4. The plots revealed that the attributes "anger", "swear", "negate", "body", "work", "negemo", "leisure", and "i" had a significant impact on the SWLS for the female users. The results of the SVM regression models for the female users were also compared with correlations determined in the second phase. The autonomous effects of independent attributes "anger", "swear", "negate", "body", "work", "negemo", and "leisure" to the dependent attribute are in parallel with the previous correlation analysis results (Fig. 4). The most significant difference between SVM models for male and female Facebook users is the correlation of SWLS with the frequency of use of body words in status updates.

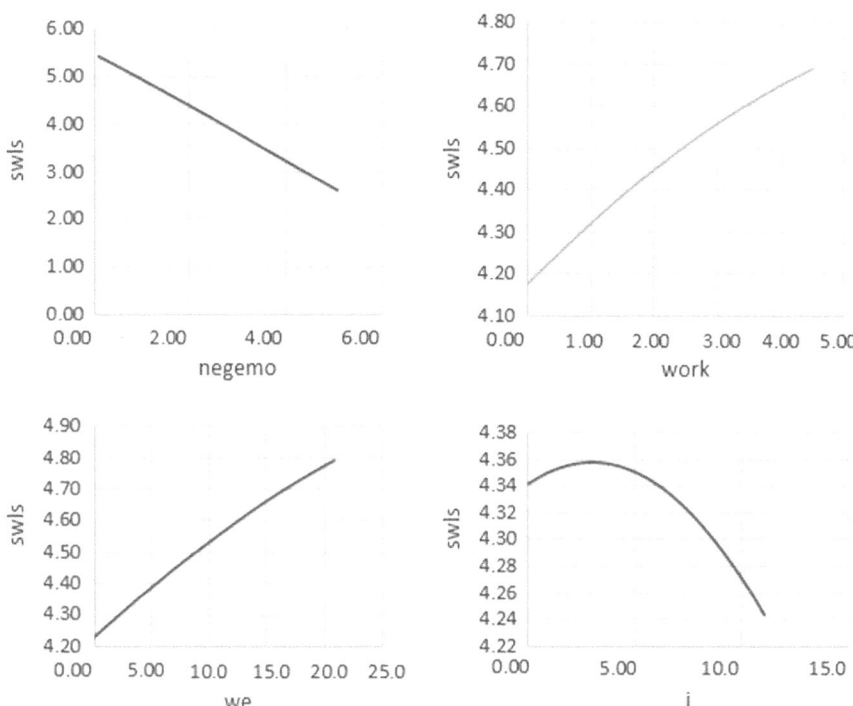

Fig. 3 Sample SVM results for male users, n_status >= 50, attributes shown: negemo, work, we, and i from left to right and from top to bottom

In the final part of SVM modeling, occurrence tests were performed and the results were compared with the findings of the SVM regression models. The occurrence test results for two satisfaction groups were provided in Table 7. The most noteworthy result is the 0.064% difference of the use of "swear" word event between users with highest satisfaction and lowest satisfaction groups. However, there is less difference for the features like "anger", "negative emotion", "work", and "negate" between the two groups. "Positive emotion" is also an intriguing word event. Slight differences in the use of positive-feeling words were identified for positive feelings such as; "love, nice, sweet" among the two groups. The second most significant contrast between the most satisfied and least satisfied group of users was specifying the "family" in any event in the status updates only once leaded to higher satisfaction. Even though family did not appear among the most significant attributes in the previous data analyses, this result, which appeared only in the occurrence tests, is indeed very logical. One other interesting finding in the occurrence tests is use of words that belong to categories of "motion" (e.g.: arrive, car, go), "space" (e.g.: down, in) and "time" (e.g.: end, until, season) did not reveal a significance difference between lowest and highest satisfaction groups, as expected. Hence, these results bolster the robustness of the occurrence test.

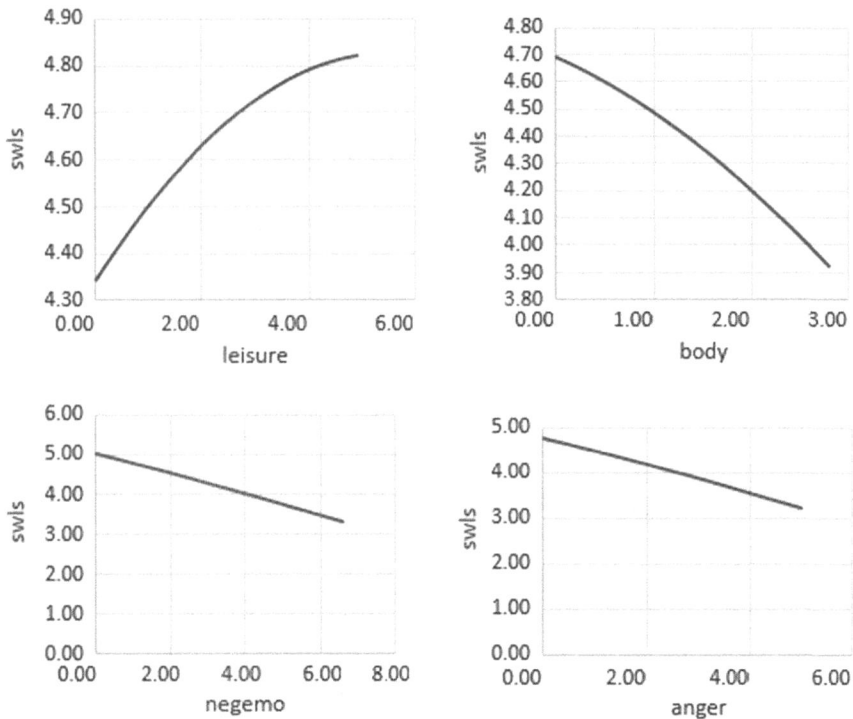

Fig. 4 Sample SVM results for female users, n_status >= 50, attributes shown: leisure, body, negemo, and anger from left to right and from top to bottom

6   Discussion and Limitations

Several techniques were used consecutively in a range of circumstances to discover logical relationships between selected attributes from myPersonality data and SWLS results. The set and order of the techniques employed were chosen by the author experimentally, after many trials. This effort resulted in clearly demonstrating relationships such as in Figs. 3 and 4. The sequential order of the used techniques is shown in Fig. 1.

In general, the results of the data analysis procedure are in parallel with the previous research, which focused on the use of first-person singular pronoun versus the use first person plural pronouns. Relationship awareness, which relates positive relationship outcomes with the couples that use "we" word more instead of "i", is pointed out in Acitelli [13]. The study revealed that positive relationship outcomes are likely to impact an individual's satisfaction in life in a positive way. Kayser et al. [40] compared the use of first-person plural pronouns with the use of first person singular pronouns between couples dealing with breast cancer. The outcomes demonstrated that couples using "we" word had better engagement with the illness, whereas, couples using "i" rather had a more avoidant manner in coping with the

Table 7 Occurrence test results for extremely satisfied and extremely dissatisfied users

LIWC category	Least SWL group % word usage	Most SWL group % word usage	Difference	LIWC category	Least SWL group % word usage	Most SWL group % word usage	Difference
swear	0.938	0.873	0.064	leisure	1.000	0.993	0.003
family	0.893	0.953	0.059	inhib	0.973	0.976	0.003
shehe	0.929	0.963	0.034	they	0.929	0.929	0.002
friend	0.876	0.906	0.030	home	0.982	0.980	0.002
humans	0.964	0.990	0.025	assent	0.982	0.980	0.002
anx	0.964	0.943	0.021	past	0.991	0.993	0.002
relig	0.991	0.970	0.021	negate	0.991	0.993	0.002
we	0.946	0.966	0.019	anger	0.991	0.993	0.002
death	0.893	0.913	0.019	excl	0.991	0.993	0.002
future	0.991	0.973	0.017	health	0.991	0.993	0.002
hear	0.991	0.973	0.017	sad	0.973	0.973	0.000
achieve	0.982	1.000	0.017	funct	1.000	1.000	0.000
nonfl	0.893	0.876	0.017	pronoun	1.000	1.000	0.000
see	1.000	0.986	0.013	ppron	1.000	1.000	0.000
body	0.982	0.993	0.011	i	1.000	1.000	0.000
ingest	0.955	0.966	0.010	article	1.000	1.000	0.000
adverb	1.000	0.990	0.01	verb	1.000	1.000	0.000
work	0.991	1.000	0.008	auxverb	1.000	1.000	0.000
You	0.982	0.990	0.007	present	1.000	1.000	0.000
money	0.973	0.966	0.006	preps	1.000	1.000	0.000
cause	1.000	0.993	0.006	quant	1.000	1.000	0.000
discrep	1.000	0.993	0.006	social	1.000	1.000	0.000
tentat	1.000	0.993	0.006	affect	1.000	1.000	0.000
number	0.973	0.980	0.006	posemo	1.000	1.000	0.000
feel	0.973	0.980	0.006	cogmech	1.000	1.000	0.000
filler	0.911	0.916	0.005	incl	1.000	1.000	0.000
sexual	0.982	0.986	0.004	percept	1.000	1.000	0.000
ipron	1.000	0.996	0.003	bio	1.000	1.000	0.000
conj	1.000	0.996	0.003	relative	1.000	1.000	0.000
negemo	1.000	0.996	0.003	motion	1.000	1.000	0.000
insight	1.000	0.996	0.003	space	1.000	1.000	0.000
certain	1.000	0.996	0.003	time	1.000	1.000	0.000

illness. Both of these research findings are in parallel with the results presented in this chapter. Other logical findings related to attributes, such as "posemo", "negemo", "family", "swear", are also explained in Results Section.

The graphical findings demonstrated in this work is unique to this study. Previous similar analysis used only numerical methods, such as regression based methods, utilizing graphical analysis during the evaluation of SVM results, provided numerous sums of graphics and tables. These graphics thus illustrate individual non-linear relationships of the selected attributes to the SWLS. If the slope functions of each attribute can be calculated, these functions can be used for the calculation of SWLS. A conceptual equation for the calculation of SWLS using individual relationships of selected attributes is described using Eqs. (7) and (8).

The use of linear projection graphs, such as in Fig. 2, to examine simultaneous effects of linguistic features to SWLS is also unique to this study. Visualization techniques utilized throughout the study helped to better understand the data and to better focus on appropriate attributes.

The research had few limitations. Explaining SWLS using Facebook data has its inherent constraints. Expression of emotions in status updates can very well be influenced by the application itself, or other irrelevant factors. Wang et al. [8] offered some clarifications for FGNH to forecast good mood or well-being of Facebook users. The relationship among Facebook status updates and the general emotion of the profile owners may not be significant, yet people can invariably hide their true feelings to paint different self-images to their Facebook friends. Another limitation is the use of LIWC, which is mainly intended for conventional types of composed dialect and does not contain modern abbreviations (e.g. LOL) or fashionably misspelled words.

Conceptual equation of SWLS based on slope functions
$$swls = \alpha * f1(att1) + \beta * f2(att2) + \gamma * f3(att3) + \theta \qquad (7)$$

Slope function for each attribute
$$fi(atti) = cn * atti^n + cn - 1 * atti^{n-1} + \cdots + c1 * atti^1 + c0 \qquad (8)$$

With regard to the Smileys dataset, which is compiled from the myPersonality application, also had certain limitations. 144 popular smileys were included in the dataset, one or two of the most expected smileys were being utilized among many users and rest of the attributes were scarcely used. Hence, neither normalization/discretization were not performed on this dataset and this data was not joined with the LIWC attributes. The techniques utilized on this dataset, such as Self Organizing Maps clustering did not lead to any significant findings.

7 Conclusions

Initially, the effects of demographic attributes on the SWLS data is inspected as proposed by Farnardi et al. [18]. To eliminate insignificant attributes, linguistic features are correlated against SWLS results. The results were pleasing as it is expected that "positive emotion", "home", "leisure", and "work" are positively correlated with SWLS, while "negative emotion", "anger", "sadness", "negation", and "swear" are negatively correlated with SWLS. It is worthy to note that the number of status updates threshold improved the correlation results. Differences between genders became more obvious after applying the threshold to the data. The main difference was, if female users are using negation, swear, and anger words, it is more likely they are to be less satisfied with their lives when compared with male users preferring the same linguistic features.

Following this, some visualization techniques are utilized to better understand the data. It is discovered that, although negation has a negative correlation with SWLS, regardless of usage of negation words, female Facebook users preferring work related features in their status updates are more likely to be satisfied with their lives. When linear models failed to explain the relation between the satisfaction index and the other attributes properly, SVM regression was logically employed as the next step. Among the results of the SVM models for both genders, the most important difference was the usage of "body" related words for female users to better explain SWLS when compared to the male counter parts. This result could support a hypothesis that female users who are obsessed with using sentences including more body-related linguistic features in their status updates are more likely to be dissatisfied with their lives.

Ultimately, the linguistic word occurrence differences between the highest satisfied user group and the least satisfied user group were examined. Although it never came up in the previous analyses, occurrence of "family" related words in status updates suggested a higher life satisfaction in this test. Although Facebook status sharings have certain limitations, the results of this study indicate that status updates can be used to successfully predict some of the people's attributes. The research findings support the fact that the linguistics styles favoured in Facebook status updates have correlations with SWLS, which was selected as a sample personal attribute. The results also reveal that the relationship between SWLS and Facebook sharing can be identified better when the datasets of male and female users are analyzed separately. Hence, the findings suggested that the online personal data should be analyzed separately for male and female genders in linguistics research considering personal attributes.

Regional factors may play an important role in linguistics studies. Future research including the effect of regional factors on the LIWC data may provide more comprehensible findings. Smileys are more universal, on the contrary, so they may not need translation and cultural or social interpretation.

The results of this study indicate that Facebook status updates have significant correlations with subjective well-being measurements. However, further research is needed to clarify the relation between the Facebook status updates and subjective well-being and to understand the contrast between these similar studies. Future social media research may also focus on real-time subjective well-being measurements across diverse populations. However, extension of linguistic dictionaries involving modern abbreviations, and region-based categories are needed to achieve real-time subjective well-being measurements. Utilization of features that do not depend on regional bias, such as smiley data can be used within this context.

References

1. Portal, T.S.-S. (2016). *Number of monthly active Twitter users worldwide from 1st quarter 2010 to 1st quarter 2016 (in millions)*. Retrieved from http://www.statista.com/statistics/282 087/number-of-monthly-active-twitter-users/.
2. Ganis, M., & Kohirkar, A. (2016). Social media analytics: Techniques and insights for extracting business value out of social media. *12*.
3. Gil de Zuniga, H. (2015). Citizenship, social media, and big data current and future research in the social sciences. *Social Science Computer Review*.
4. Ceron, A., Curini, L., & Iacus, S. M. (2015). Using sentiment analysis to monitor electoral campaigns: method matters. Evidence from the United States and Italy. *Social Science Computer Review, 33*(1).
5. Hilbert, M., Vasquez, J., Halpern, D., Valenzuela, S., & Arriagada, E. (2016). One step, two step, network step? Complementary perspectives on communication flows in twittered citizen protests. *Social Science Computer Review*.
6. Kosinski, M., Stillwell, D. J., & Graepel, T. (2013). Private traits and attributes are predictable from digital records of human behavior. *Proceedings of the National Academy of Sciences*.
7. Stillwell, D. J., & Kosinski, M. (2004). Mypersonality project: Example of successful utilization of online social networks for large scale social research. *American Psychologist, 59*(2), 93–104.
8. Wang, N., Kosinski, M., Stillwell, D. J., & Rust, J. (2014). Can well-being be measured using Facebook status updates? Validation of Facebook's gross national happiness index. *Social Indicators Research, 115*(1), 483–491.
9. Bradburn, N. M. (1969). The structure of psychological well-being.
10. Lyubomirsky, S., & Lepper, H. S. (1999). A measure of subjective happiness: Preliminary reliability and construct validation. *Social Indicators Research, 46*(2), 137–155.
11. Diener, E., Emmons, R. A., Larsen, R. J., & Griffin, S. (1985). The satisfaction with life scale. *Journal of Personality Assessment, 49*, 71–75.
12. Pavot, W., & Diener, E. (1993). Review of the satisfaction with life scale. *Psychological Assessment, 5*(2), 164.
13. Acitelli, L. K. (2002). Relationship awareness: Crossing the bridge between cognition and communication. *Communication Theory, 12*(1), 92–112.
14. Freud, D. (1901). Zur Psychopathologie des Alltagslebens (Vergessen, Versprechen, Vergreifen) nebst Bemerkungen über eine Wurzel des Aberglaubens. *European Neurology, 10*(1), 1–16.
15. Tausczik, Y. R., & Pennebaker, J. W. (2010). The psychological meaning of words: LIWC and computerized text analysis methods. *Journal of language and social psychology, 29*(1), 24–54.
16. Pennebaker, J. W., Francis, M. E., & Booth, R. J. (2001). *Linguistic inquiry and word count: LIWC 2001*. Mahwah: Lawrence Erlbaum Associates.
17. Tov, W., Ng, K. L., Lin, H., & Qiu, L. (2013). Detecting well-being via computerized content analysis of brief diary entries. *Psychological Assessment, 25*(4), 1069.
18. Farnadi, G., Geetha, S., Mehrdad, R., Michal, K., David, S., Marie-Francine, M., … Martine, D. (2014). How are you doing? Emotions and personality in Facebook. In *Proceedings of the EMPIRE Workshop of the 22nd International Conference on User Modeling, Adaptation and Personalization*.

19. Gill, A. J., French, R. M., Gergle, D., & Oberlander, J. (2008). The language of emotion in short blog texts. In *Proceedings of the 2008 ACM Conference on Computer Supported Cooperative Work* (pp. 299–302). ACM. https://doi.org/10.1145/1460563.1460612.

20. Dhir, A., Chen, S., & Nieminen, M. (2016). The effects of demographics, technology accessibility, and unwillingness to communicate in predicting internet gratifications and heavy internet use among adolescents. *Social Science Computer Review, 34*(3), 278–297.

21. Fichman, P., & Sanfilippo, M. R. (2015). Men and women react differently to online trolling, and their perceptions of the impact of trolling on online communities vary. *Social Science Computer Review, 33*(2).

22. Golder, S. A., & Macy, M. W. (2011). Diurnal and seasonal mood vary with work, sleep, and daylength across diverse cultures. *Science, 333*(6051), 1878–1881. https://doi.org/10.1126/science.1202775.

23. Coviello, L., Sohn, Y., Kramer, A. D., Marlow, C., & Franceschetti, M. (2014). Detecting emotional contagion in massive social networks. *PLoS One, 9*(3),

24. Bollen, J., Gonçalves, B., Ruan, G., & Mao, H. (2011). Happiness is assortative in online social networks. *Artificial Life*, 237–251.

25. Schwartz, H., Park, G. J., Sap, M., Weingarten, E., Eichstaedt, J., Kern, M. L., … Ungar, L. H. (2015). Extracting human temporal orientation from Facebook language. In *Proceedings of the 2015 Conference of the North American Chapter of the Association for Computational Linguistics:Human Language Technologies* (pp. 409–419). Denver, Colorado, USA: Association for Computational Linguistics.

26. Collins, S., Sun, Y., Kosinski, M., Stillwell, D., & Markuzon, N. (2015). Are you satisfied with life?: Predicting satisfaction with life from Facebook. In *International Conference on Social Computing, Behavioral-Cultural Modeling, and Prediction* (pp. 24–33). Washington, DC, USA: Springer.

27. Maurer, K. (2018). *Linguistic correlates of the quiet ego in narratives about the self*. Charleston, USA: Eastern Illinois University.

28. Khodadady, E., & Dastgahian, B. (2019). Relationship of subjective well-being and religiosity from theoretical and statistical perspectives. *SOJ Psychology, 6*(1), 1–8.

29. Wongkoblap, A., Vadillo, M. A., & Curcin, V. (2018). A multilevel predictive model for detecting social network users with depression. In *IEEE International Conference on Healthcare Informatic* (pp. 130–135). New York City, USA: IEEE.

30. Krys, K., Zelenski, J. M., Capaldi, C. A., Park, J., Tilburg, W. V., Osch, Y. V., … Uchida, Y. (2019). Putting the "We" into well-being: using collectivism-themed measures of well-being attenuates well-being's association with individualism. *Asian Journal of Social Psychology, 22*, 256–267.

31. Blais, M. R., Vallerand, R. J., Pelletier, L. G., & Briere, N. M. (1989). French-Canadian validation of the satisfaction with life scale. *Canadian Journal of Behavirol Science, 21*, 210–223.

32. Beck, A. T., Ward, C., & Mendelson, M. (1961). Beck depression inventory (BDI). *Archives of General Psychiatry, 4*(6), 561–571.

33. Stillwell, D., & Kosinski, M. (2016, 06 01). *myPersonality Project Website*. Retrieved 06 02, 2016, from myPersonality Project: http://mypersonality.org/wiki/doku.php?id=start.

34. Diener, E. (2006, February 3). *Understanding scores on the satisfaction with life scale*. Retrieved August 8, 2013, from http://internal.psychology.illinois.edu/, http://internal.psychology.illinois.edu/~ediener/Documents/Understanding%20SWLS%20Scores.pdf.

35. STATISTICA Advanced. (2016). (Dell) Retrieved September 23, 2015, from Statistica Product Web Site: http://www.statsoft.com/Products/STATISTICA/Advanced.

36. Statistica. (2015, January 1). *Statistica help*. Retrieved June 17, 2016, from Support Vector Machines—Cross-Validation Tab: http://documentation.statsoft.com/STATISTICAHelp.aspx?path=MachineLearning/MachineLearning/SupportVectorMachine/SupportVectorMachinesDialogCrossValidationTab.

37. Lin, W. F., Lin, Y. C., Huang, C. L., & Chen, L. H. (2014). We can make It better: "We" moderates the relationship between a compromising style in interpersonal conflict and well-being. *Journal of Happiness Studies*, 1–17.

38. TIBCO Software Inc. (2017). *Support Vector Machines (SVM) introductory overview*. Retrieved from Statssoft TextBook: http://www.statsoft.com/Textbook/Support-Vector-Machines#RegressionSVM.
39. Cortes, C., & Vapnik, V. (1995). Support-vector networks. *Machine Learning, 20*(3), 273–297.
40. Kayser, K., Watson, L. E., & Andrade, J. T. (2007). Cancer as a "we-disease": Examining the process of coping from a relational perspective. *Families, Systems & Health, 25*(4), 404.
41. Diener, E. (1984). Subjective well-being. *Psychological Bulletin, 95*(3), 542–575.
42. Schwartz, H., Park, G. J., Sap, M., Weingarten, E., Eichstaedt, J., Kern, M. L., … Ungar, L. H. (2015). Extracting human temporal orientation from Facebook language. In *Proceedings of the 2015 Conference of the North American Chapter of the Association for Computational Linguistics:Human Language Technologies* (pp. 409–419). Denver, Colorado, USA: Association for Computational Linguistics.

Digital Transformation to Build Smart Cities

Nasser Al Marzouqi, Cristina Bueti, and Mythili Menon

Abstract This Chapter provides insights into the smart and sustainable city concept in the international arena. Acknowledging the reality that the establishment of smart and sustainable cities is a journey and cannot be created overnight, the chapter underscores key definitions and frameworks relevant for the topic, while highlighting the important work conducted by international standards developing organizations in this domain. With the emergence of new and emerging technologies (AI, IoT, blockchain, machine learning) urban stakeholders are increasingly attempting to integrate them into their smart city plans. However, when implementing these diverse technologies within one ecosystem, it is important to be aware of the holistic interaction between them in relation to urban operations. This Chapter further delves into the interlinking of these technologies, while looking into key processes for data management. The Chapter concludes with three cases on upcoming smart cities in Asia, which have endorsed and implemented ICT-centric initiatives in various urban sectors to support their respective smart city ventures. One of the key take-aways from the cases examined in this chapter, is the moulding of the smart city plans in light of the COVID-19 pandemic by adopting AI-based applications for the monitoring of COVID-19.

Keywords Smart sustainable cities · IoT · Sustainable development goals · Big data · Artificial intelligence · Standardization · Key performance indicators · International telecommunication union

N. Al Marzouqi (✉)
Spectrum International Affairs—Telecom Regulatory Authority, Abu Dhabi, United Arab Emirates
e-mail: nasser.almarzouqi@tra.gov.ae

C. Bueti · M. Menon
International Telecommunication Union, Geneva, Switzerland
e-mail: cristina.bueti@itu.int

M. Menon
e-mail: mythili.menon@itu.int

© The Author(s), under exclusive license to Springer Nature Switzerland AG 2021
U. Ghosh et al. (eds.), *Machine Intelligence and Data Analytics for Sustainable Future Smart Cities*, Studies in Computational Intelligence 971,
https://doi.org/10.1007/978-3-030-72065-0_9

1 A Journey Towards Smart Sustainability in the Urban Context: An Introduction

With over 55% of the global population, cities form the epicentre of human settlement. Owing to the growth in population supplemented by rural–urban migration, cities are expected to house nearly 66% of the world's population by 2050 [1]. In the wake of the projected population rise, subsequent increase in demand for energy, water and other utilities could contribute to environmental degradation. Rapid growth in population will also exert pressure on the provision of basic amenities including health care, sanitation and affordable housing [2]. To cater to increasing population density in cities, several urban development concepts, including *smart cities*, *sustainable cities* and *eco-cities*, have been brought to the scene. The discussion on these concepts are primarily based on a technology- and citizen-centric approach for cities, which has been facilitated by the global discourse on "smart urbanism," encapsulating visions for a sustainable future predicated on technology-based urban infrastructures. Fragmented research on these urban development concepts has resulted in anarchy in terms of charting out their implementation in existing cities [3].

While *eco-cities* and *sustainable cities* are centred on driving urban planning to create "environmentally healthier" cities, the concept of "smart cities" relies on the incorporation of information and communication technologies (ICTs) for urban operations and on maintaining a high quality of life. The smart city concept has been researched since the 1980s. Currently, there are various definitions of what a smart city encompasses. The majority of smart city definitions focus on the integration of digital technologies into the various sectors, including health, transport, housing, manufacturing and the provision of utilities [4]. Other definitions cover data-driven processes for information and knowledge management in the urban space. Keeping these key elements in mind, and to allow for an even playing field in the smart city domain, the affordability of information and communication technologies is essential. An optimistic scenario is presented by "innovation" as technologies are becoming cheaper over time. The penetration of technologies and networks are also instrumental in assessing the potential success of a city in transitioning to a *smart city* [5].

As the smart city research zone gains traction, a common framework of understanding is required to put findings into practice. For this, a prioritization approach is needed:

- to derive best practices from existing smart city expeditions (based on simulation results); and
- to incorporate security mechanisms and ensure the provision of basic amenities for the overall safety, resilience and integrity of the smart city ecosystem.

With the inclusion of cyber-security aspects into the smart city concept, stakeholders are restructuring the smart city debate to consider additional interdependencies between socio-economic and socio-technical perspectives, and the

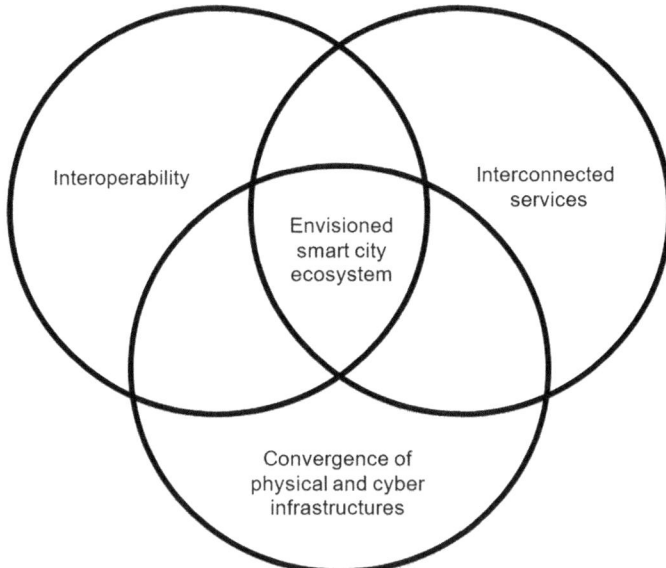

Fig. 1 Preliminary factors for smart cities [7]

physical infrastructure of cities. To this end, three factors are envisioned for the creation of smart cities in the twenty-first century (as indicated in Fig. 1) [6].

1. Convergence of cyber-physical infrastructures: This necessitates the interaction of software-controlled devices with the physical world for the provision of services/access to information.
2. Interoperability: Interoperability of the hardware and communication protocols is essential to facilitate the deployments of additional disparate devices/sensors for the delivery of services.
3. Interconnected services: The smart city ecosystem contains sensors, devices and machines that are capable of gathering and sharing data. In this context, interconnection between them is vital for the exchange of private data and for the devices to seamlessly communicate with each other, with the aim of improving the quality of urban services.

These factors bring forth the main aim for establishing smart cities—the offering of quality services to its citizens. It also highlights that the smart city domain is a heterogeneous ecosystem enveloping various sectors, devices and machines—all of which are dependent on data. In this line of thought, the concept of smart cities also finds itself to be synonymous with data hubs and knowledge to drive urban competitiveness and enhance quality of life [8].

Despite having a basic understanding of the smart city as a tech-citizen-centric urban agglomeration dependent on data, many smart city projects are unable to move beyond the pilot stage. Under these circumstances, it is essential to delve into the pitfalls encountered during the smart city transition, in order to outline the lessons learned and to prevent other urban stakeholders from following a similar unsuccessful trajectory when building their smart cities. Thus far, the importance of technologies, citizens and data has been underscored for smart cities; however, it is also important to study other interdisciplinary perspectives from smart city literature, which will help to incorporate regulatory concepts, improve stakeholder engagement and establish self-reliant smart cities—all of which will contribute to the existing smart city knowledge base [9].

Current research on smart cities does not emphasize the interaction between its different components, which can impact governance and performance. Additionally, the principle of sustainability does not take centre stage in many smart city expeditions, which can impact the components and the desired outcomes.

To reconceptualize and re-examine the research avenues on technology interactions between sectors, it is essential to map the stakeholder nexus and include sustainability as a core theme, while defining the required smart city regulatory and managerial approaches. Once established, this research stream could form the basis of an application framework for formulating policy design, analysing fieldwork and facilitating real-life smart city implementation. Accordingly, the United Nations agency for information and communication technologies, the International Telecommunication Union (ITU), laid the foundation for this work by establishing a concrete definition of the term *smart sustainable cities* in which *sustainable development*, as visualized in the Brundtland Report (1992)[1] [10], will not be undermined in the quest for smart cities. This process conducted under the umbrella of the ITU-T Focus Group on Smart Sustainable Cities (2014), analysed more than 100 existing definitions (from academia, research communities, private sector endeavours, public initiatives, NGOs and other standards-developing organizations) on smart city-related concepts. Based on this analysis, key terms were identified for inclusion in a standardized definition. These terms were categorized (see Table 1) for their easy incorporation into the definition [11].

The process concluded with the formulation of the following definition for smart sustainable cities (SSCs), together with the United Nations Economic Commission for Europe (UNECE):

> A smart sustainable city is an innovative city that uses information and communication technologies (ICTs) and other means to improve quality of life, efficiency of urban operation and services, and competitiveness, while ensuring that it meets the needs of present and future generations with respect to economic, social, environmental as well as cultural aspects.

In addition to forming part of an international standard, this definition was also embraced by the United Nations-United for Smart Sustainable Cities (U4SSC) initiative, which functions based on a unique collaboration between 17 UN agencies including ITU, UNECE and UN-Habitat. The activities of this initiative supplement the ongoing global standardization efforts for smart sustainable cities.

---

[1] Report of the World Commission on Environment and Development: Our Common Future contains the definition for sustainable development.

Table 1 Terms for inclusion in the definition for smart sustainable cities [11]

Economy	Governance	Environment	Society
Employment	Regulatory	Sustainable	People
GDP	Compliance	Renewable	Culture
Market – Global/Local	Processes	Land use	Social networks
Viability	Structure	Biodiversity	Tech Savvy
Investment	Authority	Water/Air	Demographics
PPP	Transparency	Waste	Quality of life
Value chain	Communication	Workplace	User experiences
Risk	Dialogue		Equal access
Productivity	Policies		End consumers
Innovation	Standards		Community
Compensation	Citizen services		needs
			The city as a database

The development of this standardized definition was followed by stakeholders clamouring for more guidance on the steps to be taken for smart sustainable cities along with metrics with which to evaluate the success and progress of smart sustainable city initiatives [12].

2   Future Standardization for Smart Sustainable Cities

Having a globally accepted definition of SSCs contained in an international standard, it is pertinent to understand the essence of what international standards are. Most standards developing organizations (SDOs) do not seek to define what the term "standards" entails. In the colloquial sense of the term, standards refer to the norm or requirement of a specific system. One definition put forth by the International Organization for Standardization (ISO) refers to standards as "*a document, established by a consensus of subject matter experts and approved by a recognized body that provides guidance on the design, use or performance of materials, products, processes, services, systems or persons*" [13].

> Standards can be developed by national, regional and international standards developing organizations and also by businesses or other organizations for their own internal use. They can also be developed by consortia of businesses to address a specific marketplace need, or by government departments to support regulations. [13]

International Organization for Standardization (ISO), ITU and International Electrotechnical Commission (IEC) are the three international SDOs as recognized by the World Trade Organization (WTO). All three SDOs are also actively involved in the creation of smart city-related standards. The current smart city standardization terrain is dependent on cooperation among SDOs, updating ICT standards and the inclusion of urban development notions to establish the principles of openness, interoperability, compatibility and resilience for smart cities worldwide.

The main challenges that SDOs might collectively face when developing SSC-related standards (as identified by ITU) are:

- Ensuring interoperability between city systems.
- Managing privacy, cybersecurity, resilience, and data flows on a whole-system basis.
- Understanding sector-specific needs within the smart city ecosystem and the interlinkages between sectors along with the cross-cutting impact of ICT adoption.
- Coordination with other SDOs to minimize overlap on the development of complementary SSC standards and addressing core knowledge gaps [14].

ITU has also identified a series of steps that can be followed by interested stakeholders before the start of their respective SSC expeditions (Table 2).

Table 2 Steps to initiate your smart and sustainable city journey [15]

Step 1. Setting the vision for the SSC
This step involves assessing the city's existing situation and establishing a long-term strategy for the utilization of ICTs for the delivery of services and facilitating multistakeholder involvement, including citizen engagement
Step 2. Identify your SSC targets
This second step recommends the creation of a master plan for the SSC implementation with the objectives, priorities and actions needed in the short, medium and long term. It can also include details on the adoption of emerging technologies and processes including Internet of Things (IoT), Artificial Intelligence (AI), machine learning (ML) and Big Data analytics
Step 3. Achieve political commitment
With the required political approval, local governments can commence the strategic planning for the implementation of Master Plan from Step 2
Step 4. Build your SSC
During this step, traditional infrastructures are infused with ICTs to foster the transformation to an SSC. As a part of this step, local governments are encouraged to ensure infrastructure maintenance after it is put is in place
Step 5. Measure your city-progress
After an ICT-based infrastructure is established, relevant metrics are required to assess progress against the objectives hi Step 1
Step 6. Ensure Accountability and Responsibility
This step aims to evaluate, report and flesh out best practices, which can be shared with the global smart city community

Table 3 U4SSC key performance indicators categories [16]

Key performance indicators	Categories covered
Economy	ICT infrastructure. water and sanitation, drainage, electricity supply, transport, public sector, buildings, urban planning, innovation, employment, waste
Environment	Air quality, environmental quality, energy, public space and nature
Society and culture	Education, health, housing, social inclusion, safety, food security

With reference to Step 5 (Table 2), it is essential that cities have credible indicators for systematic monitoring, management and decision-making. These indicators should also be compatible with existing city indices. Under these circumstances, cities need to make an informed choice regarding the most suitable indicators in terms of their applicability and ability to assess the extent to which a given city has become smart and sustainable. There are existing urban indicators and ICT dashboards that were utilized by cities to ascertain their smart city journey; however, standardized frameworks in this domain are limited.

While there is no one-size-fits-all model for smart and sustainable cities, the United for Smart Sustainable Cities (U4SSC) initiative has spearheaded the development of an evaluation framework consisting of key performance indicators (KPIs) to monitor the progress of smart city transitions. These KPIs have also embedded elements from the United Nations Sustainable Development Goals (SDGs)[2] in order to help cities make progress on the "smartness" as well as on sustainability fronts.

These globally agreed SSC indicators have been implemented in more than 100 cities[3] with the aim of making cities and human settlements inclusive, safe, resilient and sustainable (in line with SDG 11) by leveraging technologies. The implementation of these KPIs has facilitated knowledge sharing and dialogue between stakeholders, thereby facilitating a feedback process for the refinement of the KPIs (Table 3).

3 Smart Cities in the Advent of Emerging Technologies

There is a multitude of ICTs and processes involved in the design, development and implementation of Smart Sustainable Cities, and some of the newer and more relevant technologies are:

---

[2]The Sustainable Development Goals were adopted by the United Nations to achieve a better and more sustainable future for the current, and future, generations. These global goals cover a range of topics including poverty, world hunger, gender inequality, climate change, water management, energy consumption, innovation, land degradation, and peace and justice.

[3]Cities including Dubai, Singapore, Valencia, Pully and Moscow have implemented in the U4SSC KPIs.

- Cloud Computing: Cloud computing enables an SSC to collect, analyse and use information to be productive and viable. It allows users to obtain data with high speed at anytime and anywhere. The most notable beneficial features of the usage of cloud computing are the economic savings and the growth potential. Engaging in mega urban development via a cloud reduces the operating costs of city services [17].
- Internet of things (IoT): It is referred to as "a global infrastructure for the information society, enabling advanced services by interconnecting (physical and virtual) things based on existing and evolving interoperable information and communication technologies" [18].
- Big Data analytics: Big Data analytics refers to the utilization of massive amounts of data to support decision-making processes in smart and sustainable cities. For Big Data analytics seven Vs are kept in mind: variability, veracity, visualization, value, validity, vulnerability, and volatility [19].
- Artificial Intelligence (AI): AI can be leveraged within the smart cities to enable the urban ecosystem to learn from experiences and adapt to changes in the environment to simulate human decision-making and reasoning processes. AI-based technologies can also facilitate the automation of services, thereby reducing the need for human capital [20].

These technologies can have a crosscutting impact across sectors in a smart sustainable city. The various sectors within an SSC, which would benefit from the integration of emerging technologies include health care, waste management, education, tourism, banking etc. Each subsystem in Fig. 2 addresses a different smart sustainable city service category (Fig. 2).

IoT, predicated on the interconnection of billions of devices, allows it to share information, and coordinate operations. This ability of IoT holds the potential to transform the urban systems such that they are able to function with limited supervision. With the proliferation of smart sensors and actuators in the smart city ecosystem, voluminous streams of data are generated. These data streams can serve as the fuel for driving smart city expeditions by forming the basis of decision-making for urban planning and delivery of services [21]. Obtaining valuable insights from large data streams collected through various sources is the essence of establishing a smart sustainable city. However, the intersection of IoT and data is still to be exploited to its full potential. As such, the collection, storage and analysis of data enable problem identification, comprehension, and solution. To this end, it acts as the raw material for diagnostics and decision-making (see Fig. 3). In terms of smart cities, the intersection of IoT and data analytics can detect urban oversights, shape appropriate solutions and validate outcomes. In essence, it serves as the foundation for innovation [22].

The coupling of IoT and data analytics has replaced traditional methods of data processing which have recorded various limitations, especially when working with large amounts of data. Additionally, ML algorithms are also being employed as a means of processes large amounts of raw data in real time with great accuracy and high efficiency [23]. With more devices being added to the IoT ecosystem, it is expected that the data streams harvested will continue to grow. These data will essentially be

Fig. 2 Subsystems of SSCs [14]

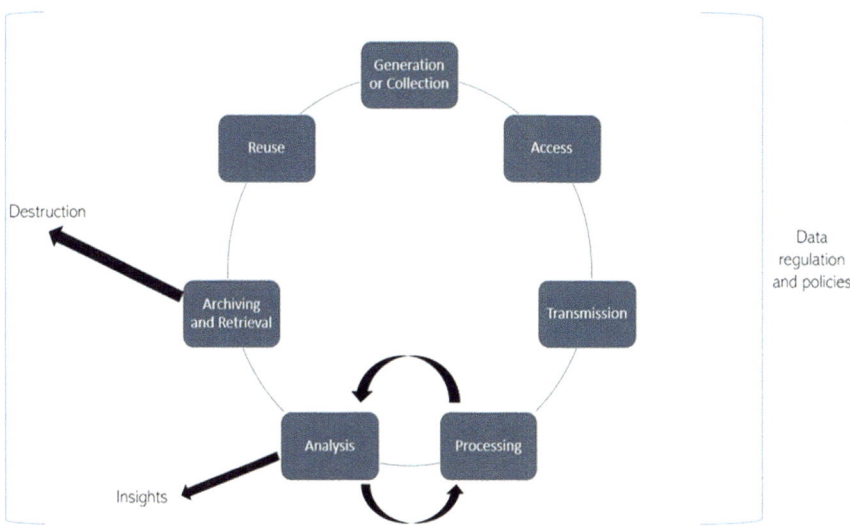

Fig. 3 Data processing cycle

gathered from devices and gateways and also reaped from other sources external to IoT (including the Web and social media). Within smart cities, information sharing through IoT facilitates decision-making in a concerted manner for urban stakeholders to base their decisions on common information, following which coordinated action can be undertaken. This data-sharing process will also strengthen the collaborative efforts between sectors. For this scenario to be made a reality, it is essential that all relevant parties have access to the data.

In this context, ITU has identified a series of requirements to ensure the usability of data [24]:

- **Accessibility to data**: Promoting openness and accessibility to data is essential for the success of any smart city endeavour. It is important to balance privacy and the accessibility of data as some sensitive data may need to be anonymized before utilization.
- **Open data**: It is suggested that basic data-sets (including utilities) are standardized against the same parameters and are made public. This is pertinent for cross-scale information sharing in smart cities. This will also improve overall operational decision-making.
- **Managing data streams**: Generation of data is only step one. It is important that cities are capable of maintaining the accuracy of the data and storing it. Analytical capabilities and data security of cities also contribute to the establishment of highly efficient databases.
- **High performance infrastructure**: For adequately utilizing the massive volumes of data, the operational capacity of the smart city infrastructure needs to be reliable, foster precise data transmission, minimize downtime, and avoid system failure.
- **Maximize efficiency**: The smart city infrastructure needs to disseminate data effectively (between sectors as and when needed), while keeping the veracity of the information intact.

To implement these requirements on a smart city-wide basis, policies, regulations and standards will need to be formulated. An appropriate governance framework for data processing and management will also have to be put in place for value creation, where stakeholders can interact during the data life-cycle. Deriving insights and contributing to the data value chain would require the underscoring of potentially commercial activities and the underlining of other actions in the move towards the establishment of smart cities. A data processing cycle (Fig. 3) within the smart city domain is incorporated to determine what insights can be garnered and what actions can be monetized. The steps within the data processing cycle can be categorized into *data core activities* and *data support activities*, which could contribute directly or indirectly to creating value from data [25]:

- Data core activities: This includes data gathering or collection, access, transmission, archiving and processing.
- Data support activities: These activities are related to the formulation and implementation of data laws, regulations and policies. Data support activities are intended to complement/supplement the core data activities and amplify the value

of the data. Within the data stakeholder nexus, an entity can conduct one or more activities in the data management cycle or several entities could carry out a single step. Figure 3 provides an overarching perspective of the main stakeholders active in the data ecosystem.

Dealing with data challenges within the IoT ecosystem is essential for smart cities to ensure adequate utilization. Smart cities with an IoT-based infrastructure need to mould their data operations in such a way that they keep the *IoT Big Data characteristics* in mind (due to the diversity of data collected from various IoT devices). The main IoT Big Data characteristics include high volume (size of data sets collected), variety (heterogeneity of data types and sources) and high velocity (speed of data streams in and out of the data sets). These IoT Big Data characteristics form the basis of identifying relevant IoT-based services for deployment (Figs. 4 and 5).

IoT, Big Data and AI can serve as the building blocks of the smart sustainable city. It is important to understand that smart sustainable cities are not just about the adoption of frontier technologies. Smart sustainable city endeavours strive to use a blend of these technologies to develop a holistic convergence architecture to offer solutions for advanced computing needs of flexibility and scalability. Before adopting these frontier technologies for urban operations within smart sustainable cities, it pertinent to understand the inter-relationship between, and the co-dependencies of, these technologies [28].

Utilizing a combination of these technologies, smart city operations across sectors (Table 4 and Fig. 6) can be envisioned for cities to function better, optimize system

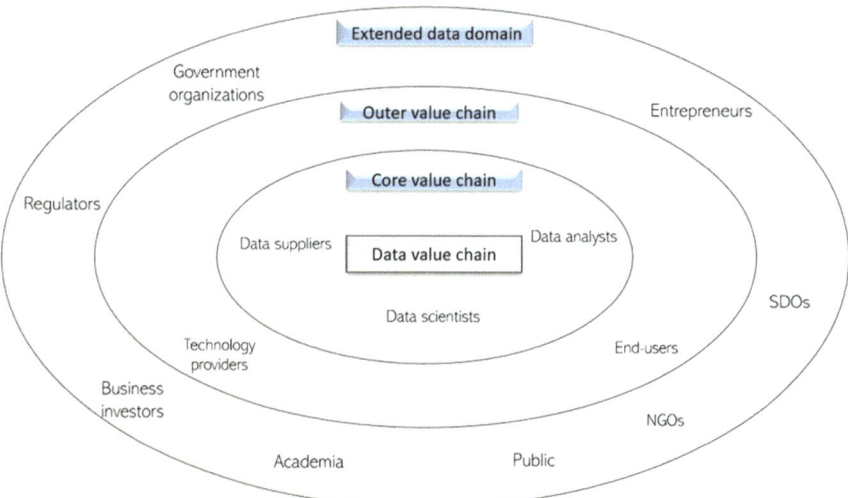

Fig. 4 Stakeholders in the data ecosystem [26]

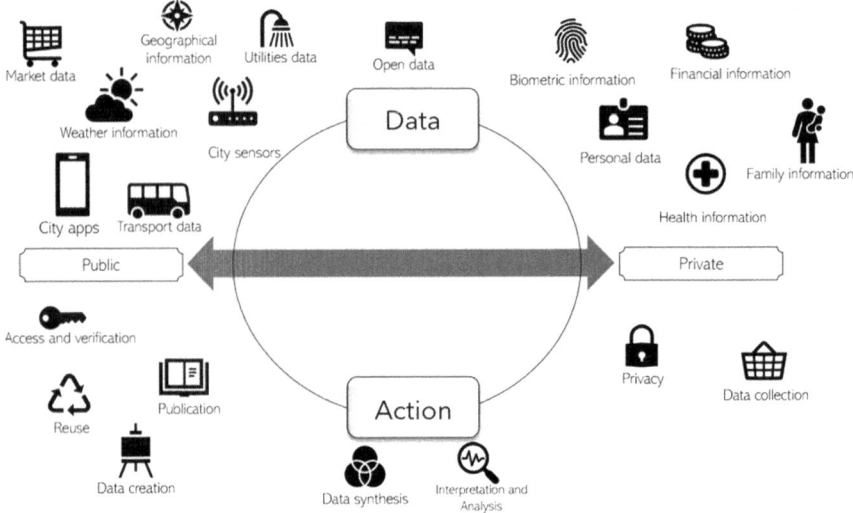

Fig. 5 Data ecosystem and processes in smart cities [27]

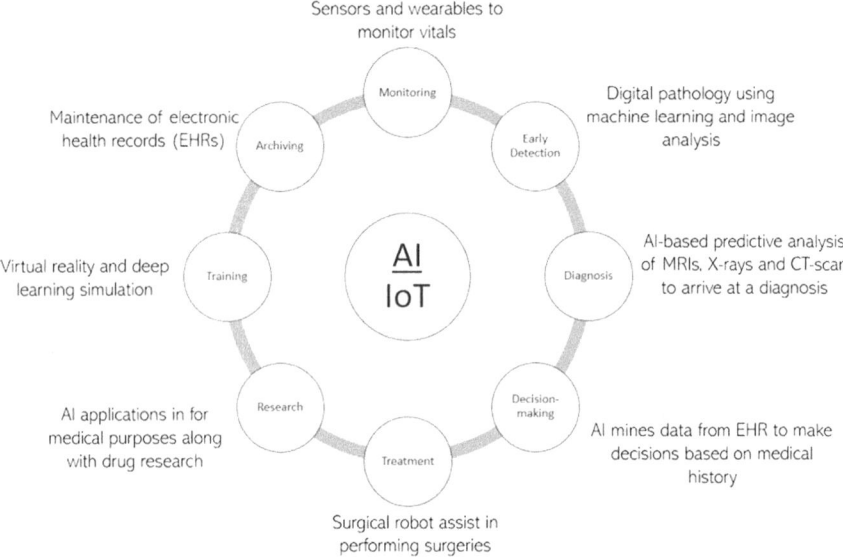

Fig. 6 IoT and AI for health care in smart sustainable cities [31]

Table 4 Utilization of IoT devices and AI for intelligent transport systems in smart cities [29]

	Channel/Device	Action/Device
	Sensors	Sensors can detect empty parking spots and transmit information via smartphone apps to the driver(s)
	Satellites	Remote sensing satellite can be utilized for vehicle detection and intermodal communication
	V2V	Vehicles can leverage V2V (communication to enable information exchange on location, traffic information and velocity) and adaptive cruise control
	Smartphone	Public transport information can be transmitted through smartphone apps
	Traffic systems	AI traffic systems to reduce wait times and traffic congestion

efficiency, improve energy efficiency, secure effective data collection and processing mechanisms, and overcome latency and bandwidth issues.

Based on the data processing cycle, IoT systems are at the forefront of connecting devices, aggregating, transmitting, archiving, analysing data for the provision of services. From the data perspective, IoT also faces a variety of challenges [32]:

- *Number of devices*:

  – IoT system needs to be equipped to deal with rapidly increasing numbers and types of devices to improve urban operations.
  – IoT system needs to manage an increasing number of devices accessing networks and transmitting large datasets.

- *Quality of service (QoS) and Quality of Experience (QoE)*: These data harnessed from the IoT ecosystem are derived from differing IoT applications, and could, therefore, have varying QoS requirements (including real-time monitoring, reducing packet loss, and latency). For data consumers, it is essential to maintain QoE in order to provide the desired data in keeping with privacy, personal data and security requirements.
- *Data transfer handling*: IoT system should be capable of avoiding network congestion from concurrent access of devices and ensuring the veracity of the transferred data.
- *Data capabilities*: Data processing capabilities need to be leveraged to ensure interoperability, integrity, privacy and security protection of data. IoT-based infrastructure needs to be scalable and flexible to accommodate diverse data sets of various sizes, formats and speeds.

To overcome challenges related to data in the IoT ecosystem, ITU specified certain requirements for Big Data relating to certain components (device, gateway, network, IoT platform and IoT application server) for IoT deployments, and to facilitate data operations and data flow.

The main standardization work on IoT and smart cities within ITU is conducted through ITU-T Study Group 20 (ITU-T SG20) on IoT and Smart Cities and Communities. In this context, SG20 aims for the coordinated development of IoT-based infrastructure, covering machine-to-machine communications and ubiquitous sensor networks, while also deploying frontier technologies/processes including AI, blockchain, robotics, Big Data analytics for various sectors (e.g. e-health, utilities, transportation systems, banking, e-commerce, future networks) in the smart city sphere. Since its establishment in 2015, ITU-T SG20 has been laying the foundation for developing a cohesive array of international standards (see Table 5) on the implementation of IoT to address urbanization problems and also achieve the key targets under the SDGs, especially SDG 11 [33].

4 Smart Cities Around the World

4.1 Case in Point: Seoul

Seoul is the capital city of South Korea and is considered to be the fourth largest metropolitan city in the world [34]. The average population of the city has been estimated to be around 10 million people. It is also known as one of the most technologically adept cities in the world [35].

The *Smart Seoul* programme was announced to overcome:

- the limitations of applying ICTs to just the city's infrastructure for economic growth; and
- the exclusion important areas like environment, safety and inhabitant satisfaction [36] (which is line with the overall aims of ITU's Smart Sustainable City definition).

Smart Seoul is projected to be one of the forerunners in the creation of smart cities and has implemented several measures along the action lines of the U4SSC KPIs. *Smart Seoul* is adopting initiatives closely in line with the indicators developed by ITU and in the future aims to establish itself as a leading Smart Sustainable City.

Seoul has been recorded as having one of the highest network capacities (20 Gb/15% usage), along with the infrastructure to deal with heavy data traffic. By 2012, 78% of Seoul's population had started using ICT-based devices. The city also plans to install free Wi-Fi networks in public spaces, which would increase the coverage to 99% [34]. Smart City Seoul utilizes location-based services that use GPS technology to promote civic integration (instead of offering just fallout services). Seoul has realized the importance of open data platforms playing a crucial role in

Table 5 ITU Standards developed on smart cities [33]

ITU-T recommendations	Specific sector (if any)	New and emerging technology/process
ITU-T Y.4461—Framework of open data in smart cities	All	Big Data, open data, IoT
ITU-T Y.4003—Overview of smart manufacturing in the context of the industrial Internet of things	Manufacturing industries	IoT, Big Data, AI
ITU-T Y.Suppl.63 to ITU-T Y.4000 series—Unlocking Internet of things with artificial intelligence	Utilities, Banking, Health, Transport	IoT, AI
ITU-T Y.4907—Reference architecture of blockchain-based unified KPI data management for smart sustainable cities	All	IoT, blockchain
ITU-T Y.4466—Framework of smart greenhouse service	Utilities, Manufacturing	IoT
ITU-T Y.4458—Requirements and functional architecture of smart street light service	City planning	IoT
ITU-T Y.4456—Requirements and Functional Architecture for Smart Parking Lot in Smart City	Transportation	IoT, Big Data
ITU-T Y.4457—Architectural framework for transportation safety services	Transportation	IoT
ITU-T Y.4200—Requirements for interoperability of smart city platforms	All	
ITU-T Y.4119—Requirements and capability framework for IoT-based automotive emergency response system	Transportation	IoT
ITU-T Y.4207—Requirements and capability framework of smart environmental monitoring	Water industry, Utilities	IoT

improving transparency in public services. Seoul's Open Data Square is a notable network, which provides interactive participatory services with uni-directional information flow [37]. This utilizes IoT, social networks for inhabitants to be able to raise issues of concern to the government [38]. Seoul foresees great potential in this system and intends to apply community mapping to address a wider range of concerns. This model also provides a platform for inhabitants to get together to develop solutions for common concerns in collaboration with the government. The Seoul Open Data Square, which was launched in 2012, aims to provide information under the following headings [34]: (i) Administrative work carried out by the central and municipal governments (ii) Welfare Culture and Tourism (iii) Environment (iv) Safety and Security (v) Health (v) The economy (vi) Transportation.

Seoul has announced a *Smart Metering project* which aims to reduce the city's total energy use and water consumption by 10%. In line with this project, it installed smart meters in more than 1000 households. This would help provide information on electricity, water and gas consumption in monetary units, accompanied by detailed information on the exact patterns of the above-mentioned resources. Sixty per cent of individuals involved in this pilot study reported that this metering system was useful, and 70% agreed to participate in similar projects in the future. With the relative success of this project, more households are to be included in the *Smart Metering Project* with the hope of achieving 100% coverage in the future [37].

Additionally, another interactive programme called *Eco-mileage* was launched in 2011. It serves 479 000 participating households in 2800 buildings and promotes green initiatives [39]. The service aims to help users save energy and cut down on carbon emissions inside and outside homes, in their domestic energy consumption and also through the purchase of certified green products and services. The service is also designed to support other spontaneous efforts [40].

South Korea as a country has been at forefront of ICT innovation, which indicates relatively high investments in ICT R&D. Smart City Seoul has the highest service diversity in terms of smart development areas. Most of the initiatives in Seoul address the areas of transportation (20%) and public facilities management (20%). The tourism sector also enjoys relatively high investments for research, which has been estimated to be around 12%. These services have largely been developed by the Seoul IT division, with assistance from the private sector. In order to prevent any duplication of services, since 2006 the implementation of these services has been coordinated by the smart city team in Seoul.

It is expected that Geographic Information System (GIS)-based applications will become integrated for different services including healthcare, transport, education tourism and manufacturing. The estimated increase in employment after the introduction of ICTs into the different sectors has not been predicted but is expected to increase over the years. The number of patent applications sent from Seoul is also not being recorded; however, considering the investment in R&D, these numbers are also expected to increase [41]. There are systems available for an integrated online reservation for public services accounts for a portion of the e-commerce transactions, which can be extended to other sectors in due course.

Mobile web technology and other mobile applications are used to provide Seoul's inhabitants with information on public services. In addition to the information on safety and security provided by the *Open Data Square*, there is the *Seoul Safety Service* (since 2008) using state-of-the-art location- based services and CCTV technologies to notify authorities and family members in case of emergencies, particularly those involving children, disabled individuals and the elderly [42]. Included in this, is the U-Children Safety System, which provides safety zones using a multiple input and output wireless infrastructure mesh network [43]. The children's smart devices assist with the location of missing children as quickly as possible. Currently, radio frequency identification (RFID) and CCTV technology within the U-Children Safety

System are being used as a sensor network to monitor the safety of primary school pupils (from kidnappers, as well as from sexual predators) [34]. RFID tags have been implemented to ensure that drivers are complying with the no-driving environmental campaigns [44].

During the COVID-19 pandemic and the subsequent lockdown, Seoul has employed voice enabled "smart" speakers, to provide care to elderly folk in the city. The target population is monitored closely for the detection of trigger words, which facilitates medical intervention, either through tele-medicine or in-person visits as and when needed. Emerging as one of the key players in the field of AI, the city of Seoul has also collaborated with IBM on its *I Care You* programme. One of the main services provided through this programme is that of an AI-based virtual assistant to decipher the social problems citizens may face during isolation. This virtual assistant is supplemented by an online survey (conducted through the webpage of the city, along with social media channels) to assess COVID-19 related changes in behaviour patterns (e.g. eating habits, sleeping patterns, mood), health, social isolation and economic factors, thereby enabling the government to be able to assess vulnerabilities [45] (Table 6).

Table 6 Initiatives and actions in Seoul [42, 46–52]

Sector	Project/Activities
Health	Ubiquitous-Health (u-Health) system allows elderly people to quickly access healthcare services, including ambulatory services, preventive treatment, and telemedicine
Agriculture	Metro Farm is Seoul's first urban smart farm, which focuses on automating the farming cycle from cultivation, irrigation and harvesting
E-government	Seoul e-government aims to provide digital benefits to all its citizens by facilitating the transition of to a smart city
Public security	Video surveillance is present throughout the city. More recently, Seoul has also commenced the utilization of AI for detection of crime through the existing surveillance devices
ITS	Vehicle-to-grid system along with self-driving cars and unmanned delivery robots arc being deployed as a part of their smart mobility project
Energy	Seoul has employed smart grids and incorporated vehicle charging sites along with smart street lighting and smart meters to ascertain energy consumption patterns
Buildings	KT Corp is working towards the development of smart building equipped with sensors for the monitoring of energy and water consumption and the detection of malfunctions
Telecommunication	Free Wi-Fi Zones arc prevalent throughout the city in public spaces

4.2 Case in Point: China—Shanghai and Wuhan

Shanghai, located in East China, is the largest city in the country, by population, with more than 27 million, and is one of the four direct-controlled municipalities [53]. The city is a global financial centre and also a transport hub with the world's busiest container port. In 2012, Shanghai launched a pilot carbon emissions trading scheme, which is the first of its kind in China, and represents a landmark step for China to control and ultimately reduce its greenhouse gas emissions [54].

Being the first in China to have a special plan in building ICTs infrastructures and setting standards for fibre-optic broadband connection for residential buildings, Shanghai is leading the country in the coverage of wireless networks—17 000 public venues are now covered by Wi-Fi and another 6.5 million households have fibre-optic broadband coverage—and in the number of actual users of fibre-optic broadband with a minimum bandwidth of 10 MB, which now amounts to more than two million users [55].

In 2015, Shanghai announced its first three-year action plan to build a Smart City. A solid foundation was laid by combining government planning coordination among different departments, and the participation of private sector companies, based domestically and overseas, such as China Telecom, China Mobile, China Unicom, Huawei, ZTE, Tencent and Datang Telecom. Under the guidance of the first-round action plan, the city has achieved remarkable progress, particularly in the development of network infrastructure.

Some of the important ICT projects listed in the three-year action plan and the 12th Five-Year Plan have been completed, such as an electronic billing platform and a digital certification platform. An urban grid management system (i.e. Smart Grid) and an electronic toll collection system (i.e. Smart Transportation) are being renovated and expanded. The building of the government service hotline "12,345" (i.e. Smart Governance) and an integrated shipping information platform (i.e. Smart Logistics) have also started in Shanghai [56]. Specifically, in terms of the Smart Transportation System in Shanghai, ICTs play a significant role with regards to the following [57]:

- Electronic ID management system: Smart ID license tags and cards were issued to the 180 000 public transit vehicles in Shanghai, 48 000 taxis and 180 000 workers in order to better monitor public transport, prevent transport license forgery and the operation of unlicensed vehicles, and to combat the rampant spread of unlicensed taxis.
- Public transport smart cards: The widely used public transport smart card is actually a smart integrated circuit card that allows the city to improve the collection of passenger flow information by using the data captured by the usage of the smart integrated circuit card. This is achieved by integrating GPS tracking to card readers thus allowing the city transportation office to understand the patterns of passenger flow.

- Bus smart information system: By installing devices on city buses, the real-time communication between vehicles, operators and bus stops is enabled, where operators can monitor vehicle position, passenger flow, as well as fuel consumption, schedule news and announce estimated arrival time.
- GPS data for each bus is fed to the operator's information system and disseminated by an exclusive digital broadcasting signal. Information screens on the buses and at bus-stop signs relay scheduling news, predicting how long a wait will be before the next bus arrives.

Wuhan, the capital of Central China's Hubei province, is a metropolis with population of 11 million and is the most populous city in central China. It is a major transportation hub for air, railway, expressways and ferry traffic. Wuhan joined the C40 Cities Climate Leadership Group, committing to reduce carbon emissions per unit of GDP by 56% (from 2005 levels) within this decade. The city has also integrated low carbon and climate change measures into the Wuhan Urban Development Strategy [64] (Table 7).

As part of the Smart Tourism plan, Wuhan will have a complete coverage of free public Wi-Fi at all the A-level scenic areas. Similarly, from the beginning of this year, Wuhan has begun to roll-out the Wuhan Bus Wi-Fi project, which aims to complete the installation of Wi-Fi devices on 5000 city buses by the end of the year, accounting for 80% of the total number of buses in the city.

The city is among the first batch of cities selected by the Ministry of Science and Technology to pilot the smart city programme; it is also the first city in the world

Table 7 Initiatives and actions in Shanghai [56–63]

Sector	Project/Activities
Health	Smart health houses have been set up in Shanghai to allow residents to acquire their medical reports without needing to go the hospital
Agriculture	The tech giant, Alibaba has been leveraging AI and Big Data for animal husbandry, forestry and fisheries
E-government	The government portal and official website has been set up to facilitate online payment for utilities, market registrations, and the provision of emergency information
Public security	Shanghai has an extensive network of AI-bascd surveillance Public security cameras to monitor crime
ITS	Shanghai has been utilizing traffic sensors, V2X communication for improving road capacities, securing traffic safety and reducing vehicular emissions. Companies like Baidu intend to roll out autonomous cars in the near future, while other companies like the technology firm AutoX seek to deploy robotaxis for public use
Energy	Shanghai has embraced smart grids and is also adding renewable forms of energy to its energy-use mix
Buildings	Tcncent has been deploying IoT in buildings to create a smart home ecosystem and monitor regular parameters, including temperature
Telecommunication	Free Wi-Fi in public transport areas and certain parks

Table 8 Initiatives and actions in Wuhan [67]

Sector	Project/Activities
Health	*Mobile hospitals* are present in Wuhan Central Hospital
Agriculture	There is a cattle breeding project in Hannan District
Government	Government has adopted a GIS-based digital decision-making assistant system
Public security	Wuhan City has an active video surveillance system
ITS	*No-waiting* toll collection systems have been adopted for roads, bridges and tunnels
Food and drug safety	To boost food safety, meat quality and source traceability is ensured by implementing blockchain
Geo-space	This includes the Wuhan geo-space information framework
Urban management	A digital grid platform for urban management has been put in place for the transition to a smart city
Telecommunication	Free Wi-Fi in "optical valley"
Smart community	Intelligent management programme in Yongqing CBD

to hold an open tender to global companies for intelligent city designs. It aims to become a leading city in terms of applying ICTs to improve people's lives and build sustainable development [65].

Based on the construction plan, 2010–2011 was the planning stage of building SSC; the construction of ICTs infrastructure was completed by 2015, and finally, by the end of 2020, a Smart Wuhan is expected to be in place. With the outbreak of the COVID-19 pandemic in Wuhan in November/December 2019, additional investments have been made in health care, AI-based surveillance and detection for COVID-19 patterns and the control of future pandemics [66]. Several programmes have already been launched to improve and provide more conveniences to people's lives. For example, a municipal administrative service centre has been set up to offer 24-h, self-service public service; a two-dimensional code food tracing system to ensure food safely and a water resource monitoring system have also been launched.

Apart from the above, various other projects have been authorized to achieve the Smart Wuhan vision, as shown in Table 8 [66].

4.3 Future Smart City Research Trajectory

While research on technological trends for smart sustainable cities continues, certain aspects and preconditions for commencing successful smart city projects are unintentionally excluded from the current literature base. A few of them have been underscored in this section based on the key features of the case studies described previously:

(A) *SSC financing*: One aspect largely omitted is the financing mechanism for related projects. The existing regulatory and standardization frameworks on SSC also do not provide guidance on funding smart city projects. This would be specifically relevant for developing countries aiming to adopt emerging technologies like AI, IoT, blockchain (for urban functions), which require substantial capital. Identifying this as a potential roadblock to initiating smart city transitions, U4SSC is developing a deliverable which will provide high-level recommendations on acquiring the required sponsorship and resources for smart city transformations and map the relevant stakeholders at the local and global level. These "Guidelines on tools and mechanisms to finance Smart Sustainable Cities project" are expected to be finalized and published in 2021 [68].

(B) *Adoption of 5G technologies*: 5G-based technologies in SSCs are still in its initial stages. The limitations of 5G roll-out (including cost, security and privacy, limited coverage in rural areas etc.), have charted out dimensions envisioned for 6G—which is already being pursued by leading players including Huawei, Apple, Samsung etc. [69].

(C) *Adapting smart cities to the COVID-19 pandemic*: The COVID-19 pandemic has resulted in worldwide economic recession. This has also impacted budgets allotted to national smart city projects. Despite this, urban stakeholders are increasingly exploring the option of utilizing frontier technologies to contain the spread of the virus, while also integrating relevant AI-based applications in healthcare to deal with the increasing pressure on the medical sector. For example, United Arab Emirates (UAE) has introduced smart helmets with thermal cameras to detect spike in body temperature for individuals who could be infected with COVID-19. More research needs to be conducted to examine additional functionalities for the employment of frontier technologies to detect and curb the spread of COVID in cities. This is essential also to improve the preparedness of cities to similar pandemics in the future [70].

5   Conclusion

While smart and sustainable cities projects are on the on rise worldwide, limited knowledge of the concept has led to urban stakeholders adopting only certain elements of a technology-centric transition. In this context, international standards have played a major role in steering stakeholders towards improving citizens' quality of life by using technology.

As research in this area continues to expand, existing smart and sustainable city KPIs are also be refined and updated to bridge knowledge gaps and support cities with the incorporation of new and emerging technologies like AI, IoT, machine learning, virtual reality, blockchain (among others). With the U4SSC KPIs, being implemented in more than 100 cities worldwide (supported by 17 United Nations agencies), this framework has also served as a credible monitoring tool to measure progress towards the city's own goals, as well as the Sustainable Development Goals. The occurrence of the COVID-19 pandemic has stymied global smart city transitions. It has also highlighted the drastic improvements needed in the health infrastructure of aspiring smart cities. Artificial intelligence, leveraging on the available data ecosystem can play a central role here in not just bolstering the overall medical sector but also contribute to the research base being built on for COVID-19 [2, 30].

References

1. United Nations. (2018). The World's Cities in 2018. Retrieved October 28, 2020, from https://www.un.org/en/development/desa/population/publications/pdf/urbanization/the_worlds_cities_in_2018_data_booklet.pdf.
2. Aliyu, A., & Amadu, L. (2017). Urbanization, cities, and health: The challenges to Nigeria—A review. Retrieved October 28, 2020, from https://www.ncbi.nlm.nih.gov/pmc/articles/PMC5676403/.
3. Sepasgozar, S., Hawken, S., Sargolzaei, S., & Foroozanfa, M. (2018). Implementing citizen centric technology in developing smart cities: A model for predicting the acceptance of urban technologies. *Technological Forecasting and Social Change, 142,* 105–116. https://doi.org/10.1016/j.techfore.2018.09.012.
4. Lu, H., de Jong, M., & ten Heuvelhof, E. (2018). Explaining the variety in smart eco city development in China-What policy network theory can teach us about overcoming barriers in implementation? *Journal of Cleaner Production, 196,* 135–149. https://doi.org/10.1016/j.jclepro.2018.05.266.
5. Yang, C. (2020). Historicizing the smart cities: Genealogy as a method of critique for smart urbanism. *Telematics and Informatics, 55,*. https://doi.org/10.1016/j.tele.2020.101438.
6. Yao, T., Huang, Z., & Zhao, W. (2020). Are smart cities more ecologically efficient? Evidence from China: Sustainable Cities and Society. https://doi.org/10.1016/j.scs.2019.102008.
7. Sokolov, A., Veselitskaya, N., Carabias, V., & Yildirim, O. (2019). Scenario-based identification of key factors for smart cities development policies. *Technology Forecasting and Social Change, 148,*. https://doi.org/10.1016/j.techfore.2019.119729.
8. Saif Almuraqab, N. (2020). Introduction to the critical success factors of E-government adoption of the utilization of emerging smart cities technologies. *Solving Urban Infrastructure Problems Using Smart City Technologies, 3–15,*. https://doi.org/10.1016/B978-0-12-816816-5.00001-2.
9. Monfaredzadeh, T., & Krueger, R. (2015). Investigating social factors of sustainability in a smart city. *Proceedia Engineering, 118,* 1112–1118. https://doi.org/10.1016/j.proeng.2015.08.452.
10. United Nations. (1992). Report of the World Commission on Environment and Development: Our Common Future.
11. ITU. (2016). Shaping smarter and more sustainable cities: striving for SDGs. Focus Group on Smart Sustainable Cities. Retrieved October 28, 2020, from https://www.itu.int/en/publications/Documents/tsb/2016Shaping-smarter-and-more-sustainable-cities/index.html.
12. ITU. (2014). Smart sustainable cities: An analysis of definitions. Focus Group on Smart Sustainable Cities.
13. ISO (2020) Standards in our world. Retrieved October 28, 2020, from https://www.iso.org/sites/ConsumersStandards/1_standards.html.
14. ITU (2014). Overview of smart sustainable cities infrastructure. Focus Group on Smart Sustainable Cities.
15. Recommendation ITU-T Y.Sup32 (2016) Smart sustainable cities—A guide for city leaders.
16. ITU (2017) Collection Methodology for Key Performance Indicators for Smart Sustainable Cities. Retrieved October 28, 2020, from https://www.itu.int/en/publications/Documents/tsb/2017-U4SSC-CollectionMethodology/index.html.

17. Jiang, D. (2020). The construction of smart city information system based on the Internet of Things and cloud computing. *Computer Communications, 150,* 158–166. https://doi.org/10.1016/j.comcom.2019.10.035.
18. ITU. (2012). Overview of the Internet of things. Retrieved October 28, 2020, from https://handle.itu.int/11.1002/1000/11559.
19. Munawar, H., Quyyum, S., Sepasgozar, S. (2020). Big Data and Its Applications in Smart Real Estate and the Disaster Management Life Cycle: A Systematic Analysis. Retrieved October 28, 2020, from https://www.mdpi.com/2504-2289/4/2/4.
20. Ullah, Z., Al-Turjman, F., Mostarda, L., & Gagliardi, R. (2020). Applications of Artificial Intelligence and Machine learning in smart cities. *Computer Communications, 154,* 313–323. https://doi.org/10.1016/j.comcom.2020.02.069.
21. ITU. (2019). Data processing and management framework for IoT and smart cities and communities.
22. Ben Atitallah, S., Driss, M., Boulila, W., & Ben, H. (2020). Ghezala leveraging deep learning and IoT big data analytics to support the smart cities development: Review and future directions. *Computer Science Review, 38,.* https://doi.org/10.1016/j.cosrev.2020.100303.
23. Zamanifar, A. (2021). Data Analytics in IOT-Based Health Care. https://doi.org/10.1016/B978-0-12-819314-3.00007-0.
24. ITU. (2014). Anonymization infrastructure and open data in smart sustainable cities. Focus Group on Smart Sustainable Cities.
25. Rahul, K., & Banyal, R. K. (2020). Data life cycle management in Big Data analytics. *Procedia Computer Science, 173,* 364–371. https://doi.org/10.1016/j.procs.2020.06.042.
26. Data Economy. (2020). Big data ecosystem stakeholders. Retrieved November 11, 2020, from https://www.dataeconomy.eu/stakeholders/#page-content.
27. IBM. (2014). Going Beyond Data Science Toward an Analytics Ecosystem: Part 2. Retrieved November 2, 2020, from https://www.ibmbigdatahub.com/.
28. Chen, Y. (2020). IoT, cloud, big data and AI in interdisciplinary domains. *Simulation Modelling Practice and Theory, 102,.* https://doi.org/10.1016/j.simpat.2020.102070.
29. Vanitha, N., Karthikeyan, J., Kavitha, G., & Radhika, K. (2020). Modelling of Intelligent Transportation System for Human Safety Using IoT. https://doi.org/10.1016/j.matpr.2020.06.421.
30. Reuters. (2018). China's Baidu gets green light for self-driving vehicle tests in Beijing. Retrieved November 2, 2020, from https://br.reuters.com/article/autos-selfdriving-baidu-idUSL3N1R51A5.
31. Straw, I. (2020). The automation of bias in medical Artificial Intelligence (AI): Decoding the past to create a better future. *Artificial Intelligence in Medicine, 110,.* https://doi.org/10.1016/j.artmed.2020.101965.
32. ITU. (2019). Framework to support data interoperability in IoT environments. Retrieved October 28, 2020, from https://www.itu.int/pub/T-FG/publications.aspx?lang=en&parent=T-FG-DPM-2019-3.3.
33. SG20: Internet of things (IoT) and smart cities and communities (SCC). https://www.itu.int/en/ITU-T/studygroups/2017-2020/20/Pages/default.aspx.
34. ITU. (2013). Smart Cities Seoul: A case study. ITU-T Technology Watch Report.
35. Embassy of India. (2018). Factsheet and Countrynote. Retrieved November 2, 2020, from https://www.indembassyseoul.gov.in/page/korea-country-note/.
36. Kim, B., Yoo, M., Chul Park, K., Lee, K. R., & Kim, J. (2020). A value of civic voices for smart city: A Big Data analysis of civic queries posed by Seoul Citizens. https://doi.org/10.1016/j.cities.2020.102941.
37. Lee, J., Gong, M., & Hu, M. -C. (2013). Towards an effective framework for building smart cities: Lessons from Seoul and San Francisco. https://doi.org/10.1016/j.techfore.2013.08.033.
38. Yi, C., Park, & J. (2015), Design and Implementation of an End-of-Life Vehicle Recycling Center Based on IoT (Internet of Things) in Korea. https://doi.org/10.1016/j.procir.2015.02.007.

39. Eco-mileage. (2020). What is Eco-mileage. Retrieved November 2, 2020, from https://ecomil eage.seoul.go.kr/home/index.do.
40. Lee, H., Park, H., Kho, S. Y., & Kim, D. Y. (2019). Assessing transit competitiveness in Seoul considering actual transit travel times based on smart card data. *Journal of Transport Geography, 80,*. https://doi.org/10.1016/j.jtrangeo.2019.102546.
41. Kim, B. K., & Hwang, J. (2019). Longitudinal small and medium enterprise (SME) data on survival, research and development (R&D) investment, and patent applications in Korea's innovation clusters from 2008 to 2014. *Data in Brief, 25,*. https://doi.org/10.1016/j.dib.2019.103967.
42. Seoul Metropolitan Government. (2017). Seoul e-Government. Retrieved November 2, 2020, from https://www.metropolis.org/sites/default/files/seoul_e-government_english.pdf.
43. Choi, J., & Kim, H. M (2021). State-of-the-art of Korean smart cities: A critical review of the Sejong smart city plan. In *Smart Cities for Technological and Social Innovation: Case Studies, Current Trends, and Future Steps* (pp. 51–72), 1st edn, Academic Press.
44. Escolar, S., Villanueva, F., Santofimia, M. J., Villa, D., del Toro, G. X., & Carlos L´opez, J. (2018). A multiple-attribute decision making-based approach for smart city rankings design. *Technological Forecasting and Social Change, 142,* 42–55. https://doi.org/10.1016/j.techfore.2018.07.024.
45. IBM. (2020). Collaborated with the City of Seoul, South Korea to Develop an AI-based Virtual Assistant for its Citizens using IBM Watson. Retrieved November 4, 2011, from https://new sroom.ibm.com/2020-08-20-IBM-Collaborated-with-the-City-of-Seoul-South-Korea-to-Dev elop-an-AIbased-Virtual-Assistant-for-its-Citizens-using-IBM-Watson.
46. Moon, G., Lim, K.W., Yoo, Y.M., An, H.M., Lee, K.S., & Szu, H. (2011). Ubiquitous-Health (U-Health) monitoring systems for elders and caregivers. In *Proceedings of SPIE—The International Society for Optical Engineering*.
47. BBC. (2020). Is underground farming the future of food. Retrieved November 5, 2011, from https://www.bbc.com/travel/story/20200723-is-underground-farming-the-future-of-food.
48. ZDNet. (2020). Seoul to install AI cameras for crime detection. Retrieved November 2, 2011, from https://www.zdnet.com/article/seoul-to-install-ai-cameras-for-crime-detection/.
49. Maeng, K., Jeon, S. R., Park, T., & Cho, Y. (2020). Network effects of connected and autonomous vehicles in South Korea: A consumer preference approach. *Research in Transportation Economics*. https://doi.org/10.1016/j.retrec.2020.100998.
50. Choi, J., & Do, P. D. (2020). Process and features of Smart Grid, Micro Grid and Super Grid in South Korea. *IFAC-PapersOnLine, 49,* 218–223. https://doi.org/10.1016/j.ifacol.2016.10.686.
51. Business Korea. (2020). KT Launches New Smart Building Management Service. Retrieved November 4, 2011, from https://www.businesskorea.co.kr/news/articleView.html?idxno=49928. Accessed 4.11.2020.
52. Wiman. (2020). Where may I connect to free WiFi in South Korea? Retrieved November 4, 2011, from https://www.wiman.me/south-korea.
53. World Population Review. (2020). Retrieved November 4, 2011, from https://worldpopulation review.com/world-cities/shanghai-population.
54. Wang, Y., Ren, H., Dong, L., Park, H. S., Zhang, Y., & Xu, Y. (2019). Smart solutions shape for sustainable low-carbon future: A review on smart cities and industrial parks in China. *Technological Forecasting and Social Change, 144,* 103–117. https://doi.org/10.1016/j.techfore.2019.04.014.
55. Zhu, S., Li, D., & Feng, H. (2019). Is smart city resilient? Evidence from China. *Sustainable Cities and Society, 50,*. https://doi.org/10.1016/j.scs.2019.101636.
56. Jian, L., Yongqiang, Z., & Hyoungmi, K. (2018). The potential and economics of EV smart charging: A case study in Shanghai. *Energy Policy, 119,* 206–214. https://doi.org/10.1016/j.enpol.2018.04.037.
57. Shmelev, S. E., Shmeleva, I. A. (2019). Multidimensional sustainability benchmarking for smart megacities. *Cities, 92,* 134–163. https://doi.org/10.1016/j.cities.2019.03.015.
58. CGTN. (2019). Visiting a smart health house in Shanghai. Retrieved November 2, 2011, from https://news.cgtn.com/news/08-21/Live-Visiting-a-Smart-Health-House-in-Sha nghai-Jlh5MxwxEY/index.html.

59. China Daily. (2018). AI tech boosting smart agriculture. Retrieved November 2, 2011, from https://www.chinadaily.com.cn/a/201808/07/WS5b68f3f3a3100d951b8c8f62.html.

60. Shanghai Municipality. (2020). Retrieved November 4, 2011, from https://en.shio.gov.cn/.

61. China Tech. (2019). Shanghai district installscomprehensivesurveillance system. Retrieved November 2, 2011, from https://asia.nikkei.com/Business/China-tech/Shanghai-district-instal lscomprehensive-surveillance-system.

62. TechCrunch. (2020). AutoX launches its RoboTaxi service in Shanghai, competing with Didi's pilot program. Retrieved November 2, 2011, from https://techcrunch.com/2020/08/16/autox-launches-its-robotaxiservice-in-shanghai-competing-with-didis-pilot-program/.

63. Atha, K., Callahan, J., Chen, J., Drun, J., Green, K., Lafferty, B., McReynolds, J., Mulvenon, J., Rosen, B., & Walz, E. (2020). SOSi. Retrieved November 2, 2011, from https://www.uscc.gov/sites/default/files/2020-04/China_Smart_Cities_Development.pdf. Accessed 2.11.2020.

64. WiFiMap. (2020). Retrieved November 2, 2011, from https://www.wifimap.io/639-shanghai-free-wifi.

65. Rameri, D., Benevolo, C., Veglianti, E., & Li, Y. (2018). Understanding smart cities as a glocal strategy: A comparison between Italy and China. *Technological Forecasting and Social Change, 142,* 26–41. https://doi.org/10.1016/j.techfore.2018.07.025.

66. Cowley, R., Caprotti, F., Ferretti, M., & Zhong, C. (2018). Ordinary chinese smart cities: The Case of Wuhan. Inside smart cities, In book: Inside Smart Cities: Place, Politics and Urban Innovation, Routledge, https://doi.org/10.4324/9781351166201-4.

67. Allam, Z., Dey, G., & Jones, D. (2020). Artificial Intelligence (AI) provided early detection of the Coronavirus (COVID-19) in China and will influence future urban health policy internationally. *AI, 1*(2), 156–165. https://doi.org/10.3390/ai1020009.

68. Helsinki Times. (2020). China's mobile cabin hospitals serve as 'cozy home' for patients. Retrieved November 2, 2011, from https://www.helsinkitimes.fi/china-news/17432-china-s-mobile-cabin-hospitalsserve-as-cozy-home-for-patients.html. Accessed 2.11.2020.

69. United for Smart Sustainable Cities. U4SSC Thematic Groups. Retrieved November 2, 2011, from https://www.itu.int/en/ITU-T/ssc/united/Pages/thematic-groups.aspx.

70. Jian, Y. (2021). Economic development of smart city industry based on 5G network and wireless sensors. *Microprocessers and Microsystems, 80,*. https://doi.org/10.1016/j.micpro.2020.103563.

71. Kim, H. (2021). Smart cities beyond COVID-19. *Technological Forecasting and Social Change, 142,* 105–116. https://doi.org/10.1016/j.techfore.2018.09.012.

Applying IoT and Data Analytics to Thermal Comfort: A Review

Maysaa Khalil, Moez Esseghir, and Leila Merghem-Boulahia

Abstract The widespread popularity of Internet of things (IoT) enables a huge amount of fine-grained thermal comfort and, implicitly, energy efficiency data to be assembled. Meanwhile, the movement toward a human-centric approach in the domain of smart homes has constantly been advancing worldwide. How to collect, prepare, store and employ the huge amount of data collected by the IoT devices to promote and enhance the thermal comfort inside buildings while preserving energy constraints is an important issue. To date, different IoT applications have been introduced in the thermal comfort domain. To provide a thorough overview of the prevailing research and to identify challenges for the upcoming research, this paper conducts a review on the IoT thermal comfort applications and thermal comfort analytics in buildings. Following the three stages of analytics: descriptive, predictive and prescriptive, we propose the basic application areas as thermal comfort analysis, thermal comfort prediction and thermal environment control. We also review the novel techniques endorsed by each application. In addition, we discuss some research trends associated with the topic like machine learning techniques, data privacy and security and federated learning.

Keywords Thermal comfort · IoT · Machine learning · E-health · Deep learning · Data analytics

M. Khalil · M. Esseghir (✉) · L. Merghem-Boulahia
ICD/ERA, University of Technology of Troyes, 12 Rue Marie Curie, 10300 Troyes, France
e-mail: moez.esseghir@utt.fr

M. Khalil
e-mail: maysaa.khalil@utt.fr

L. Merghem-Boulahia
e-mail: leila.merghem@utt.fr

1 Introduction

Nowadays, the Internet of things (IoT) applications have become an essential part in our life. The term is based on the concept that everything is connected everywhere [42]. The new pattern aims at switching real-world objects into smart objects [75] that are able to communicate via the Internet to generate a shared infrastructure that is able to connect human to human, things to things and human to things [1]. In fact, IoT has speedily engaged in different critical fields in human life such as e-health [59], home automation [72], thermal comfort [62], energy management [68], and so forth [21].

On the other hand, global warming and its impending follow-ups have become a major global issue. Scientists and governments have agreed that cleaner and sustainable solutions can help in reducing the impact of this phenomenon. Thus, governments have implemented measures to reduce greenhouse gases emissions, the major catalyst for global warming [63, 67]. Studies [12, 61] have agreed that the energy sector, specifically the electric one has a vital impact on emissions and can be regulated by public policies to reduce greenhouse gases emissions.

According to EU statistics [7], buildings represent 40% of all energy consumption and 36% of CO_2 emissions in Europe due to the age of buildings. More specifically, 50% of building's energy consumption is due to Heating Refrigeration and Air Conditioning (HVAC) energy use. Studies [2] reveal that if the current energy consumption pattern persists, the world energy consumption will increase more than 50% before 2030. The concept of smart building has been introduced to address problems implied by this observation. The principle is to integrate "datification" into buildings to optimize their usage in terms of comfort and energy. When looking at the use of HVACs, occupant's thermal satisfaction plays a vital role. Therefore, it is important to rely on an IoT environment that is able to provide the balance between energy consumption and thermal comfort [78].

Different researchers have developed several mechanisms to apply IoT to the issue of thermal comfort. Some has focused on the analysis of the data such as revealing some thermal comfort insights or the effects of different input parameters on the thermal sensation. Others tried to predict the thermal sensation using different machine learning models. There are also some researchers that focused on the IoT architecture: the sensing layer, the wearable devices and network layer. Finally, some moved toward controlling the thermal environment in order to preserve energy. Consequently, our purpose is to propose and classify the most influential scientific publications that mix the fields of data analytics, IoT and thermal comfort.

1.1 Bibliometric Analysis

In order to maintain an overview of the ongoing research in thermal comfort analytics, a bibliometric analysis was conducted on 16 September 2020 using the acknowledged database, Scopus. The query for Scopus is as follows: (("IoT" OR "sensor"

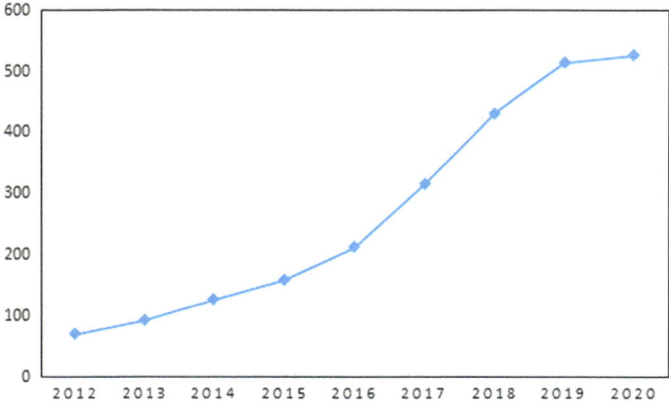

Fig. 1 Number of publication indexed by Scopus

OR "Internet of things") AND "thermal comfort" AND ("household" OR "resident" OR "residential" OR "building" OR "smart building" OR "industrial" OR "individual" OR "occupant" OR "inhabitant") AND ("energy efficiency" OR "prediction" OR "clustering" OR "forecasting" OR "profiling" OR "classification" OR "machine learning" OR "environment control" OR "image processing")).

Figure 1 presents the number of publications indexed by Scopus from 2012 to 2020. There are 2451 publications in total in the Scopus library. The sum of publication before 2012 was under a low level and it started to advance gradually after the year of 2012. However, during 2017 and later on the research publications increased rapidly. The result is not surprising at all. The IoT or the smart building term started appearing in the 1999. It requires years to collect data and then bring research verdicts to publications.

Smart home IoT applications allow the automatic control inside houses including thermal comfort, energy management, security and connectivity. Smart homes market sector has doubled its sales in the last three years and is expected to increase profits by 60% by 2024 [71]. Sine 2013, smart home has become a trend but it has reached the summit in the last few years as a result of the success of commercial devices like Amazon Alexa. This was considered as a revolution as IoT evolved from a scientific interest to a fashionable user trend. Therefore, another query was created on 16 September 2020 using Google trends data in order to visualize people's interest in "Smart Home", "Alexa" and "Nest" trends.

In Fig. 2, the comparison incorporates the trend of Google Nest as well. The graph aims at revealing the commercial phenomenon behind IoT in smart homes. Nest is a smart thermostat that was bought by Google in 2014 where it became more popular. In 2018, smart meters started to be mandatory in developed countries which explains the peak happened. Moreover, at the beginning of every year since 2017, Amazon releases the Christmas advertisement related to Alexa resulting in a peak.

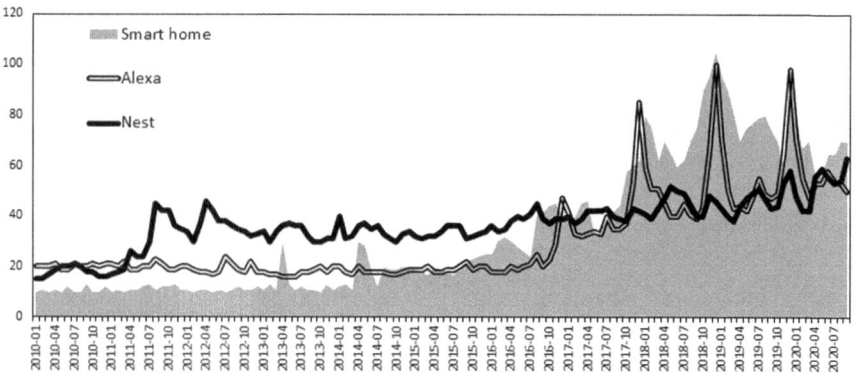

Fig. 2 Normalized Google trends of the terms "Smart home", "Alexa" and "Nest"

1.2   Contributions and Organization

The following paper seeks a comprehensive review of ongoing research and identify challenges for the application of IoT in thermal comfort. The contribution of this paper is as follows:

– Providing a comprehensive detailed study of the IoT architecture based on four major components which are IoT devices, communication technologies, IoT middlewares, and data storage and processing.
– Conducting a well-designed taxonomy for IoT application in thermal comfort from the perspective of sensing layer.
– Conducting a well-designed taxonomy for IoT data analytics in thermal comfort from the perspective of thermal comfort analysis, thermal comfort prediction and environmental control.
– Examining open research challenges for forthcoming research orientations, including machine learning techniques, data privacy and security and federated learning.

The remainder of the paper is organized as follows. A brief description of thermal comfort and open thermal comfort datasets are presented in Sect. 2. Section 3 presents the IoT ecosystem related to thermal comfort including all major components. Section 4 conducts the survey on IoT applications in thermal comfort depending on different compositions of the sensing layer. Section 5 discusses some applications on data analytics for thermal comfort analysis, prediction and control. Several future research issues are considered in Sect. 6. Section 7 draws the conclusions.

2   Thermal Comfort

In 2001, the widely referenced National Human Activity Pattern Survey stated that people spend 87% of their times indoors [35]. The problem of identifying the thermal comfort of occupants inside buildings has been a challenging issue through out decades. Since the 1970s, two different schools of thought emerged to solve the problem: the steady-state model and the adaptive model.

The steady-state model or Fanger's [22] model was introduced by Fanger. The model relates the thermal sensation to subjective parameters such as the metabolic rate and clothing insulation. Fanger derived his model in climate chambers where the model requests four environmental variables: indoor air temperature, mean radiant temperature, indoor relative humidity and indoor air velocity. The model also necessitates the presence of two personal variables: metabolic rate and clothing insulation. After feeding these parameters into the model, a Predicted Mean Vote (PMV) index is calculated. The PMV index aims at predicting and evaluating the thermal sensation of the individuals. The model is based on heat balance equation as explained in [28] taking into consideration both the physical heat transfer process and other thermo-physiological processes such as sweating, etc. Fanger also related the PMV index to a subjective parameter, the Predicted Percentage of Dissatisfied (PPD). PPD measures the percentage of thermally dissatisfied occupants. However, studies have revealed the failure of PMV model outside the exact conditions required by Fanger as it does not take into consideration the effect of occupant adaptation. Therefore, Fanger and Toftum [23] elaborated another model which includes an expectancy factor.

On the other hand, Nicol and Humphreys [53] claimed that experiments in climate chamber miss some fundamental parameters in thermal comfort analysis such as time sequence and social factors leading to the creation of the Adaptive Model. De Dear and Brager [17] stated that occupant's thermal comfort in naturally ventilated buildings depends in great part on the external temperature and occupant's adaptability. In recent studies, people in mixed environments can adjust or change their thermal environment with operable windows or through metabolic alterations. Therefore, the PMV model is considered inadequate [70, 80]. Moreover, the adaptive model standard specifies thermal design guidance by indicating the range of accepted indoor temperature depending on the climate zone [15, 16].

In fact, both models are regulated by ISO 7730:2005 [30], EN 16798-1:2019 [9] and ASHRAE 55:2017 [5] regulations and are globally used and accepted but with some significant differences among them [26]. Moreover, several studies illustrate that thermal comfort has an effect on occupant's health and performance. Lan et al. [38] were able to derive a quantitative relationship between thermal comfort and productivity losses. They concluded that in both cold and hot situations, thermal discomfort results in less performance.

In this framework, IoT represents a solution for the issue of thermal comfort control inside buildings. Recently, IoT developers have been proposing plenty of IoT based systems for smart homes to provide comfort inside buildings while retaining energy consumption [43]. Furthermore, IoT applied to thermal comfort consider the energy matter from the user's point of view not only passively from the building leading to better comfort.

2.1 Open Thermal Comfort Datasets

Many companies and researchers are hesitant to upload their thermal comfort data globally due to different matters including privacy and security. This has been a challenge for conducting research in thermal comfort data analytics and its applications. However, some datasets at household and office levels have been made public of which are summarized in Table 1.

Table 1 Basic information of different open thermal comfort datasets

Name	Brief description	Number of records	Frequency	Duration
ASHRAE global thermal comfort database II [46]	– Collection and normalization of row data from last two decades – Set of indoor climatic observations – Intended to support inquires about thermal comfort	107584	No freq.	Between 1979 and 2016
The scales project [64]	– Sensor data and survey data from 30 countries – Intended to understand the causes of thermal discomfort	8226	No freq.	Between 1986 and 2010
Langevin Longitudinal Dataset [39]	– Indoor and outdoor sensor data – Questionnaire related to thermal sensation and other personal factors	678621	Each 15 min	Between July 20112 and August 2013
ERA5-Heat [20]	– Dataset completed after reanalysis from European Center of Medium Range Forecast	–	Each hour	Between 1979 and 2020
Winter Thermal Comfort and health for the elderly [29]	– Datasets from 43 houses in Bath,UK – Elderly Occupants – Sensor data and questionnaire about thermal comfort and health	424	Every 90 min	Between 2016 and 2018

- **ASHRAE Global Database II**: In 2014, the ASHRAE Global Thermal Comfort Database II project was launched under the leadership of University of California at Berkeley's Center for the Built Environment and The University of Sydney's Indoor Environmental Quality (IEQ) Laboratory. The database is intended to support diverse inquiries about thermal comfort in field settings.
- **The Scales Project**: The study aims at gaining knowledge of the thermal perception scales by survey participants. Questionnaires were derived in 21 language versions and then conducted in 57 cities of 30 countries.
- **Lagevin Longitudinal Dataset**: The dataset adopts findings from a one-year longitudinal case study of occupants thermal comfort in an air-conditioned office building in Center City Philadelphia. It is the most utilisable dataset in the machine learning domain related to thermal comfort. It has gained attention due to the large amount of sensed data related to outdoor and indoor environment. Moreover, questionnaires made it possible to gather personal information about the occupants.

3 IoT Ecosystem

In the following section, an overview of the IoT architecture for thermal comfort is presented. It incorporates four major components, which are the IoT devices, communication technologies, IoT middlewares and data storage and processing [77]. Figure 3 illustrates the IoT ecosystem related to thermal comfort. The four essential components presented are the core of every IoT environment as presented in the figure. The description of the components in relation to thermal comfort is presented as follows.

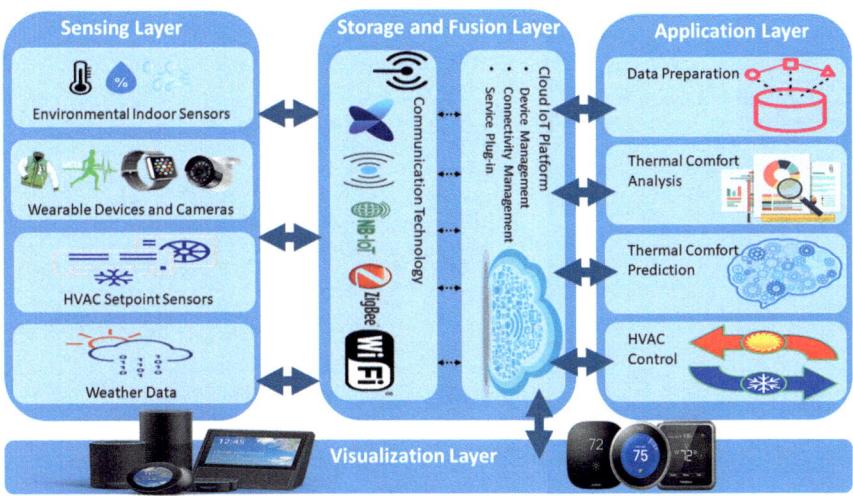

Fig. 3 Illustration of IoT ecosystem in thermal comfort

3.1 IoT Devices

IoT devices form the sensing layer provided in Fig. 3. These devices involve embedded systems that communicate with sensors and actuators with wireless connections. Figure 4 reflects the typical IoT devices for thermal comfort. The microprocessors, memory, communication module and input/output interfaces form the embedded system. The sensors are used to monitor and measure different personal and environmental attributes related to thermal comfort. The sensors can be classified into indoor environmental sensors, outdoor environmental sensors, wearable sensors and HVAC sensors. These sensors are used to gather information, such as air temperature, relative humidity, cooling set-point, solar radiation and metabolic rate. Actuators, on the other hand, are the devices that manipulate the physical environment. They transform electrical signals into tangible actions. IoT devices are suitable for thermal comfort measurements because of the different key characteristics they possess: memory, durability, power efficiency, coverage, reliability, cost and computational efficiency.

3.2 Communication Technologies

The communication technology assures the successful deployment of IoT systems. Existing technologies can be categorized according to four factors: standard, spectrum, application scenario, topology and power consumption [52].

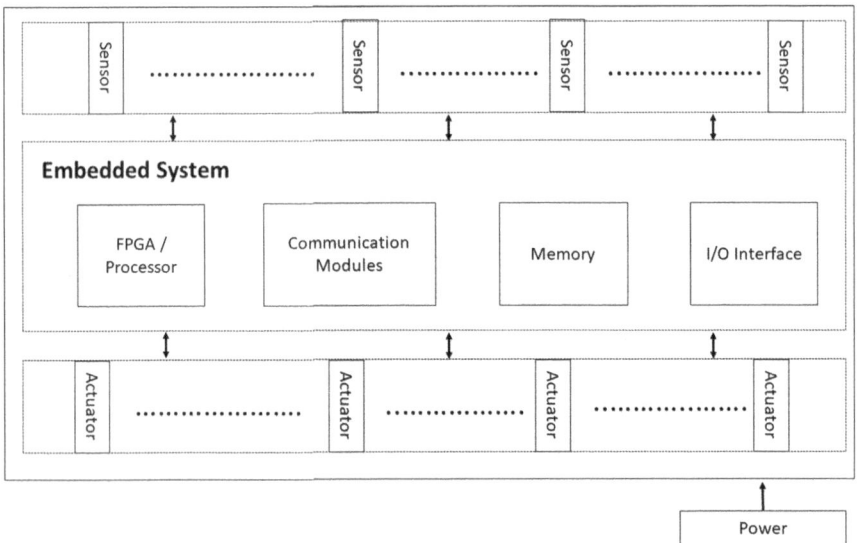

Fig. 4 Architecture of IoT devices

- *Standard*: The communication standard is divided into short-range communication standard and long-range communication standard. In fact, there are many existing wireless communication standards of which are listed in Table 2. Short-range standards are able to cover distance withing 100 m. Examples on short-range standards are ZigBee, Bluetooth, Wi-Fi, Z-Wave, active radio frequency identification (RFID) and passive systems. Long-range are able to cover distances up to 10s of kilometers. Long-range communication standards are considered as the low power wide area (LPWA). Examples on long-range standards are LoRa, Nb-IoT and Sigfox [58, 66].
- *Spectrum*: The communication spectrum is divided into licensed spectrum and unlicensed spectrum. The licensed spectrum is usually allocated to cellular network. Licensed spectrum provides an efficient traffic management, better reliability, higher quality of service (QoS), less interference, higher security levels, low infrastructure cost for users and wider coverage. The weaknesses of the licensed spectrum are related to the cost of subscription for data transmission. The unlicensed spectrum exploits the Industrial scientific medical (ISM) band, which is an industrial scientific and medical radio frequency (RF) band. The weaknesses of the unlicensed spectrum are related to security and infrastructure cost. The latter is related to the difficulty and the cost of installing the equipment necessary to run this unlicensed spectrum to satisfy utilities business.
- *Application Scenario*: IoT devices can act as nodes or as backhaul networks depending on the IoT application. Nodes are able to transmit low data with low power consumption covering very short distances, whereas backhaul network sets up data transmission on high rates over long distances. Besides, there are communication technologies that allow bi-directional link which allows forwarding error correction, handshaking, data encryption and communication between devices. Some studies [19] have showed that each communication technology has its own

Table 2 Communication technology

Type	Network	Speed	Bi-directional link	Spectrum	Range	Topology
RFID	PAN	400 kbps	Yes	Unlicensed	<3 m	P2P
NFC	PAN	400 kbps	No	Unlicensed	<0.1 m	P2P
Wi-Fi	LAN	11 Mbps	Yes	Unlicensed	1 km	Star
ZigBee	LAN	250 kbps	Yes	Unlicensed	10–100 m	Mesh, star, tree
Bluetooth	PAN	700 kbps	Yes	Unlicensed	<30 m	Star
SigFox	LPWAN	100 bps	Yes	Licensed	<50 km	Star
LoRaWan	LPWAN	50 kbps	Yes	Licensed	<20 km	Star
Nb-IoT	LPWAN	200 kbps	Yes	Licensed	<10 km	Star
Z-Wave	LAN	100 kbps	Yes	Unlicensed	<100 m	Mesh

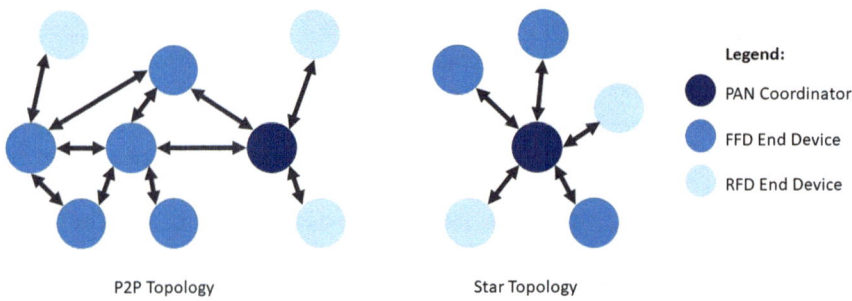

P2P Topology   Star Topology

Fig. 5 Communication technology topology

advantages and disadvantages. Therefore, the most suitable technology depends on the application itself.

– *Topology*: There are different kinds of topology like peer-to-peer (P2P), star, mesh, tree, ring and bus topology. For example in the IEEE 802.15.4 network standard [51], IoT devices play two different roles: personal area coordinator (PAN) or an end device. Also, an IoT device has two different functions: full function device (FFD) or reduced function device (RFD). Figure 5 presents the P2P and star topologies in the IEEE 802.15.4 standard. In P2P topology, the PAN is an FFD while other devices can be FFD or RFD. FFD can connect to any other devices while the RFD can only connect to one FFD and establishes no connection with other RFD. Note that only PAN can start the communication. On the other hand, PAN initiates the communication in the star topology and connects with other devices. However, the end devices can only establish connection with the PAN.

– *Power Consumption*: A limited power supply problem comes up with wireless connectivity between sensors [47]. In an ideal situation, a system where a sensor running on a single AAA battery for years is the most preferable in an IoT application. The IEEE 802.15.4 standard solves the issue of limited power supply. It specifies a wireless ink for low-power personal area networks (LowPANs).

3.3   IoT Middlewares

The transfer of data via the Internet necessitates the presence of adequate security, support of real-time data and the ease of accessibility. This requires the presence of the IoT middleware which assures the mentioned functionalities. Existing architecture of IoT middleware are grouped into 3 class types according to [52].

– *Service-based solution*: it generally adopts the service-oriented architecture (SOA) [36]. The mentioned architecture permits users to deploy a diverse range of IoT devices as services. It is a high-performing heavyweight middleware usually deployed on a set of nodes running in the cloud or on a powerful gateway between IoT device and the applications.

– *Cloud-based solution*: This solution limits the type and the number of IoT devices the users can deploy. However, it allows users to collect and interpret collected data. The functional components of the middleware are limited to what is available on the cloud. Available services of IoT devices on the cloud are only accessed or controlled by cloud supported RESTful APIs or vendor's provided application.
– *Actor-based solution*: It was first discussed by TerraSwam [74]. The architecture is based on three concentric circles. The outermost circle contains sensors and actuators. The middle consists of gateways such as Arduino, RaspberryPi, smartphone and laptop. The inner most circle is the cloud. For more details on actor-based framework, please refer to [52].

Most of the middlewares are service-based. Service description has to be comprehensive but energy efficient to become suitable for resource constrained devices. Also, service discovery and composition should be autonomous and dynamically adaptive. The key difference among the three solutions are the openness of the solution in supporting a new IoT device type, the type of middleware services and where the IoT middleware can be embedded.

3.4 Data Storage and Processing

Data driven thermal comfort requires the collection of huge, dynamic, spatial data and image data that requires both storage and processing [44]. Data may be historical data, sensor data and real-time data. Data complexity ranges from structured to non-structured data [69] that can be a video, image, text, audio, digit and binary number. The IoT platform allows the storage of sensor's collected data. Nowadays, edge or fog computing is advocated. In edge computing, the IoT devices carry out computations and analysis in order to decrease latency of critical applications and reduce costs [45, 81].

There exist plenty of thermal comfort management systems in smart buildings. Those systems have been developed to deploy various forms of data to generate insights that provides thermal control inside houses. Examples of some commercially available platforms: Alexa, Nest, Ecobee, Honeywell, Samsung Home and Siemens Home.

3.5 Summary

There are some technical parameters that need to be taken into account when deploying the IoT device. In wireless connectivity, the following most be taken into account: range, speed, latency, mobility, sensor degradation, security and gateway cost. In

communication technology, technologies with LPWAN are gaining attention specially because of the NB-IoT technology. The latter technology includes interesting
capabilities: low device power consumption, long distance coverage, bidirectional
data flow, low device cost and huge amount of low-throughput devices. There are
plenty of IoT applications in thermal comfort which have adopted the previous architecture explained in this section. Some of these applications are presented in the
following Section.

4   IoT Application in Thermal Comfort

In this section, we are going to discuss how IoT smart objects are used to study both
thermal comfort and energy efficiency. Smart objects are acknowledged as a part of
the IoT application for their inter-connectivity. The following review presents several
approaches and solutions for thermal comfort IoT environment. After achieving a
comprehensive study of the available literature, it has been chosen to group the
research done based on the sensing layer. For a solution strategy grouping, please
refer to Sect. 5. Among the literature, we found some studies that employ sensors to
capture environmental data and questionnaires to identify personal factors. Besides,
some studies employ wearable devices to capture personal factors. Among others,
we found studies that employ smart HVAC sensors. Finally, there exist some studies
that used infrared sensing system and video cameras to capture personal factors: the
metabolic rate and clothing insulation.

4.1   Applications Employing Environmental Sensors and Questionnaires

It is quite common in the literature to measure Fanger's parameters (humidity, air
temperature, air velocity and mean radiant temperature) using indoor equipped sensors. However, some researchers go beyond these parameters giving the fact that his
model shows weaknesses in naturally ventilated buildings and with adapted occupants. Other extracted parameters include outer environmental data and more indoor
environmental data. Langevin et al. [39] presented a longitudinal field study of occupant behaviour in an air-conditioned building. Over the course of one year, continuous
measurements were made for the weather, the local indoor environment and certain
behavioral actions using different equipped sensors. Behavior parameters, including
personal fan use and personal heater use, were measured using the WattsUp meters.
WattsUp meter is a power Analyzer/Watt Meter with computer interface through
which a data logger function records all data into non-volatile memory. Window
state was measured by using HOBO UX-90 state loggers. Ambient temperature and

relative humidity were measured using HOBO U12 base sensors. Daily surveys were conducted including questions related to work flow, thermal sensation, recent activity and recent clothing level.

Hassan et al. [27] suggested an IoT system that merges together data captured from sensors with the Building Information Modeling. They targeted the ability to monitor elderly thermal comfort in eldercare centers. Along with sensor data, they have conducted surveys and site management to gather information from the elderly including activity, cognitive perception and age. Al-Kuwari et al. [3] proposed an IoT environment for smart home automation with thermal comfort considerations. The presented design exploits the EmonCMS platform in collecting data. The sensing of the different environmental variables inside the building is handled by NodeMCU-ESP8266 microcontroller board, which makes it possible to process and upload/download data to/from the EmonCMS cloud server. However, this proposal neglects the relation between personal factors and thermal comfort.

Zhao et al. [86] designed an IoT control system for thermal comfort in intelligent buildings. The system includes a sensor module that groups a temperature sensor, humidity sensor and a wind speed sensor. Metabolic rate and clothing insulation were collected upon surveys provided to the occupants. Rajith et al. [60] built a real-time optimized HVAC control system on the top of IoT framework. Environmental attributes were collected by different equipped sensors where MQTT protocol establishes the connection between the IoT gatway and the cloud. Moreover, user feedback information regarding their thermal sensation are collected for real-time processing in the distributed cloud environment.

4.2 Applications Employing Environmental Sensors and Wearable Devices

The applications presented in this section adopt wearable technologies aiming at measuring personal metabolic rate and clothing insulation. Salamone et al. [62] tried to demonstrate the reliability of IoT applications in assessing thermal comfort. Therefore, they designed an IoT system based on two principal systems for data harvesting: the nearable system and the wearable system. The nearable system consists of low-cost sensors able to measure indoor environmental variables. The wearable system consists of a wearable device: the Empatica E4 wristband (Empatica Inc., Cambridge, MA, USA) which allows the measurement of the heart rate, the activity level and the skin temperature.

Laftchiev and Nikovski [37] proposed an IoT based system for personal thermal comfort. The model was created after collecting data from different equipped sensors for environmental variables whereas biometric readings were measured by the Microsoft Band. Ciabattoni et al. [14] presented a low-cost IoT based system for thermal comfort. Different ambient sensors were used to provide the requested functionality including humidity and temperature sensors. Besides, the integration of the device with a smartwatch allowed the analysis of personal comfort parameters. ALsaleem et al. [4] used the Microsoft Smart Band 2 with Hobo Data Logger UX100 in order to collect both personal and environmental data in order to build the IoT environment for thermal comfort monitoring.

4.3 Applications Employing Smart HVAC System

The main focus of the following applications is on the HVAC system and the IoT environments designed to fulfill the less energy consumption target. Marche and Nitti [48] proposed an IoT smart system incorporating five HVAC in five different rooms with an aim to propose a user-centric IoT approach. The main point is to figure out a trade-off between occupant's thermal comfort and cost. That is, the system will be able to find the most appropriate working time for the 5 HVAC systems based on room temperature and occupancy. En et al. [18] presented a model predictive control (MPC) based thermal comfort controller for cyber-physical home systems. The house is featured with more than 300 sensors that are connected with ECHONET Lite version 1.1. Ali et al. [2] developed an IoT-based smart framework to control the Air Conditioning (AC) to provide a thermally comfortable environment. User's feeling toward the environment and sensor data are integrated with an enhanced PMV index.

4.4 Applications Employing Infrared Sensing System and Video Cameras

Different methods have been adopted to measure personal factors. However, surveys are far from the concept of IoT and wearable devices may generate inconvenience. This has lead to the penetration of researches that use video cameras or infrared sensing system to evaluate certain parameters.

Zang et al. [82] proposed a thermal comfort IoT platform with automatic regulation. Continuous data are collected via environment sensors such as humidity sensor. The special thing is that they have integrated video cameras to the IoT platform to capture occupant's activity and clothing insulation. The images generated can be mapped into different metabolic rate and clothing insulation by machine learning classification algorithms. Besides, Ghahramani et al. [25] presented an IoT application to calculate personal thermal comfort. They used data from the occupants face using infrared thermography. They used the face because it has high density of blood vessels and it is not covered. Jazizadeh et al. [32] designed a novel approach for enabling RGB video cameras as sensors for measuring personalized thermoregulation states, an indicator of thermal comfort.

5 IoT and Data Analytics in Thermal Comfort

The scientific process of transforming IoT data into powerful insights used to make better decisions is known as data analytics. It is usually anatomized into three phases: descriptive analytics (how do the data look like), predictive analytics (what is going to be found from the data) and prescriptive analytics (what decisions can be made from the data). This section of thermal comfort IoT data analytics is conducted from three aspects: analyzing thermal comfort, predicting thermal comfort and controlling the indoor thermal environment.

5.1 Approaches Involving IoT Environments to Analyze Thermal Comfort

In this section, the works on monitoring thermal comfort are reviewed based on bad data detection and thermal comfort analysis. Bad data detection is crucial because training a forecast or clustering model on the collected dataset with bad data may arise bias and failure in parameter estimation and model creation. In fact, no work was found on bad data detection related to thermal comfort, therefore, works related to anomaly detection in smart buildings were added. For thermal comfort analysis, studies aiming at analyzing situations in terms of thermal sensation without proposing improvements to the environment were presented. Both works are presented in Table 3 through which the smart objects used, the model formulated, the type of building, the location of the experiment and the accuracy of the methods are presented.

5.1.1 Anomaly Detection

Bad data as presented in the following section can represent unusual patterns or missing data generated from data collection failure, unplanned events or communication failure. A data mining approach to solve energy inefficiencies detection problem in smart buildings is presented in [56]. The data mining system is developed after extracting knowledge from different sensors equipped inside the building. These sets of ZigBee sensors are strategically located in the BlueNet building. A master sensor coordinates and manages all sensors sending the data metering to a device that aims at collecting sensor information every 30 s. In [8], the authors proposed a new mechanism for the characterisation of time series data generated by smart buildings and identification of unusual data patterns. They have used an enhanced Symbolic Aggregate approXimation (SAX) process which incorporates an optimised tuning of the time window width and the symbol intervals depending on the building energy and comfort behaviour. The mechanism can be used to characterise building behaviour, define appropriate energy management strategies and send timely alerts based on anomaly detection outcomes.

Table 3 Main contribution to analyze thermal comfort using IoT data

Source	Smart objects	Model	Context	Location	Accuracy
Pena et al. [56]	Indoor ZigBee sensors and outdoor sensors	Data mining techniques	Controlled office building	BlueNet Building, Seville, Spain	Not mentioned
Capozzoli et al. [8]	Indoor sensors	SAX process and classification trees	Six-storey glazed building	Sant Cugat del Valles, Spain (exp. 1) and Torino, Italy (exp. 2)	Ranged between 80 and 90%
Choi et al. [13]	Thermistor-type skin sensor	Stepwise regression	Experimental chamber at University of Southern California	Los Angeles, California, USA	Ranged between 94 and 95%
Ploennings et al. [57]	Thermostat and occupancy sensors	Not used	IBM Research Living Laboratory Building	Dublin, Ireland	Not mentioned
Park et al. [55]	Indoor sensors	Dynamic and static models using Mat-lab/Simulink	Residential Building	Not mentioned	Not mentioned

5.1.2 Thermal Comfort Monitoring

As mentioned before, these studies only conduct experimentation to test the proposed framework. Besides, they do not possess material consequences on energy efficiency nor on occupant's satisfaction. An example is a study in [13], where the authors performed a set of experiments on 18 parameters to explore the correlation between thermal sensation and skin temperature. In order to collect user preferences and environmental data, a data acquisition system was adopted. For skin temperature, a thermistor-type sensor was equipped in 8 different body locations. The 7-point scale of ASHRAE standard was used to detect thermal sensation. The scale ranges between -3 for hot sensation, 0 for neutral and +3 for cold sensation. A stepwise regression analysis was conducted to cast the predictive parameters. Results showed that thermal sensation is highly correlated with the skin temperature. Moreover, authors claimed that data obtained from the waist, arm and wrist results in the maximum accuracy (95.97%). Another study that is worth mentioning is the demo by Ploennigs et al. [57] that provided a combination between virtual reality and thermal comfort concept. The demo allowed users to visualize real-time data related to environmental parameters collected by sensors deployed inside the building. This will allow occupants to better understand the interaction between their thermal sensation and the collected parameters. In another study [55], a complete smart building system was created. The sensing layer of the IoT system includes electrical devices, personal wristband, smartphone application and a smart HVAC system. Two models were adopted: a static model to measure the heating system and a dynamic model to measure user's thermal comfort. The optimal PMV value ranges between -0.2 to 0.2. Therefore, when the PMV out ranges the optimal values, a cooling or heating system will be requested.

5.2 Approaches Involving IoT Environments to Predict Thermal Comfort Using Machine Learning

Thermal comfort prediction, in most cases, implies the action of predicting the PMV index developed by Fanger but using different input parameters. Figure 6 presents the number of publications for which a machine learning model has been used to forecast thermal comfort. Among the algorithms, neural networks, Support Vector Machines (SVM), Random Forest and fuzzy logic ranked at top choice. Because there is no final decision on algorithm selection, the most adequate strategy in current researches is adopting different algorithms to test a single dataset aiming at minimizing the prediction biases. Table 4 gives a brief summary of the contributions below.

Zhange et al. [84] used IoT senor's data from a building to derive an accurate deep neural network (DNN) thermal comfort prediction model. A fine grained approach was proposed to predict the PMV index using different comfort factors. The model was trained for each factor through which the corresponding PMV index was evaluated. Results revealed the accuracy of the proposed DNN model and the

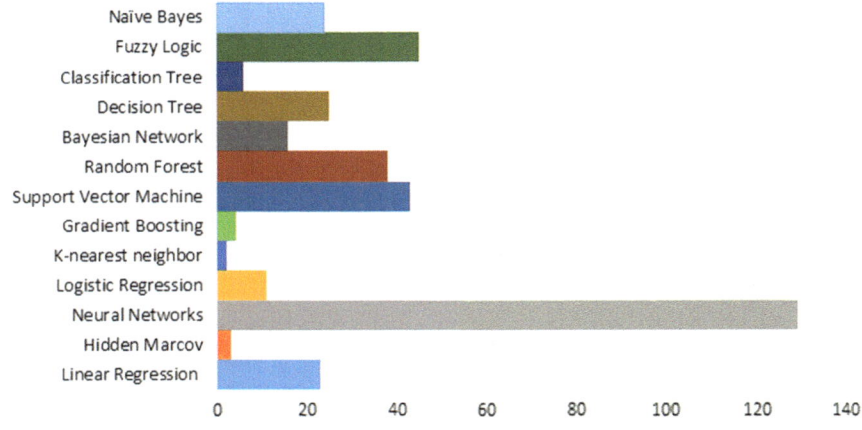

Fig. 6 Machine Learning Models deployed in thermal comfort prediction according to Scorpus

Table 4 Main contribution to predict thermal comfort using IoT data

Source	Smart Objects	Model	Context	Location	Accuracy
Zhange et al. [84]	Indoor and outdoor sensors with question-naires	Deep learning model	Controlled office building	Center City Philadelphia, USA	90%
Megri et al. [50]	Comfort meter and a Portable Heat Stress Monitor	SVM machine	North Carolina A&T State University building	Greensboro, USA	93%
Chaudhuri et al. [10]	Sensing devices	Random forest	Regulated laboratory environment	Singapore	92.86% for males and 94.29% for females
Lee et al. [40]	Indoor sensors	Bayesian Network	Center of High Performance Buildings	Purdue University, West Lafayette, Indiana	Not mentioned
Khalil et al. [33]	Indoor sensors and question-naires	Autoregressive with exogenous variables (ARX) model	Controlled buildings	Philadeplphia, USA	93%

out-performance of the used model in comparison to other machine learning models. Besides, the model showed higher performance with deeper network topology and more neurons. Another study in [50] adopted the SVM machine learning model where more adequate nonlinear kernels had been used to predict thermal comfort index. The study focused on supervised machine learning through which an instructor provides the output samples while training. Experimental factors include air temperature, mean radiant temperature, relative humidity, air velocity, metabolic rate and clothing insulation. The mentioned factors were used for training the SVM machine. Results demonstrate a high correlation between the thermal index resulted from the SVM machine with a polynomial kernel of the second order and the thermal index calculated using Fanger's model.

Another study [10] investigated the possibility to predict occupant's thermal sensation, either comfort or discomfort, using physiological parameter information. Five physiological parameters were recorded including hand skin temperature, hand skin conductance, pulse rate, blood oxygen saturation and blood pressure. Four subjective responses were measured including thermal comfort, thermal preference, humidity sensation, and airflow sensation. Besides, thermal sensation survey was recorded along the experiment. Statistical analysis and predictive modeling using Random Forest were implemented. Results show a high accuracy in thermal state prediction equal to 92% and 94% in males and females respectively. Moreover, Lee et al. [40] proposed a general thermal comfort model under the hypothesis "Different people prefer different thermal conditions" with a Bayesian approach. The authors claimed that thermal sensation is based on the overall thermal stress and personal thermal characteristics. The thermal stress can be derived using physical equations and few parameters whereas the thermal characteristic is considered as a hidden random variable. Then, occupants are clustered based on this hidden variable. Results demonstrate a clear evidence of the existence of the multi-clusters and show accurate model predictions.

In [33], an accurate thermal comfort model for smart building control in real time is derived from an IoT generated building data. Auto-regressive with exogenous variables (ARX) model is used to determine the relation between thermal comfort and different personal, outdoor and indoor attributes. As thermal comfort model may derive from different attributes, a sensitivity analysis using Morris method is used to figure out the best model input parameters that results in lower error and cost. After that, an ARX model using the chosen parameters was trained. Experiments showed that Fanger's parameters are the best combination of parameters used to estimate the PMV index. Results demonstrated the efficiency of ARX model in terms of prediction accuracy and complexity. Besides, results revealed that the ARX model improves the mean absolute error and complexity of the thermal prediction when compared to other machine learning models.

5.3 Approaches Involving IoT Environments to Control the Indoor Thermal Environment

In this section, we will focus on occupant-centric building control (OOC) which reflects control systems that uses data from occupants, indoor environment and weather conditions to regulate the indoor environment. The regulations target the indoor temperature, humidity, CO_2 level and/or illuminance via different actuators and devices including thermostat, windows, fans, etc. in order to provide thermal comfort for occupants with energy savings. Table 5 presents a brief summary of the work presented below.

Feldemier and Paradiso [24] designed a personalized HVAC system for building comfort control that is focusing on the occupant. A central network hub, room nodes, portable nodes and control nodes form the system. User's contribution is nothing to mention. That is, users will only have to press a button on the wrist they are wearing whenever they feel uncomfortable. These uncomfortable signals are sent to the designed system that controls the air-conditioning system to direct the airflow. The system showed prosperous advantages in terms of comfortable and energy savings with an 80% comfort goal and 24% energy savings.

In another work presented in [73], the authors tried to control thermal comfort inside a building. They presented a complete IoT architecture including sensors and

Table 5 Occupant-centric thermal comfort control

Source	Smart objects	Model	Heating verso Cooling	Comfort improvement	Energy savings
Feldmeier and Paradiso [24]	Smart HVAC, wearable devices, sensors and control nodes	Hybridized control system	Cooling	More than 80%	Up to 24%
Sung et al. [73]	Smart HVAC and sensors	Machine learning	Cooling	83%	6-11.3%
Zhange et al. [83]	Foot warmers	Berkeley simple measurement and actuation pofile	Heating	80–100%	37–75% depending on the weather
Chaudhuri et al. [11]	Wearable devices and indoor sensors	Neural networks	Both	19.8–21.2%	36.5%

smart terminals. Indoor environmental parameters were analyzed and simulated on MATLAB to derive their impact on thermal comfort index. Three control modes were adopted: Comfort mode, Energy-saving mode and General model with a 0, 0.7 and 0.5 targeted thermal sensation respectively. Human interface was used for monitoring and controlling the load. Moreover, data were stored in the database for analysis. Results show that the expected thermal sensation is achieved in the three modes.

Zhang et al. [83] equipped an office with foot warmers with indoor temperature less than normal in winter and claimed no effect on occupants thermal comfort with great energy savings. The foot warmers were connected to the internet. Energy savings were up to 75% depending on the weather. The astonishing result was the 100% satisfaction rate, which is the optimal target in comfort field. However, the disadvantages of this study is that it only considers the heating system.

In [11], the authors present a feedforward neural network based control framework for thermal comfort and energy saving inside buildings. Explicitly, the framework contains two main components: a prediction model and an optimization algorithm. These components will act between the occupant and the Air Conditioning and Mechanical Ventilation (ACMV) system. The prediction model estimates the thermal sensation index. The ACMV energy consumption is modelled using feedforward neural network with three input variables: air temperature and cooling operating frequencies. Depending on the predicted value, the optimization method locates the optimal operating state. Results demonstrate a 36.5% energy saving and 19.8% - 21.2% comfort improving.

6 Challenges

6.1 Machine Learning Models

Figure 7 presents the predicting accuracy of thermal comfort index in different machine learning models discussed in [79]. These models include Linear regression, Tree regression, Classification tree, Linear discriminant analysis, Logistic regression, Decision tree, Boosted trees, Bayesian network, Bayesian modeling, Naive Bayes, Artificial neural networks, K nearest neighbors, Adaboost, Gradient boosting machine, Support vector machine, Random forests, Gaussian process classifier, Rule-based classifier, Fuzzy logic, Extra tree, and Hidden Markov model. The accuracy varied in a wide range with a standard deviation of around 15%. Different reasons cause this variance including the input variables, the sampled dataset and the model itself. However, median predicting accuracy of 84% is highly encouraging. Therefore, machine learning has offered an alternative approach for thermal comfort prediction. Directions toward the future include shifting towards personal comfort models [76]. This will encourage the fundamental of occupant-centric strategies [34]. But this is totally a modern trend that needs further investigations. Another future focus might be on balancing between model accuracy and input parameters. This will result in a cost-efficient model considering sensing cost, computing cost and data management. In fact, some studies [41] tried to cover this balance but more research is needed. The final future focus is related to public accessibility. The available resources can boost a joint effort among concerned researchers. Unfortunately, only few data-driven thermal comfort studies publicly published their code, web tool, model or data [85, 87]. Intentions must be made to make these models publicly accessible in a way to avoid work duplication.

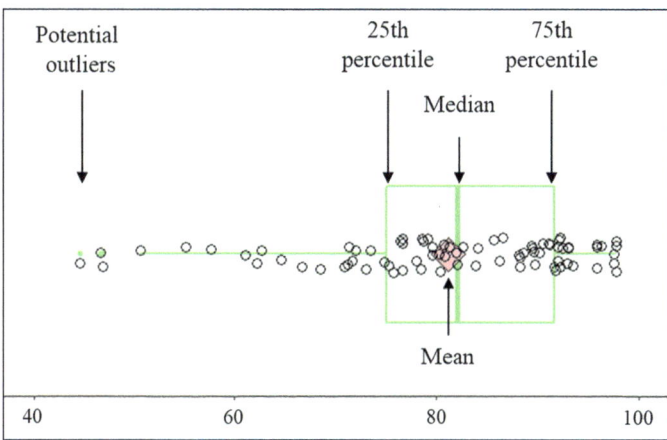

Fig. 7 Predicting accuracy of different machine learning models

6.2 Data Privacy and Security

In fact, the concern regarding occupant's data privacy and security is one of the fundamental barrier to the privilege of smart IoT control systems inside buildings. Several existing works related to data privacy and security issues in buildings focus on the data communication architecture and physical systems [54]. But, the mechanism of studying data privacy and security from the data analytics aspect is still limited. To solve the issue related to data privacy, end-to-end privacy-preservation methods that permit the knowledge extraction from data have been proposed [31, 65]. The application of privacy-preservation method and how the IoT application in thermal comfort would benefit from it can be further studied.

On the other hand, the security of IoT devices should get more attention between researchers. More studies are needed in the development of IoT devices that are able to adopt new security schemes such as the signcryption algorithm [6]. The algorithm blocks eavesdropping and unauthorized data modification by combining data encryption and digital signature. Indeed, in order to prevent physical intruders and attackers from the IoT devices, security measures need to be more investigated.

6.3 Federated Learning

The rich datasets collected from buildings related to thermal comfort and energy consumption are often privacy sensitive as mentioned before and/or large in quantity. This might preclude logging to the data center and training there using conventional approaches. Therefore, a distributed approach most be adopted in thermal comfort prediction that will allow training on distributed buildings through learning a shared model by aggregating locally-computed updates. This decentralized approach is considered as Federated Learning [49] and its architecture is presented in Fig. 8, where the central server chooses among a set of houses to which the current model will be sent. For each selected house, the provided model is updated based on local data by performing iterations of a machine-learning algorithm and then sent back to the server. On the server, the models are aggregated leading to a construction of the new model. However, until now, very few work focused on integrating federated learning in building's management systems. Hence, the integration of federated learning in smart buildings must be further studied.

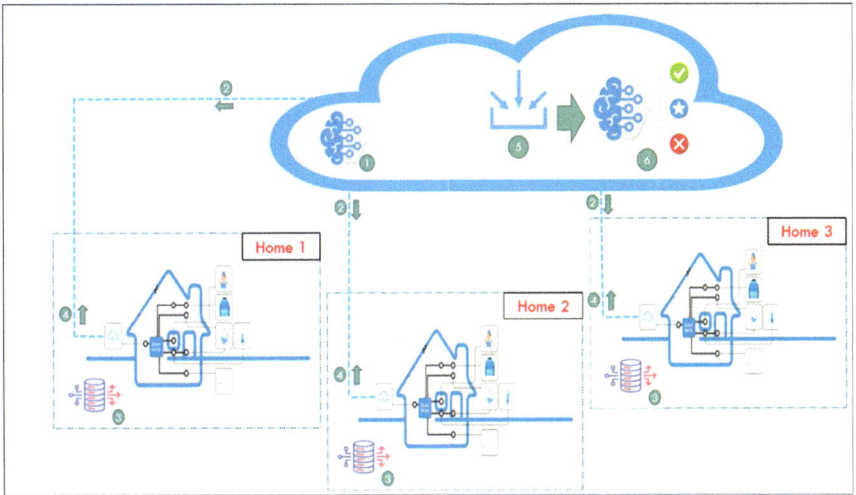

Fig. 8 Federated learning architecture for thermal comfort prediction

7   Conclusion

As the march to a smart building continues with the implementation of foundational smart building management capabilities, it is vital that utilities develop and execute an analytics strategy to profit from the additional information at their fingertips. Thermal comfort control is becoming increasingly complex. Occupant's expectations continue to sour and developers have new alternatives to meet occupant's comfort needs. Simply put, analytics excellence is a strategic imperative. A comprehensive IoT analytics vision and strategy is a most for utilities to maximize the potential of data analytics. The first step is to recognize that data is a strategic asset and should be managed as such. IoT-generated data analytics are key enablers for enhanced thermal comfort control.

Therefore, we have provided a comprehensive review of IoT and data analytics in thermal comfort, including the applications in thermal comfort prediction, anomaly detection, monitoring and thermal environment control. The current developments in this field have been encapsulated and discussed. In addition, we have proposed future research orientations from the prospective of machine learning, data privacy and federated learning. Thermal comfort IoT data analytics in smart buildings is still an emerging research area. We hope that this review can answer reader's questions related to all IoT aspects in thermal comfort control.

Acknowledgements This work is partly supported by grants from Troyes Champagne métropole and the Conseil Départemental de l'Aube.

References

1. Aggarwal, R., & Das, M. L. (2012). Rfid security in the context of internet of things. In *Proceedings of the First International Conference on Security of Internet of Things* (pp. 51–56)
2. Ali, A., Shukor, S. A., Rahim, N., Razlan, Z., Jamal, Z., & Kohlhof, K. (2019). IoT-based smart air conditioning control for thermal comfort. In *2019 IEEE International Conference on Automatic Control and Intelligent Systems (I2CACIS)* (pp. 289–294). IEEE
3. Al-Kuwari, M., Ramadan, A., Ismael, Y., Al-Sughair, L., Gastli, A., & Benammar, M. (2018). Smart-home automation using iot-based sensing and monitoring platform. In *2018 IEEE 12th International Conference on Compatibility, Power Electronics and Power Engineering (CPE-POWERENG 2018)* (pp. 1–6). IEEE
4. Alsaleem, F., Tesfay, M. K., Rafaie, M., Sinkar, K., Besarla, D., & Arunasalam, P. (2020). An IoT framework for modeling and controlling thermal comfort in buildings
5. ANSI/ASHRAE Standard 55-2013, A. (2013). Thermal environmental conditions for human occupancy
6. Barbosa, M., & Farshim, P. (2008). Certificateless signcryption. In *Proceedings of the 2008 ACM symposium on Information, computer and communications security* (pp. 369–372)
7. Burman, E., Mumovic, D., & Kimpian, J. (2014). Towards measurement and verification of energy performance under the framework of the european directive for energy performance of buildings. *Energy, 77,* 153–163.
8. Capozzoli, A., Piscitelli, M. S., Brandi, S., Grassi, D., & Chicco, G. (2018). Automated load pattern learning and anomaly detection for enhancing energy management in smart buildings. *Energy, 157,* 336–352.
9. CEN-CENELEC Management Center, E. (2019). Energy performance of buildings—ventilation for buildings—part 1: Indoor environmental input parameters for design and assessment of energy performance of buildings addressing indoor air quality, thermal environment, lighting and acoustics—module m1-6. CEN-CENELEC Management Centre
10. Chaudhuri, T., Zhai, D., Soh, Y. C., Li, H., & Xie, L. (2018). Random forest based thermal comfort prediction from gender-specific physiological parameters using wearable sensing technology. *Energy and Buildings, 166,* 391–406.
11. Chaudhuri, T., Soh, Y. C., Li, H., & Xie, L. (2019). A feedforward neural network based indoor-climate control framework for thermal comfort and energy saving in buildings. *Applied Energy, 248,* 44–53.
12. Cherubini, F., Bird, N. D., Cowie, A., Jungmeier, G., Schlamadinger, B., & Woess-Gallasch, S. (2009). Energy-and greenhouse gas-based LCA of biofuel and bioenergy systems: Key issues, ranges and recommendations. *Resources, Conservation and Recycling, 53*(8), 434–447.
13. Choi, J. H., & Yeom, D. (2017). Study of data-driven thermal sensation prediction model as a function of local body skin temperatures in a built environment. *Building and Environment, 121,* 130–147.
14. Ciabattoni, L., Ferracuti, F., Ippoliti, G., Longhi, S., & Turri, G. (2016). IoT based indoor personal comfort levels monitoring. In *2016 IEEE International Conference on Consumer Electronics (ICCE)* (pp. 125–126). IEEE
15. De Dear, R., & Brager, G. S. (2001). The adaptive model of thermal comfort and energy conservation in the built environment. *International Journal of Biometeorology, 45*(2), 100–108.
16. De Dear, R. J., & Brager, G. S. (2002). Thermal comfort in naturally ventilated buildings: revisions to Ashrae standard 55. *Energy and Buildings, 34*(6), 549–561.
17. De Dear, R., & Brager, G. S. (1998). Developing an adaptive model of thermal comfort and preference.
18. En, O. S., Yoshiki, M., Lim, Y., & Tan, Y. (2018). Predictive thermal comfort control for cyber-physical home systems. In *2018 13th Annual Conference on System of Systems Engineering (SoSE)* (pp. 444–451). IEEE

19. Erlinghagen, S., Lichtensteiger, B., & Markard, J. (2015). Smart meter communication standards in Europe-a comparison. *Renewable and Sustainable Energy Reviews, 43*, 1249–1262.
20. European centre for medium-range weather forecasts (2020). Retrieved Sept 16, 2020, from https://cds.climate.copernicus.eu/cdsapp#!/dataset/derived-utci-historical?tab=form.
21. Evangelos, A. K., Nikolaos, D. T., & Anthony, C. B. (2011). Integrating rfids and smart objects into a unified internet of things architecture. *Advances in Internet of Things*.
22. Fanger, P. O., et al. (1970). Thermal comfort. analysis and applications in environmental engineering. Thermal comfort. *Analysis and Applications in Environmental Engineering*.
23. Fanger, P. O., & Toftum, J. (2002). Extension of the PMV model to non-air-conditioned buildings in warm climates. *Energy and Buildings, 34*(6), 533–536.
24. Feldmeier, M., & Paradiso, J. A. (2010). Personalized HVAC control system. In *2010 Internet of Things (IOT)* (pp. 1–8). IEEE
25. Ghahramani, A., Castro, G., Karvigh, S. A., & Becerik-Gerber, B. (2018). Towards unsupervised learning of thermal comfort using infrared thermography. *Applied Energy, 211*, 41–49.
26. Grignon-Massé, L., Adnot, J., & Rivière, P. (2008). A preliminary attempt to unify the different approaches of summer comfort evaluation in the European context. Citeseer
27. Hassan, M. T., Yu, J., Zhu, W., Liu, F., Liu, J., & An, N. (2018). Monitoring thermal comfort with IoT technologies: a pilot study in Chinese eldercare centers. In *International Conference on Human Aspects of IT for the Aged Population* (pp. 303–314). Springer
28. Höppe, P. R. (1993). Heat balance modelling. *Experientia, 49*(9), 741–746.
29. Hughes, C., Natarajan, S., Liu, C., Chung, W. J., & Herrera, M. (2019). Winter thermal comfort and health in the elderly. *Energy Policy, 134*, 110954.
30. ISO, I. (2005). 7730: Ergonomics of the thermal environment–analytical determination and interpretation of thermal comfort using calculation of the PMV and PPD indices and local thermal comfort criteria. Management **3**(605), e615.
31. Jayaraman, P. P., Yang, X., Yavari, A., Georgakopoulos, D., & Yi, X. (2017). Privacy preserving internet of things: From privacy techniques to a blueprint architecture and efficient implementation. *Future Generation Computer Systems, 76*, 540–549.
32. Jazizadeh, F., & Jung, W. (2018). Personalized thermal comfort inference using RGB video images for distributed HVAC control. *Applied Energy, 220*, 829–841.
33. Khalil, M., Esseghir, M., & Merghem-Boulahia, L. (2020). An IoT environment for estimating occupants' thermal comfort. In *2020 IEEE 31st Annual International Symposium on Personal, Indoor and Mobile Radio Communications (PIMRC)* (pp. 1–6). IEEE.
34. Kim, J., Schiavon, S., & Brager, G. (2018). Personal comfort models-a new paradigm in thermal comfort for occupant-centric environmental control. *Building and Environment, 132*, 114–124.
35. Klepeis, N. E., Nelson, W. C., Ott, W. R., Robinson, J. P., Tsang, A. M., Switzer, P., et al. (2001). The national human activity pattern survey (nhaps): A resource for assessing exposure to environmental pollutants. *Journal of Exposure Science & Environmental Epidemiology, 11*(3), 231–252.
36. Krafzig, D., Banke, K., & Slama, D. (2005). *Enterprise SOA: Service-oriented architecture best practices*. Prentice Hall Professional
37. Laftchiev, E., & Nikovski, D. (2016). An IoT system to estimate personal thermal comfort. In *2016 IEEE 3rd World Forum on Internet of Things (WF-IoT)* (pp. 672–677). IEEE.
38. Lan, L., Wargocki, P., & Lian, Z. (2011). Quantitative measurement of productivity loss due to thermal discomfort. *Energy and Buildings, 43*(5), 1057–1062.
39. Langevin, J., Gurian, P. L., & Wen, J. (2015). Tracking the human-building interaction: A longitudinal field study of occupant behavior in air-conditioned offices. *Journal of Environmental Psychology, 42*, 94–115.
40. Lee, S., Bilionis, I., Karava, P., & Tzempelikos, A. (2017). A Bayesian approach for probabilistic classification and inference of occupant thermal preferences in office buildings. *Building and Environment, 118*, 323–343.
41. Lee, S., Karava, P., Tzempelikos, A., & Bilionis, I. (2019). Inference of thermal preference profiles for personalized thermal environments with actual building occupants. *Building and Environment, 148*, 714–729.

42. Leminen, S., Westerlund, M., Rajahonka, M., & Siuruainen, R. (2012). Towards IoT ecosystems and business models. In *Internet of things, smart spaces, and next generation networking* (pp. 15–26). Springer.

43. Li, X., Lu, R., Liang, X., Shen, X., Chen, J., & Lin, X. (2011). Smart community: An internet of things application. *IEEE Communications Magazine, 49*(11), 68–75.

44. Li, S., Da Xu, L., & Zhao, S. (2015). The internet of things: A survey. *Information Systems Frontiers, 17*(2), 243–259.

45. Li, H., Ota, K., & Dong, M. (2018). Learning IoT in edge: Deep learning for the internet of things with edge computing. *IEEE Network, 32*(1), 96–101.

46. Ličina, V. F., Cheung, T., Zhang, H., De Dear, R., Parkinson, T., Arens, E., et al. (2018). Development of the Ashrae global thermal comfort database II. *Building and Environment, 142*, 502–512.

47. Mahmoud, M. S., & Mohamad, A. A. (2016). A study of efficient power consumption wireless communication techniques/modules for internet of things (IoT) applications.

48. Marche, C., & Nitti, M. (2019). IoT for the users: Thermal comfort and cost saving. In *Proceedings of the ACM MobiHoc Workshop on Pervasive Systems in the IoT Era* (pp. 55–60).

49. McMahan, B., Moore, E., Ramage, D., Hampson, S., & Arcas, B. A. (2017). Communication-efficient learning of deep networks from decentralized data. In *Artificial intelligence and statistics* (pp. 1273–1282). PMLR.

50. Megri, A. C., & El Naqa, I. (2016). Prediction of the thermal comfort indices using improved support vector machine classifiers and nonlinear kernel functions. *Indoor and Built Environment, 25*(1), 6–16.

51. Molisch, A. F., Balakrishnan, K., Chong, C. C., Emami, S., Fort, A., Karedal, J., Kunisch, J., Schantz, H., Schuster, U., & Siwiak, K. (2004). IEEE 802.15. 4a channel model–final report. *IEEE P802 , 15*(04), 0662.

52. Ngu, A. H., Gutierrez, M., Metsis, V., Nepal, S., & Sheng, Q. Z. (2016). IoT middleware: A survey on issues and enabling technologies. *IEEE Internet of Things Journal, 4*(1), 1–20.

53. Nicol, J. F., & Humphreys, M. A. (1973). Thermal comfort as part of a self-regulating system.

54. Pappachan, P., Degeling, M., Yus, R., Das, A., Bhagavatula, S., Melicher, W., Naeini, P.E., Zhang, S., Bauer, L., Kobsa, A., et al. (2017). Towards privacy-aware smart buildings: Capturing, communicating, and enforcing privacy policies and preferences. In *2017 IEEE 37th International Conference on Distributed Computing Systems Workshops (ICDCSW)* (pp. 193–198). IEEE.

55. Park, H., & Rhee, S. B. (2018). Iot-based smart building environment service for occupants' thermal comfort. *Journal of Sensors, 2018*.

56. Peña, M., Biscarri, F., Guerrero, J. I., Monedero, I., & León, C. (2016). Rule-based system to detect energy efficiency anomalies in smart buildings, a data mining approach. *Expert Systems with Applications, 56*, 242–255.

57. Ploennigs, J., Clement, J., Pietropaoli, B. (2015). Demo abstract: The immersive reality of building data. In *Proceedings of the 2nd ACM International Conference on Embedded Systems for Energy-Efficient Built Environments* (pp. 99–100).

58. Porkodi, R., & Bhuvaneswari, V. (2014). The internet of things (IoT) applications and communication enabling technology standards: An overview. In *2014 International Conference on Intelligent Computing Applications* (pp. 324–329). IEEE

59. Rahmani, A. M., Thanigaivelan, N. K., Gia, T. N., Granados, J., Negash, B., Liljeberg, P., & Tenhunen, H. (2015). Smart e-health gateway: Bringing intelligence to internet-of-things based ubiquitous healthcare systems. In *2015 12th Annual IEEE Consumer Communications and Networking Conference (CCNC)* (pp. 826–834). IEEE.

60. Rajith, A., Soki, S., & Hiroshi, M. (2018). Real-time optimized hvac control system on top of an IoT framework. In *2018 Third International Conference on Fog and Mobile Edge Computing (FMEC)* (pp. 181–186). IEEE.

61. Ramseur, J. L. (2017). Us carbon dioxide emissions trends and projections: Role of the clean power plan and other factors. US Congressional Research Service.

62. Salamone, F., Belussi, L., Currò, C., Danza, L., Ghellere, M., Guazzi, G., et al. (2018). Integrated method for personal thermal comfort assessment and optimization through users' feedback, iot and machine learning: a case study. *Sensors*, *18*(5), 1602.

63. Samli, A. C. (1994). Toward a model of international consumer behavior: Key considerations and research avenues. *Journal of International Consumer Marketing*, *7*(1), 63–84.

64. Schweiker, M., Abdul-Zahra, A., André, M., Al-Atrash, F., Al-Khatri, H., Alprianti, R. R., et al. (2019). The scales project, a cross-national dataset on the interpretation of thermal perception scales. *Scientific Data*, *6*(1), 1–10.

65. Sharma, S., Chen, K., & Sheth, A. (2018). Toward practical privacy-preserving analytics for IoT and cloud-based healthcare systems. *IEEE Internet Computing*, *22*(2), 42–51.

66. Sheng, Z., Yang, S., Yu, Y., Vasilakos, A. V., McCann, J. A., & Leung, K. K. (2013). A survey on the IETF protocol suite for the internet of things: Standards, challenges, and opportunities. *IEEE Wireless Communications*, *20*(6), 91–98.

67. Shove, E. (2010). Beyond the ABC: Climate change policy and theories of social change. *Environment and Planning A*, *42*(6), 1273–1285.

68. Shrouf, F., Ordieres, J., & Miragliotta, G. (2014). Smart factories in industry 4.0: A review of the concept and of energy management approached in production based on the internet of things paradigm. In *2014 IEEE International Conference on Industrial Engineering and Engineering Management* (pp. 697–701). IEEE.

69. Smith, M. R., Martinez, T., & Giraud-Carrier, C. (2014). An instance level analysis of data complexity. *Machine Learning*, *95*(2), 225–256.

70. Spindler, H. C., & Norford, L. K. (2009). Naturally ventilated and mixed-mode buildings-part ii: Optimal control. *Building and Environment*, *44*(4), 750–761.

71. Statista/smart home (2020). Retrieved Sept 16, 2020, from https://www.statista.com/outlook/279/100/smart-home/worldwide.

72. Stojkoska, B. L. R., & Trivodaliev, K. V. (2017). A review of internet of things for smart home: Challenges and solutions. *Journal of Cleaner Production*, *140*, 1454–1464.

73. Sung, W. T., Hsiao, S. J., & Shih, J. A. (2019). Construction of indoor thermal comfort environmental monitoring system based on the IoT architecture. *Journal of Sensors*, *2019*.

74. The terraswarm research center (2013). Retrieved Sept 17, 2020, from https://ptolemy.berkeley.edu/projects/terraswarm/

75. Tomat, V., Ramallo-González, A. P., & Skarmeta Gómez, A. F. (2020). A comprehensive survey about thermal comfort under the IoT paradigm: Is crowdsensing the new horizon? *Sensors*, *20*(16), 4647.

76. Vissers, D. (2012). The human body as sensor for thermal comfort control. Eindhoven University of Technology.

77. Wei, C., & Li, Y. (2011). Design of energy consumption monitoring and energy-saving management system of intelligent building based on the internet of things. In *2011 International Conference on Electronics, Communications and Control (ICECC)* (pp. 3650–3652). IEEE.

78. Wei, S., Jones, R., & De Wilde, P. (2014). Driving factors for occupant-controlled space heating in residential buildings. *Energy and Buildings*, *70*, 36–44.

79. Xie, J., Li, H., Li, C., Zhang, J., & Luo, M. (2020). Review on occupant-centric thermal comfort sensing, predicting, and controlling. *Energy and Buildings*, 110392.

80. Yang, L., Yan, H., & Lam, J. C. (2014). Thermal comfort and building energy consumption implications-a review. *Applied Energy*, *115*, 164–173.

81. Yu, W., Liang, F., He, X., Hatcher, W. G., Lu, C., Lin, J., et al. (2017). A survey on the edge computing for the internet of things. *IEEE Access*, *6*, 6900–6919.

82. Zang, M., Xing, Z., & Tan, Y. (2019). Iot-based personal thermal comfort control for livable environment. *International Journal of Distributed Sensor Networks*, *15*(7), 1550147719865506.

83. Zhang, H., Arens, E., Taub, M., Dickerhoff, D., Bauman, F., Fountain, M., et al. (2015). Using footwarmers in offices for thermal comfort and energy savings. *Energy and Buildings*, *104*, 233–243.

84. Zhang, W., Hu, W., & Wen, Y. (2018). Thermal comfort modeling for smart buildings: A fine-grained deep learning approach. *IEEE Internet of Things Journal, 6*(2), 2540–2549.
85. Zhao, Q., Zhao, Y., Wang, F., Wang, J., Jiang, Y., & Zhang, F. (2014). A data-driven method to describe the personalized dynamic thermal comfort in ordinary office environment: From model to application. *Building and Environment, 72*, 309–318.
86. Zhao, Y., Genovese, P. V., & Li, Z. (2020). Intelligent thermal comfort controlling system for buildings based On IoT and AI. *Future Internet, 12*(2), 30.
87. Zhou, X., Xu, L., Zhang, J., Niu, B., Luo, M., Zhou, G., et al. (2020). Data-driven thermal comfort model via support vector machine algorithms: Insights from ashrae rp-884 database. *Energy and Buildings, 211*, 109795.

Rapid IoT Prototyping

Della Krachai Mohamed and Melouk Kheira

Abstract IoT nodes are spread out everywhere in the world, collecting data and transmitting it to local or cloud storage for further analysis and processing using machine learning or artificial intelligence algorithms. These nodes are the base elements of IoT networks infrastructures that constructs the foundation of smart cities. Actually, IoT nodes, are installed to perform two essential tasks: measurements and data processing and broadcasting. This is achieved through a programmable hardware that is made mainly of a processing unit and a wireless communication stack. Full disclosure, back in 2005 or so when iphone got into market, everyone heared about smartphone. Fast forward 15 years and well, a lot has changed. There is a lot more hardware (Enhanced CPUs and sensors), operating systems, applications, and everything in-between out there now that wasn't so available before. As the demand of this technology is exponentially rising, developing tools and frameworks for rapid production is a challenge to allow rapid prototyping and access to market. Through this chapter, we will use Simulink to develop a code generator for a dedicated hardware IoT node based on espressif ESP8266 microcontroller. Simulink gives the IoT nodes not only the ability to accomplish measurements, but also the power to make decisions and performs actions using various algorithms, such as PID control, fuzzy logic and neural networks.

Keywords IoT · Simulink · Rapid prototyping · ESP8266 · Smart cities · Software framework

D. K. Mohamed (✉) · M. Kheira
Automatics Department, University of Science and Technologies of Oran,
Mohammed Boudiaf BP1500, El Menaouar, Oran, Algeria
e-mail: mohamed.dellakrachai@univ-usto.dz

1   Introduction

The Internet of Things [1–3] connects billions of things and billions of people. It can now be considered as one of the most powerful tools for creating, modifying and sharing countless information. It promises to be the engine of major transformations in the lives of individuals by democratizing new facilities and services in the mobility sector. Smart cities, smart grids, smart lighting systems, smart transportations, smart homes, smart industries, smart agrigulture, clouds and IoT [4–6], it all began with the smart phone, a revolutionary device that changed the history of the technolgy and human being. Starting by integrating operating systems, applications, internet and services on the phones, allowed the development of another vision of market, business and communucation.

In the business to business (B2B) sector, the fields of application are multiplying, particularly in energy management in buildings, in the fields of industry, agriculture, security, transport or still health. Industry is being upgraded to the 4th generation, where IIoT(Industrial Internet of Things) will take place. Interconnecting nodes with reliable and secure communication will, no doubt, make industry smart and productive.

In industry 4, enginners need IoT for sensor data acquisition, processing, performing control algorithms and actuating systems, such as robots, machines, ...etc. The challenge is to start designing a dedicated framework from scratch, which requires to write thousands of lines of code [7–9], master sensors/actuators hardware and control engineering algorithms. This will take obviously a huge time to the final framework. The second approach is to take profit from existing IoT platforms. This approach suffers from the heterogeneous nature of these IoT platforms [10]. In order to shorten the development time, while taking into account the aforementioned facts, the contribution outlined in this chapter, consists of developing a framework on top of an existing engineering software which is intensively applied in industry. The choice is pointed to Matlab/Simulink due to the facts that it implements thousands of multiphysics/multidomain optimised and ready to use toolboxes. The developed framework will serve as a deployment platform for rapid prototyping of IoT projects, including control, local and cloud based storage and visualizing telemetry using a third party tools.

The outline of this chapter is organised as follows: in Sect. 2, a brief overview of internet of things is given. Simulink software is defined in Sect. 3, while Sect. 4 details the framework development stages. Code generation and demonstrations are given in Sect. 5 for validation of the framework, using real world applications.

2   Internet of Things Architecture

By 2020, the Gartner Institute [11] estimates that more than 65% of enterprises (up from 30% today) will adopt IoT products. This says that we are witnessing a real digital revolution that will radically change our lifestyles. At the heart of the

Internet of Things is the object's ability to interconnect and interact with its physical environment. It therefore includes [12]:

- Objects connected directly to the internet,
- Machine to machine M2M: communication between machines and access to the information system without human intervention, through Bluetooth, RFID, Wifi, 4G and soon 5G ...
- "Smart connected devices" such as tablets or smartphones.

The Internet of Things works mainly with sensors and connected objects installed in physical infrastructures. Connected objects interact with their environment through sensors: temperature, speed, humidity, vibration ... In the Internet of Things, an object can be a vehicle, an industrial machine or even a parking space. Sensors will then emit data that will be fed back using a wireless network on IoT platforms. They can thus be analyzed and pre-processed to get the most out of them. These data management and data visualization platforms are the new IoT solutions allowing governments, companies or even users to analyze data and draw conclusions in order to be able to adapt practices and behaviors.

This notion of the Internet of Things can be explained by the concept of Service oriented Architecture, ch11Georgios broken down into four different layers as shown in Fig. 1.

- Sensing layer that collects and process data. The analysis of raw data, temperature, vibration, humidity ... aims to make it usable. With this in mind, connected objects are:

  - Instrumented: they collect and store data in real time,
  - Interconnected: they are shared using a wireless network with other information systems
  - Intelligent: they are analyzed and enriched to help users, companies and institutions in their decision-making.

- Network layer used for data transmission. The Internet of Things is made up of a heterogeneous set of networks that allow these objects to communicate. Among the best known are the cellular networks of historical telecoms operators which allow objects equipped with an M2M SIM card to upload and send data. Actually, LPWA networks, notably with LoRa and Sigfox. Long-range low-speed networks are protocols entirely dedicated to communication between objects.

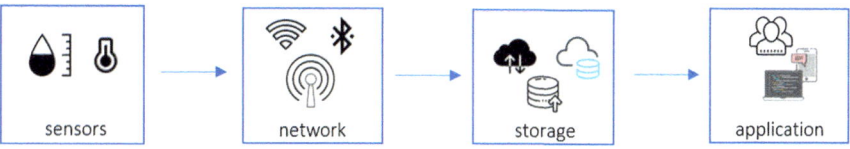

Fig. 1 IoT architecture

- `Platform layer` manages device communication, data storage and services. In the industrial sector, for example, we can now monitor machines remotely, do predictive maintenance of equipment, or improve product traceability.
- `Application layer` presents data and interacts with customers.

3 Simulink—A Complete Multi-domain Engineering Software

Several companies [13, 14] use Matlab/Simulink in their research and develoment departments to model, simulate and deploy code to their hardware.

`MATLAB®/Simulink`[1] is a functional diagram environment for multi-domain simulation and the Model-Based Design approach. It supports system-level design and simulation, automatic code generation, and continuous testing and verification of on-board systems. `Simulink` supports the following features:

- Used to model, analyze and simulate dynamic systems using block diagrams.
- Fully integrated with MATLAB,[2] easy and fast to learn and flexible.
- It has comprehensive block library which can be used to simulate linear, non-linear or discrete systems.
- C codes can be generated from Simulink models for embedded applications and rapid prototyping of control systems.

`MATLAB®/Simulink` provides libraries of components. Assembling these components with connections produces a so-called "Simulink model". The resulting model is easier to interpret and reuse because it reflects the structure of the system to be modeled, rather than a purely mathematical depiction of the system. Models can also integrate control and signal processing algorithms, state machines from Stateflow, neural networks, fuzzy logic, ...etc. Working in a common environment helps different teams developing models and other complex systems avoid the pain of cosimulation and uncover integration issues earlier in the design process.

Simulink uses the V-Model for system development. The V-model (Fig. 2) is a representation of system development that highlights verification and validation steps in the system development process. The left side of the 'V' identifies steps that lead to code generation, including system specification and detailed software design. The right side of the 'V' focuses on the verification and validation of steps cited on the left side, including software and system integration.

Code generation technology and related products provide tooling that can be applied to the V-model for system development. Simulink and Real-Time Embedded Coder will be used to perform code generation for IoT applications.

---

[1] Simulink is a trademark of Mathworks company.

[2] www.mathworks.com.

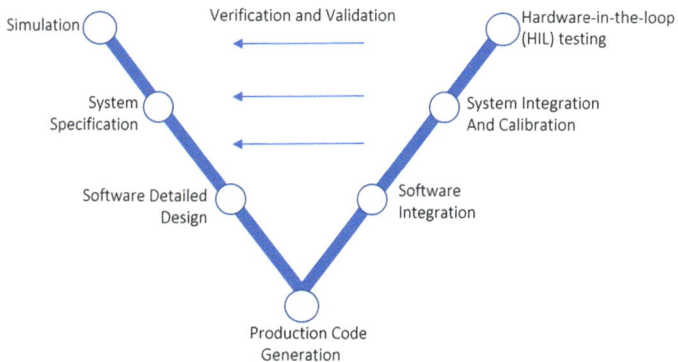

Fig. 2 V-Model for System Development

4 IoT Framework Development

The development of the framework was divided into two parts:

- A hardware node—based on `Espressif` ESP8266 microcontroller [15–20],
- Software package, that automatically compiles and upload a the executable to the node hardware.

4.1 IoT—Hardware Node

The IoT developed node in build on top of ESP8266 microcontroller. `ESP8266` is a cost-effective and highly integrated Wi-Fi `MCU` for `IoT` applications. It has the following features[3]:

- **High Durability**: ESP8266EX is capable of functioning consistently in industrial environments, due to its wide operating temperature range. With highly-integrated on-chip features and minimal external discrete component count, the chip offers reliability, compactness and robustness.
- **Power-Saving Architecture**: Engineered for mobile devices, wearable electronics and IoT applications, ESP8266EX achieves low power consumption with a combination of several proprietary technologies. The power-saving architecture features three modes of operation: active mode, sleep mode and deep sleep mode. This allows battery-powered designs to run longer.
- **Compactness**: ESP8266EX is integrated with a 32-bit Tensilica processor, standard digital peripheral interfaces, antenna switches, RF balun, power amplifier, low noise receive amplifier, filters and power management modules.

---

[3]https://www.espressif.com/files/documentation.

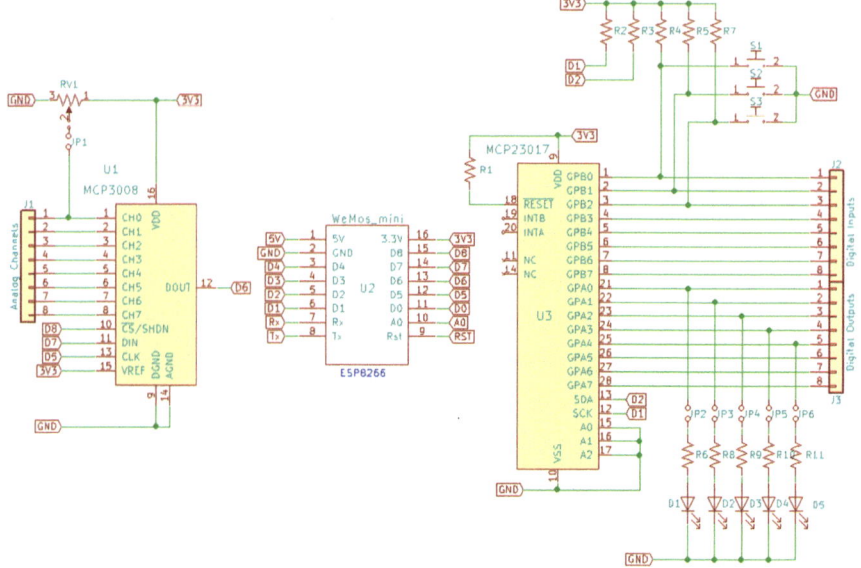

Fig. 3 Node electronics schematic

- **32-bit Tensilica Processor**: The ESP8266EX microcontroller integrates a Tensilica L106 32-bit RISC processor, which achieves extra-low power consumption and reaches a maximum clock speed of 160 MHz. The Real-Time Operating System (RTOS) and Wi-Fi stack allow about 80% of the processing power to be available for user application programming and development.

The ESP8266 has a low number of GPI/O and one ADC input reducing thus its interfacing ability with other components. Ports expansion circuits have been used to extend the number of General Purpose Inputs (GPI) to eight(8), the General Purpose Outputs (GPO) to eight(8) using the MCP23017 IC, Analog Inputs to eight (8) using the MCP3008 IC and PWM Outputs to sixteen (16) using the PCA9685 IC. The schematic of the board has been developed under KiCAD[4] software and is given in Fig. 3.

The corresponding board is given in Fig. 4. All the circuits are connected on an I2C bus, master slave architecture. Where ESP8266 is the master and the rest of circuits are addressed slaves. The master choose the slave to communicate with by issuing its specific address. The concerned slave responds to the master request. A set of LEDs and push-buttons can be used as inputs/outputs for applications purposes.

---

[4]https://kicad-pcb.org/.

Fig. 4 Node PCB and
components

4.2 IoT—Software Framework

In this section, we are going to develop a library of blocs (Fig. 5) that implements:

- Board peripherals,
- Data relatives,
- Common things.

Each bloc in the library is composed of three components as depicted in Fig. 6.

S-Function C-file that implements the anatomy of the block: the number of input/output ports, their size and width, the type of parameters of the block, the sample time and the interface that reports these parameters to the TLC file.

Block Mask Front-end user interface. It is a dialog box where the user can configure the block, using the parameters given in the S-Function, e.g. pin number, sample time, topic, ...etc.

TLC file Target Language Compiler file: automate how the code will be generated from the bloc. This file implements the initialization and the behavior of the block in the model for each time step of the execution.

The S-Function file defines the list of block parameters and implements functions for block initialization, setting properties of input/ouput ports, their data types, the sample time and the interface to code generation. The target language compiler (tlc) file contains instructions that serve for generating code for the target hardware. From the simulink model (Fig. 7), the code generator produces a set of source files (*.c and *.h) and a Makefile that is responsible of compiling and uploading the produced binary file to the hardware target.

  Here is an example of blinking a LED on GPO0 each 1s as depicted in Fig. 8.

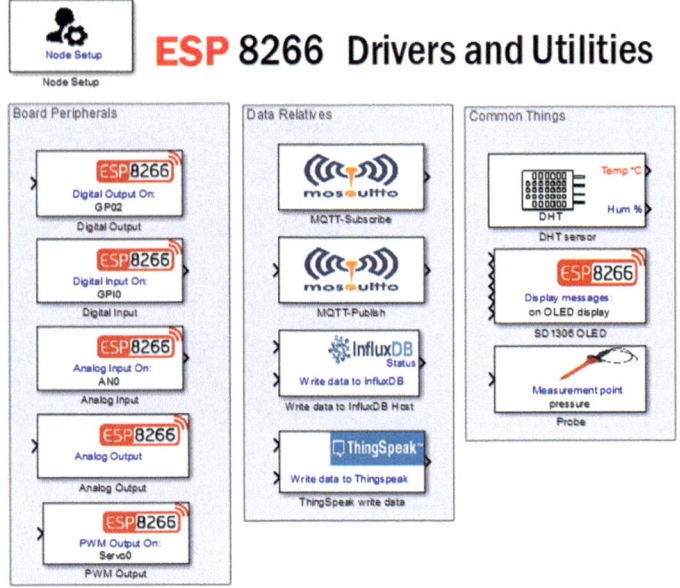

Fig. 5 The developed library

Fig. 6 Block development
workflow

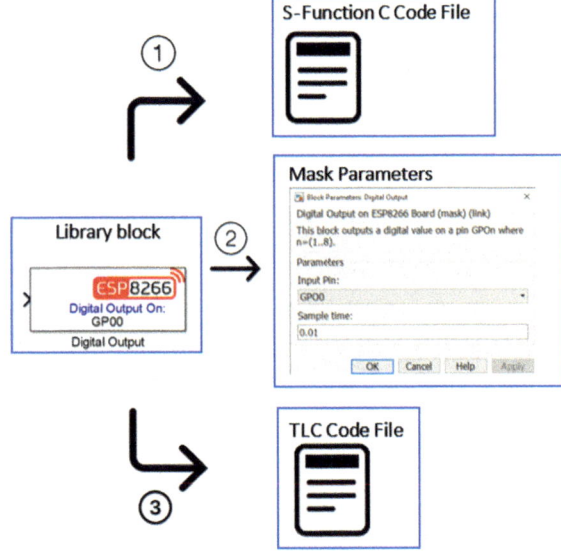

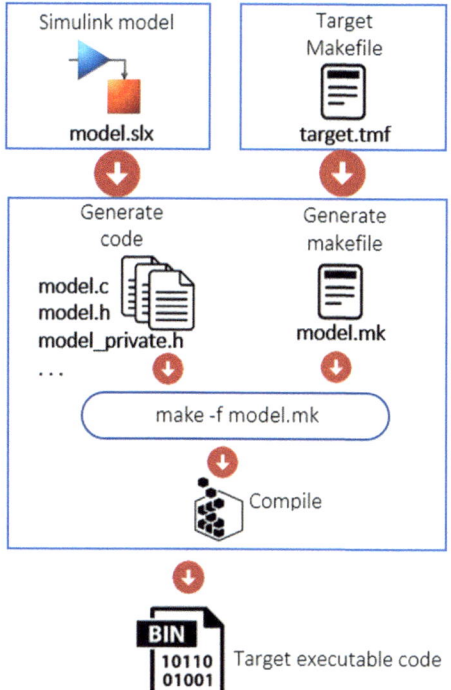

Fig. 7 Code generation workflow

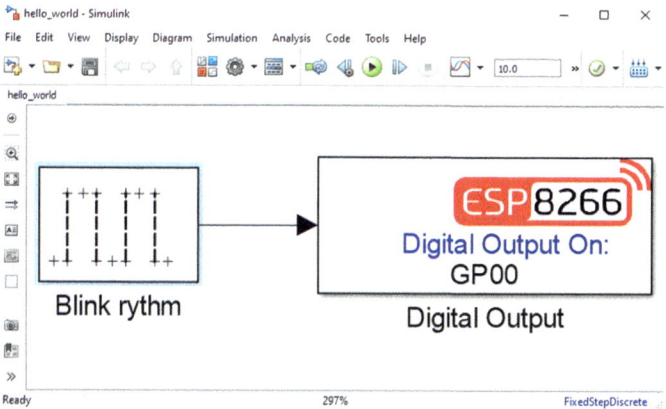

Fig. 8 An example of library usage—blinking a LED

Fig. 9 Scope view of the pin GPO0

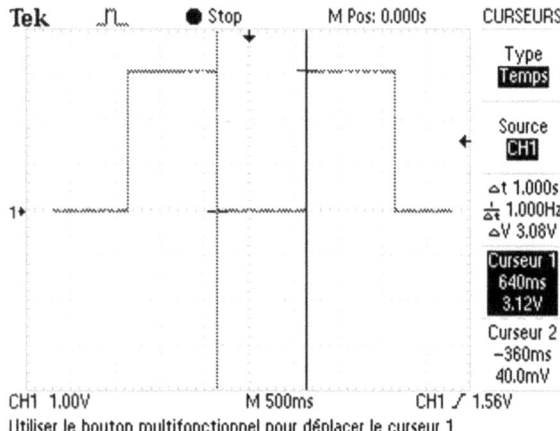

In order to compile the files, PlatformIO[5] is used as a framework for that purpose.

- PlatformIO is suitable for larger projects.
- PlatformIO offers better control of project dependencies. Each project has its own dependencies, and the tool manages them independently of other projects.
- PlatformIO, the created projects are self-contained. To compile the project, everything is stored inside the project directory.
- PlatfromIO integrates all of the toolchain components.

The compilation produces object (`.o`) files from sources (`.cpp`) files, and the libraries (`.a`). If the overall process is done without errors, the building system of PlatformIO links all the objects together to produce a binary file `firmware.bin` which is downloaded to the IoT node.

Figure 9 gives a scope view of GPO0 pin on the board. It is observed that the LED connected on this pin is blinking with a rythm of 1s ($\triangle t = 1s$). As the blink program is considered as a kind of "Hello, world!" program for testing every compiler, interpreter or framework, an advanced example in which all aspects of IoT are integrated is given in the next section.

5 Micro-climate Monitoring IoT Project

To validate the developed code generator, an example of IoT humidity and temperature monitoring system is designed (Fig. 10). This system is based on DHT22 sensor which measures humidity and temperature. These measurements are then stored

---

[5]https://platformio.org/.

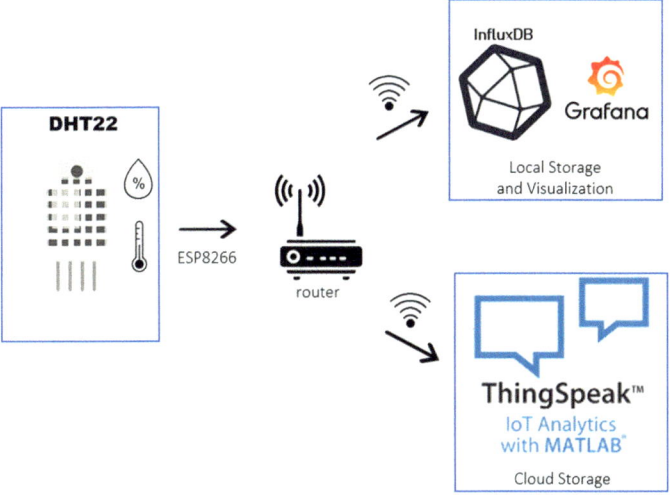

Fig. 10 IoT monitoring of temperature and humidity

locally on an InfluxDB[6] database and visualized via Grafana[7] software. On the other side, a cloud storage is also integrated to the example, to alow remote storage on Thingspeak.[8]

5.1 The Sensor

The DHT22 is a commonly used temperature and humidity sensor. The sensor comes with a dedicated NTC thermistor to measure temperature and an internal circuit to output the values of temperature and humidity as serial data. The sensor is also factory calibrated. The sensor can measure temperature from −40 °C to 80 °C and humidity from 0% to 100% with an accuracy of ± 1°C and ± 1%.

For measuring humidity they use the humidity sensing component which has two electrodes with moisture holding substrate between them. So as the humidity changes, the conductivity of the substrate changes or the resistance between these electrodes changes. This change in resistance is measured and processed by the IC which makes it ready to be read by a microcontroller.

On the other hand, for measuring temperature these sensors use a NTC temperature sensor or a thermistor. A thermistor is actually a variable resistor that changes its resistance with change of the temperature. The term NTC means "Negative Temperature Coefficient", which means that the resistance decreases with increase of

---

[6]https://www.influxdata.com/.

[7]https://grafana.com/.

[8]https://thingspeak.com/.

the temperature. These measurements make it possible to construct graphs in order to visualize performance and see trends, to detect or even anticipate anomalies or failures and to alert in the event of a malfunction.

5.2 Data Storage and Visualization

In this application, we will set up InfluxDB and Grafana where InfluxDB will be used as a storage solution and Grafana as a data formatting and visualization tool.

Grafana [21] is one of the benchmarks for visualizing and formatting metric data. It allows to create dashboards and charts from a multitude of data sources (e.g. InfluxDB) and also issue alerts according to thresholds.

ThingSpeak is an open source "Internet of Things" application and API to store and retrieve data from things using HTTP over the Internet or via a Local Area Network. With ThingSpeak, one can create sensor logging applications, location tracking applications, and a social network of things with status updates.

In addition to storing and retrieving numeric and alphanumeric data, the ThingSpeak API allows for numeric data processing such as timescaling, averaging, median, summing, and rounding. Each ThingSpeak Channel supports data entries of up to 8 data fields. In this example, we will be using ThingSpeak to visualize data collected in real time.

5.3 MQTT IoT Protocol

MQTT[9] (Message Queuing Telemetry Transport) [22] is an ISO standard publish-subscribe-based messaging protocol. It was originally created by Dr. Andy Stanford-Clark and Arlen Nipper in 1999. It works on top of the TCP/IP protocol. It is designed to be lightweight and uses only low network bandwidth. The publish/subscribe messaging pattern requires a message broker (a server). MQTT has become the messaging standard for IoT (Internet of Things) solutions and M2M (machine-to-machine) connectivity (Fig. 11).

Messages within MQTT are published as topics. Topics are structures in a hierarchy using the slash (/) character as delimiter. A structure such as sensors/ Pressure/ allows a subscriber to specify that it should only be sent data from clients that publish to the Pressure topic.

All the blocks in the library are coded to publish their states via mqqt broker. Mosquitto[10] is used as a broker to manage the communication exchange.

---

[9]https://mqtt.org/.

[10]https://mosquitto.org/.

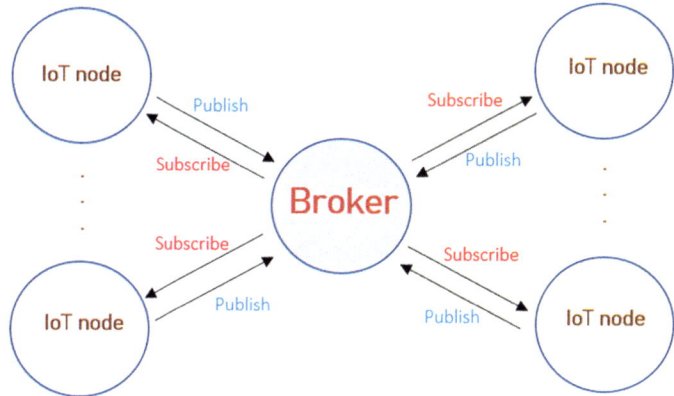

Fig. 11 MQTT architecture

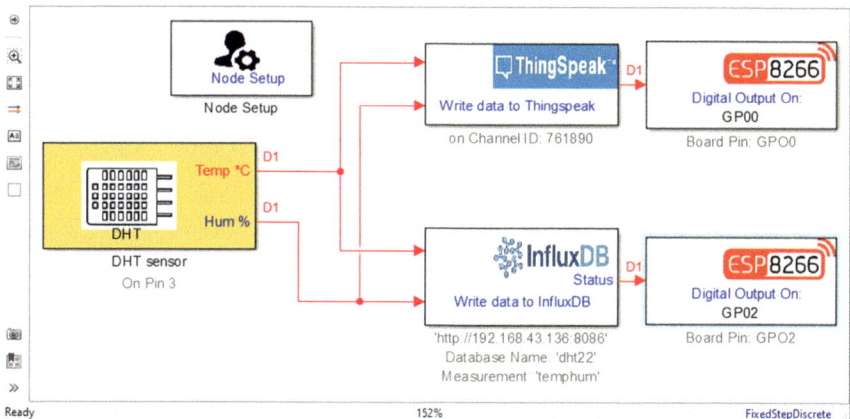

Fig. 12 IoT microclimate monitoring example

5.4 Building the Application

In this example (Fig. 12), the sensor is connected on pin 3 of the ESP8266. Temperature and humidity are sent to InfluxDB and Thingspeak for storage.

The model is first configured via the block "Node Setup", where the user defines WiFi credentials as well as mqtt broker `IP address` and its `port`. The DHT22 block is designed to generate code that outputs the measured temperature and humidity and outputs them on its corresponding ports. These two measurements are fed as inputs to the blocks `Write data to InfluxDB` and `Write data to Thingspeak` for local and remote storage. `GPO0` and `GPO1` are used as indicators for success of writing operations.

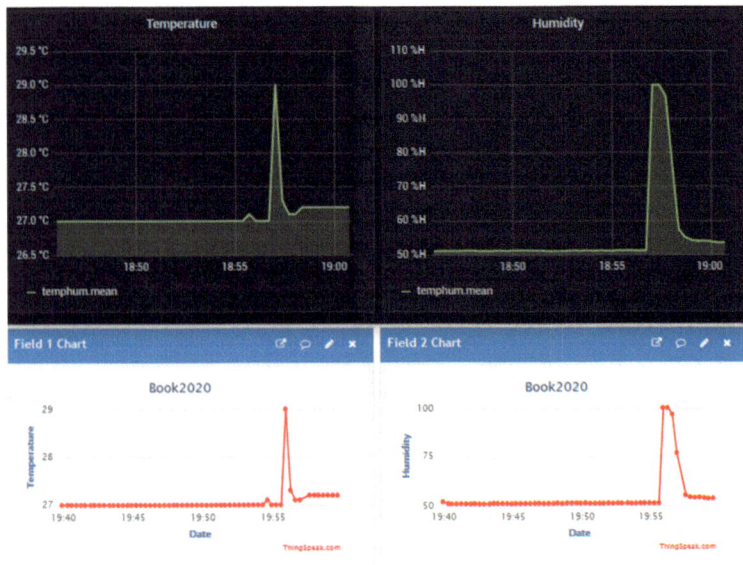

Fig. 13 Results visualization on Grafana(top) and Thingspeak(bottom)

Figure 13 shows the results. In the upper side, temperature and humidity data that are stored in Influx database are plotted using grafana. On the bottom side, the two graphs are issued from Thingspeak cloud storage.

For the sake of conformity, Fig. 14 shows the debugging values issued from the serial port of the ESP8266.

All these steps where carried out in order to check the full and sure operation of the node. This example has been given for a demonstration purposes. It can be used as a node for smart city environment monitoring, a smart home comfort management or in a smart greenhouse project. More functionalities are implemented in the framework in order to develop large scale IoT applications.

6   Conclusion

Every day, connected objects will generate billions of pieces of information that will enable companies to create new services. Several architectures and frameworks have been developed for that purpose. IoT frameworks developed for industry do not allow multi-domain/multiphysics modeling and design in the same mainframe, they need intervention and collaboration of multiple software in each field to accomplish the task. The framework presented in this chapter, has been developed based on diagram modeling, integration of powerful signal processing, control algorithms and artificial intelligence on top of Simulink product. The framework was also designed

Fig. 14 Debugging on the serial port

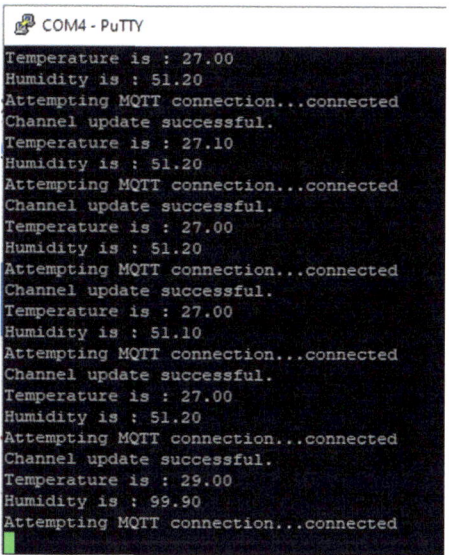

to be extensible to include future functionalities. It was also designed for boosting existing capabilities of Simulink by extending its native library with new out-of-the-box features relative to IoT and rapid prototyping. Cloud storage as well as Local storage has been added to the blockset for saving data for post-processing and decision making as well as visualization. Examples given for demonstartion purposes, show how fast to design an IoT applications and deploy them to real world nodes.

References

1. Hany F. A., Robert J. W., Gary B. W. (2018). Internet of things: State-of-the-art, challenges, applications, and open issues. *International Journal of Intelligent Computing Research (IJICR), 9*(3). https://doi.org/10.20533/ijicr.2042.4655.2018.0112.
2. Takahiro, I., Masaki, K., Shusuke, O., & Masaru, K. (2017). IoT technologies: State of the art and a software development framework. *Smart Sensors Networks,*. https://doi.org/10.1016/B978-0-12-809859-2.00002-4. Elsevier.
3. Rajkumar, B. (2016). *Amir Vahid Dastjerdi, "Internet of Things Principles and Paradigms".* Elsevier.
4. Vangelis, A., Elias, T., Henrich, C. P., Adam, K., & Alessandro, B. (2017). *Designing, developing, and facilitating smart cities urban.* Springer.
5. Sayed O. F. T. (2018). *Building Smart Drones withESP8266 and Arduino.* Packt Publishing
6. Catalin B. (2017). *ESP8266 home automation projects.* Packt Publishing.
7. Anand T. (2019). *Build your own IoT platform.* Apress.
8. Bin C., Gurkan S., Flavio C., Erno K., Kazuyuki T., & Atsushi K., FogFlow: Easy programming of IoT services over cloud and edges for smart cities. *IEEE Internet of Things Journal.* https://doi.org/10.1109/JIOT.2017.2747214.

9. Edel, S., Ileana, O., Emmanuel, G., Pau, F. I C., & Finn K. (2015). SDL—The IoT language. Springer International Publishing Switzerland 2015, J. Fischer et al. (Eds.): SDL 2015, LNCS 9369, pp. 27–41, 2015. https://doi.org/10.1007/978-3-319-24912-4_3.
10. Guth, J., Breitenbücher, U., Falkenthal, M., Fremantle, P.,Kopp, O., Leymann, F., & Reinfurt, L. (2018). A detailed analysis of IoT platform architectures: concepts, similarities, and differences. In *Book "Internet of Everything: Algorithms, Methodologies, Technologies and Perspectives.* Springer (2018).
11. Mark H. (2017). Leading the IoT, Gartner Insights on How to Lead in a Connected World. Gartner, Inc. www.gartner.com/imagesrv/books/iot/iotEbook_digital.pdf.
12. Georgios, K., Nikolaos, V., & Michael, H. (2017). *Components and services for IoT platforms.* Springer.
13. Jéphane F., Florent F., Matthieu F., Ursula G., FranôSois G., Thierry H., Florian L., Stéphane L., Patrick M., Pierre-Nicolas P., & Alain S., Model quality objectives for embedded software development with MATLAB and simulink. *9th European Congress, Embedded Real Time Software and Systems, ERTS2018.*
14. Sébastien D. (2019). Industry 4.0 and digital twins. Matlab Expo.
15. Marco S. (2017). *ESP8266 internet of things cookbook.* Packt Publishing.
16. Daniel E. (2017). *ESP8266 weather station.* Leanpub (2017).
17. Peter H., & Lizzie P. (2020). *IoT Development for ESP32 and ESP8266 with Javascript.* Apress.
18. Thankur, M. R., NodeMCU ESP8266 Communication Methods and Protocols Programming with Arduino IDE.
19. Kolban, N. (2016). *Kolban's book on ESP8266.* Leanpub.
20. Marco, S. (2106). *Internet of things with ESP8266.* Packt Publishing.
21. Eric, S. (2020). *Learn Grafana 7.0, A beginner's guide to getting well versed in analytics, interactive dashboards, and monitoring.* Packt Publishing.
22. Gastón, C. H. (2017). *MQTT Essentials—A Lightweight IoT protocol.* Packt Publishing.

Prototype of a BCI-Based Autonomous Communicating Robot for Disability Assessment

Amin Zammouri and Soufian Zerouali

Abstract Designing intelligent and assistive devices for disabled persons' mobility requires appropriate assistive tools and techniques in order to insure efficiency and security in use. In this work we present a new architecture of an autonomous and communicating mobile robot for disability assistance. This architecture represents a first step towards an implementation to an intelligent wheelchair for the mobility of individuals with motor impairment. The proposed design provides two use modules namely the navigation based only on the brain signals and the intramural mapping of buildings. For the first use module, our approach exploits the statistical characteristics of the brain signal to extract eye movements and transform them into commands to control the robot. Concerning the intramural mapping module, our proposed solution is based on an obstacle detection algorithm implemented on a Raspberry PI 3 board using the ROS system.

Keywords EEG · BCI · Eye mouvements · Intramural mapping · Autonomous mobile robot

1 Introduction

In the field of robotics a booming market is installed. This is a direct consequence of the decline in costs since the 90s and which covered all types of robots namely domiciliary, industrial and leisure robots. This decrease in robot costs has made it possible to extend applications of this field to the military, industrial and medical areas. In the context of locomotion, the first designed robots were limited by the constraint of stability. This, because it is a discipline that requires skills in robotics, artificial intelligence (AI), industrial and computer engineering [7].

A. Zammouri (✉)
EPF Gradute School of Engineering, Sceaux, France
e-mail: amin.zammouri@epf.fr

S. Zerouali
Kubota Europe, Beauvais, France

© The Author(s), under exclusive license to Springer Nature Switzerland AG 2021
U. Ghosh et al. (eds.), *Machine Intelligence and Data Analytics for Sustainable Future Smart Cities*, Studies in Computational Intelligence 971,
https://doi.org/10.1007/978-3-030-72065-0_12

Nowadays, increasing the autonomy of people in situations of disability is in an increasing interest [12]. This may be justified at a first time by the phenomenon of aging but also accidents and chronic diseases take part in this situation. As an example, the legal definition of disability according to the French law of 2005, it covers any limitation of activity to the restriction of participation in life. According to this example, more than two million individuals are administratively recognized as disabled in France, and 60% of whom have motor impairment. In this French example, 12% of the reported impairments are attributed to accidents. On another hand, 10% are due to "early" causes such as complexity of pregnancy, congenital malformation and even hereditary diseases. In this context, assistance robotic has become one of the most invested sectors [6]. This covers sectors from helping robots to auxiliary robots in life.

With the advent of artificial intelligence, rapid advances have been made in robotics. These advances have made it possible to increase the degree of interaction between humans and robots. The refinement of such interaction has allowed the appearance of what is called "Assistive Robotic" (AR) [2]. The aim of such a discipline is to improve the autonomy and life quality of persons with disabilities. In this context, understanding the needs of disabled persons and adapting these needs to robotic services is the serious challenge of any assistive robotic system. Given this constraint, assistive robotic has been regrouped in different axes in order to offer a diversity of choices in disability assistance.

In terms of physical assistance, which is considered to be the direct application of AR, designed systems aim to increase the individual's independence during a physical task. For example in the case of muscle degeneration in the upper limb, self-feeding becomes a tedious, or even an impossible task. To overcome this deficiency, works from assistive robotics propose the use of a robotic arm. In this context, the Assistive Robotic Manipulation (ARM), known as MANUS [5], represents one of the most commonly studied robots in literature works. Based on six degrees of freedom and an end effector, the MANUS is intended to be mounted on a wheelchair for a general handling. This robotic device has been designed for clinical cases where there is a quadriplegic patient or suffering from a neuromuscular disease. In addition to self-feeding, the MANUS offers its users simple aids such as opening the door of a room or wearing glasses. As MANUS, several approaches are proposed in the literature and are based on the residual motor abilities of the individual. However, in the case where no motor ability is available, these robotic systems are of no use. In such a situation, the use of other approaches is necessary. These approaches can rely on voice recognition or eye-tracking in order to have a controlling tool of a robotic device. However, in cases of extreme disability, i.e. no standard neuromuscular peripheral is functional, the use of brain signals represents the only overcoming reliable approach to increase the dependence of the patient [6].

In the case of a total motor deficiency, i.e. no residual motor ability is present; a possibility to overcome this isolation can be implemented using paradigms based on non-muscular information pipes. These information channels could be like the Heart Rate Variation (HRV), the Galvanic Skin Response (GSR), eye movements recorded on an eye tracker and brain signals which result in respect to the electrical brain activities, i.e. Electroencephalogram signals (EEG) [4, 8]. However, taking into account the inter-individual variability, brain signals-based paradigms represent the most reliable solution for creating communication systems. The use and the analysis of brain activity in relation to intelligent systems has always been referred to the concept of Brain–Computer Interfaces (BCI) [1, 3, 10, 14]. These interfaces allow communication which does not require any muscular ability and the user intentions are mediated through brain signals. Related to the mobility assistance, Rebsamen [9] seeks to generate, in the user's brain, a response following the presentation of a stimulus. In this paradigm, randomly flashed choices are presented to the user. When the user's intended selected choice is flashed, a cerebral response is generated at the central brain area. This response is a large positive deflection in the brain signal after the reception of a stimulus. This deflection is known as the P300 Event Related Potential (ERP) component [13]. This component is generated at the central brain area 300 ms after the reception of the stimulus. To allow navigation in a typical building, the user selects his desired direction on a GUI. Apart from the ERP paradigm, BCI can be designed through exploiting the asynchronous brain signals. Unlike ERPs, these signals are voluntarily generated by the user and no stimulation is used. In such a context, the imagination of movements is the most used approach through literature works. In this realm, Tonin [11] designed control architecture for a telepresence robot based on asynchronous BCI. This system, based on the imagination of hand movements, allows patients in clinics to control a telepresence robot situated from more than 100 km. In order to focus patients' attention on movements imagination, shared control is introduced and is based on automatic obstacle avoidance.

In this work, we present a new architecture of a communicating autonomous robot. The proposed design provides two modules of the robot use, namely (1) the control of the robot using only EEG signals and (2) the intramural mapping of a building. In the first module, our control approach is based on the detection of eye movements on the EEG signal. This approach exploits the statistical characteristics of the EEG signal. On a graphical user interface (GUI), and based on his eye movements, the user moves a cursor to select the robot movement direction. For the test and the validation of this module, our experimentations were conducted using a reference EEG assembly with a dry active electrode placed at the Fp1 position according to the 10–20 international system. On another hand, and for a first validation of the proposed use mode, this experimental protocol included a mal participant with any eye or cognitive impairment. As a test task, the participant was asked to move the robot from a departure point and going through three stations before reaching the arrival point. Experimental results from this first module show that the precision of the proposed approach reached 86%. This result allows us to postulate that such an approach could be one of the alternatives for the control of a wheelchair by persons with severe motor disability. For the second module, the intramural mapping represents a useful feature for patients with disabilities. To do this, our adopted approach is based on obstacle avoidance and wall tracking. In order to demonstrate and validate the capabilities of this module, a case study was discussed. In this case, the robot simultaneously locates, browses and builds a map of its environment.

The remain of this chapter is organized as follow. In Sect. 2 we give a presentation on our adopted model. We present each of experimental setups and the adopted approaches for both use modules. In Sect. 3, we present the obtained results in relation to each use module. These results are then discussed in the same section. Finally, perspectives and future works are introduced in the conclusion.

2 Adopted Model

2.1 Experimental Setup

Since in this work we propose a system with two modules of use, two versions of architecture have been proposed. The "light" version implements only the intramural mapping module. In the "complete" version we have implemented the two modules. In the two proposed versions the logical processes are carried out based on a Raspberry PI embedded system which has an ability to implement the Robot Operating System (ROS). For interfacing the different actuators, the two versions are based on Arduino boards. In order to enhance the robot autonomy in terms of obstacle avoidance and wall tracking, we relied on an ultra-sound sensor. The two designed robots are presented on Fig. 1. The common architecture of the two versions is presented in Fig. 2. Concerning the navigation module using the brain signals, our experimental protocol consists in using a single active EEG electrode (Ag/Cl electrode) placed at the Fp1 position according to the 10–20 international system.

For the validation of the navigation module, the experimental protocol was tested on a volunteer subject from the project team. The subject was trained in the method with which he has to use the system for experimentation. The experimentation consists in placing the participant in front of a GUI on which we present a cursor and the possible directions for moving the robot on the plan. The participant moves his eyes to select the desired direction for moving the robot.

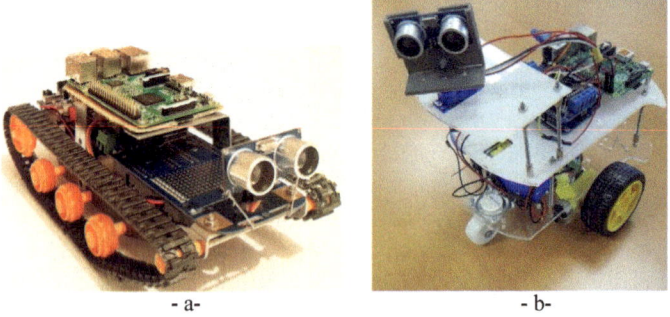

- a-   - b-

Fig. 1 "Complete" version of the designed robot **a** and "Light" version of the designed robot **b**

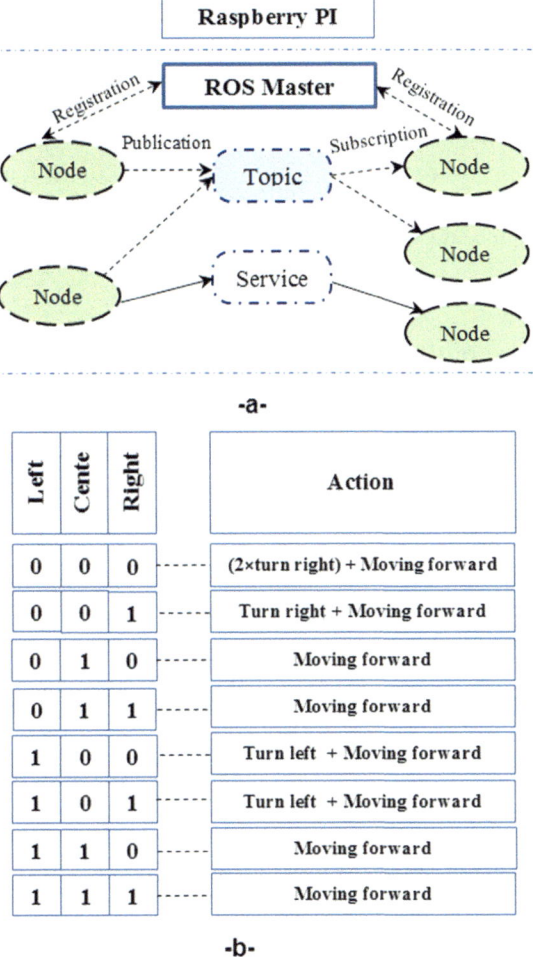

-a-

-b-

Fig. 2 a ROS communication structure. ROS starts by executing the Master which allows to all the ROS executables (Nodes) to meet and inter-communicate. **b** Robot states and actions diagrams. On the left side of the Fig. 1 represents a distance greater than the safety distance regarding the obstacle

2.2 *Intramural Mapping Module*

The robot autonomy in the mapping module mainly concerns obstacle avoidance and wall tracking using an ultrasonic distance sensor. In our approach, we chose to connect the different sensors and actuators to the Raspberry PI which is considered as the logic processing unit in our architecture. We chose an implementation using the ROS. The choice of such a system is justified by its fundamental concept which offers parallel operations of a large number of executables (called Nodes) in order to make them exchanging information in a synchronous way (called Service) or asynchronous (called Topic) (Fig. 3).

At the ROSSERIAL stack, communication between ROS on the Raspberry PI and the Arduino board can be established. To allow the estimation of the position of the robot during a movement, the ultrasonic sensor allows a measurement starting from a known departure position. For such an estimation of the position and the trajectory based on the odometric measurements, we propose the models, presented on Fig. 4, for the robot displacement.

2.3 Eye Movements-Based Navigation Module

Our approach for controlling the mobile robot based only on brain signals consists on using and improving an eye movement detection algorithm from our previous works [15]. This improvement consists in taking into account a new parameter which is the power spectral density (PSD) of the alpha brain wave [8–12 Hz] during the production of an eye movement. For the sake of a lack understanding, we give a brief description of this algorithm. Our approach is based on detecting eye movement. Such an approach requires a calibration session in which we try to measure a typical eye movement of the user. During this session, the user moves his eyes in different times. The importance of the typical eye movement, which is specific to the user, lies in the refinement of the detection process during experiments through parametric comparisons. For the validation of our proposed method, an existing dataset of eye movements recorded on ten subjects is used.

The proposed eye movement detection process is based on statistical characteristics of the EEG signal. Let $s(t_i)$ be the measurement of the EEG signal by the electrode Fp1 at the instant t_i. We demonstrate that $s(t)$ follows a Normal Gaussian Law, i.e. $s(t) \sim \mathcal{N}(\mu_s, \sigma_s^2)$ where μ_s and σ_s^2 represent the EEG signal mean and variance respectively. The EEG data normality leads us to postulate that the pure EEG data, which are not contaminated by other external signals, are within the Gauss' confidence interval. Taking into-account the inter-individual variability, this confidence interval, denoted by I_C, is defined as presented in Eq. (1). In our experiments, we chose to vary the parameter p from 1 to 5 so as to exclude the least percentage of pure EEG data from the interval I_c.

$$I_c = [\mu_s - p * \sigma_s, \mu_s + p * \sigma_s]. \tag{1}$$

The next step in a eye movement identifying process is seeking a point from the signal which verifies $(t_i) \notin I_c$. Once this point is identified we seek to determine the first $s(t_j)$ which verifies $s(t_j) \approx 0$ and $t_j < t_i$. On the other hand, we seek to identify the third point verifying $s(t_{j'}) \approx 0$ and $t_{j'} > t_i$. The instants t_j and $t_{j'}$ represent the estimated times of starting and ending of the eye movement. A graphical representation of these parameters is given on Fig. 3. As a third step in the identification process, we apply a Fisher-Senedecor test for a comparison based on

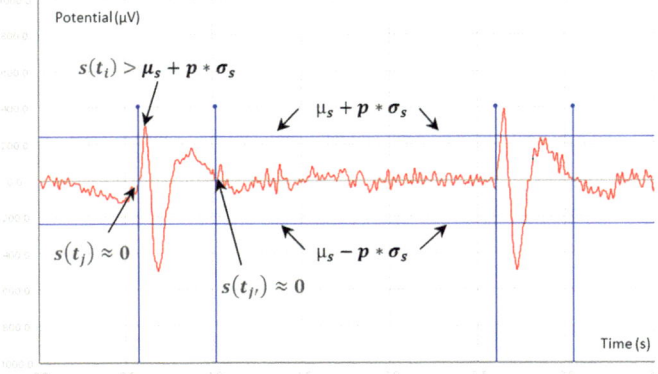

Fig. 3 Detection of an eye blink using the signal statistical parameters

the typical eye movement and the blink supposed to be identified between instants t_j and $t_{j'}$.

The use of the Fisher-Snedecor test aims at comparing if two samples belong to the same population. In our case, the two samples are the typical eye movement and the estimated one. Let s_e be the EEG signal delimited by instants t_j and $t_{j'}$. And let s_b be the EEG signal which composes the typical movement. We denote by n_{s_e} and n_{s_b} sizes of s_e and s_b respectively. The experimental variances of s_e and s_b are, respectively $S_{s_e}^2$ and $S_{s_b}^2$ as shown in Eqs. (2) and (3). μ_{s_e} and μ_{s_b} are means of s_e and s_b respectively.

$$S_{s_e}^2 = \frac{1}{n_{s_e}} \sum_{i=1}^{n_{s_e}} \left(s_e(t_i) - \mu_{s_e}\right)^2, \tag{2}$$

$$S_{s_b}^2 = \frac{1}{n_{s_b}} \sum_{i=1}^{n_{s_b}} \left(s_b(t_i) - \mu_{s_b}\right)^2. \tag{3}$$

The two equations lead us to postulate two hypotheses. Null Hypothesis (H_0) corresponds to the case where $S_{s_e}^2 = S_{s_b}^2$. The Alternative Hypothesis (H_A) represents the case of $S_{s_e}^2 \neq S_{s_b}^2$. The statistic of the Fisher-Snedecor test is defined as follow:

$$F = (S_{(s_e)}^2)/(S_{(s_b)}^2) \tag{4}$$

Based on a bilateral test with a specific confidence threshold λ. The test can rejects the H_0 hypothesis with a ρ risk if $F > \lambda$. In the case where $F \leq \lambda$, H_0 is accepted.

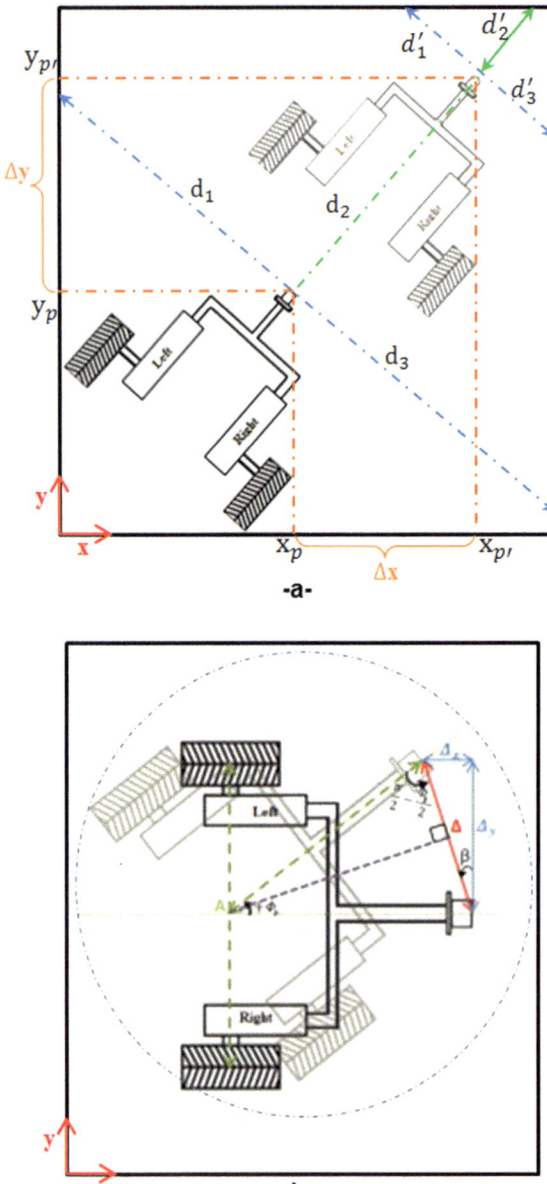

Fig. 4 **a** Translation model. **b** Rotation model

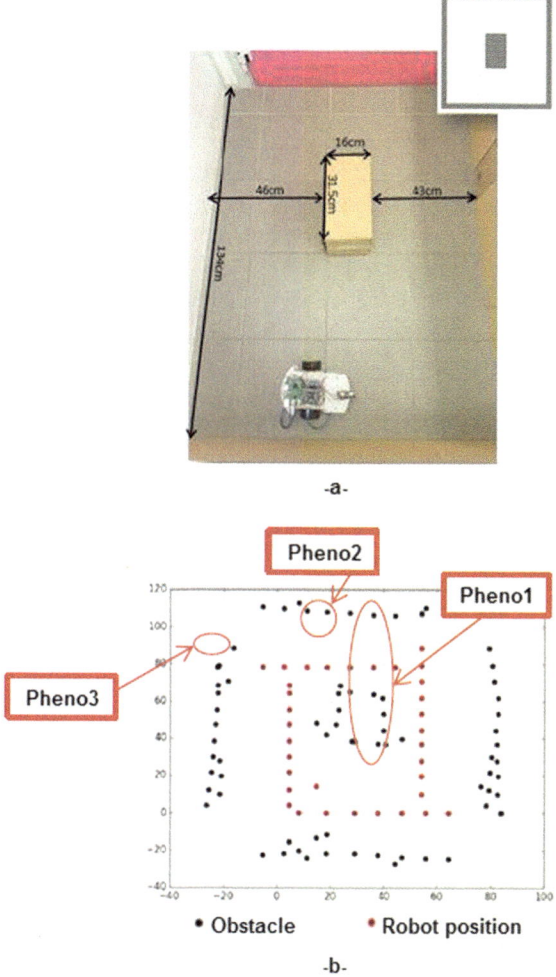

Fig. 5 **a** Test environment for the mapping module. **b** Results of the mapping module in terms of estimating obstacles and the robot positions

3   Results and Discussions

In order to test and validate the capabilities of the mapping module, we discussed the case study described in Fig. 5. In this case study, the robot locates tracks and simultaneously builds a map of its environment. Such a test can be considered as a first step in the SLAM machine learning. In Fig. 5 we present each of the environments where the robot performs its movements and the results obtained in terms of mapping and trajectory estimation without going through the wheels encoders. Figure 5 shows that the robot is able to map the entire test environment and to accurately estimate its positions. On another hand, results from Fig. 5 present a limitation in mapping. This limitation is translated by the robot's inability to accurately map the corners of the environment. This is shown on Fig. 5 by phenomena "Pheno1" and "Pheno2". Indeed, the principle of the robot navigation consists of two steps. The first consists of a 180° scan of the environment with the ultrasonic sensor through sampling three points (0°, 90°, 180°). Subsequently, the robot moves according to what it has just detected based on the algorithm of Fig. 2. Thereafter, it returns to scanning. The "Pheno2" is justified by the servomotor pitch that we have fixed to 90° and also by the ultrasonic sensor field which detects a wall on the side despite having already passed it. On the other hand, Fig. 5 shows an approximation between the robot placements ("Pheno3"). This is due a purely technical problem related to the battery. Indeed, when the battery charge decreases, the motor lacks power and this affects its displacement, including the rotation which requires more power.

For the navigation module, the used eye movement approach was trained on an existing dataset. The validation process of controlling the robot was performed based on a mal participant. This validation process focused on measuring the performance, in terms of accuracy, we used metrics of Sensitivity (Sen) and Specificity (Spe) as defined in the following equations:

$$Sen = \frac{TP}{TP + FN}, \tag{5}$$

$$Spe = \frac{FP}{FP + TP}, \tag{6}$$

where TP represents the number of real detection, done by the algorithm, and which are also annotated by a human expert. FN, represents the number of detections made by the human expert which were not made by the algorithm. Finally, FP represents the number of detections made by the algorithm and which are not made by the human expert.

In Table 1 we report results of our eye movement detection algorithm implemented on the used existing dataset. The used dataset contains expert's annotation. Decisions made from our algorithm are compared to the expert's annotations. The results presented in Table 1 give reasonably sensitivity greater than 85.71%, while the specificity reached more than 99.48%. Since the decisions of our algorithm are compared to the expert's annotations. The agreement in this context is evaluated meaning the Cohen's Kappa test. The Kappa coefficient is computed as follow:

$$k = \frac{P_r(a) - P_r(e)}{1 - P_r(e)} \tag{7}$$

where $P_r(a)$ represents the existing agreement between the detection algorithm and the expert, and $P_r(e)$ is the probability of a random agreement.

Table 1 Performance metrics of the eye movement detecting approach trained on the existing dataset

Subjects	Performance metrics		
	Sensitivity (%)	Specificity (%)	Kappa's coefficient
Subject 1	97.29	100	0.98
Subject 2	98.47	100	0.99
Subject 3	100	100	1
Subject 4	96	100	0.97
Subject 5	100	100	1
Subject 6	97.14	100	0.98
Subject 7	92.10	100	0.95
Subject 8	95.61	100	0.97
Subject 9	85.71	99.48	0.88
Subject 10	93.73	100	0.98
Average	95.6	99.94	0.97

In order to give additional quantitative performance measure of our eye movement detection method, we computed signal-to-artifact ratio (SAR). In this case, an eye movement is considered as an artifact on the EEG signal. Results are compared to those obtained using Single-Channel ICA (SC-ICA) method. Figure 6 shows that the SAR outcomes obtained frome our eye movement detection algorithm are close to those obtained based on the SC-ICA.

Next in Table 2 we present the results of the chosen evaluation process. For the selected representative participant, the obtained results represent a sensitivity rate of 86% and a specificity of 69%. The first metric value reflects the precision of the proposed algorithm to detect true eye movements. However, concerning the sec ond metric value, we explain it by the test environment influence on the EEG acquisition system. Indeed, since the acquisition system uses a Bluetooth-based information transmission, this transmission channel could be disturbed by the magnetic waves which could be f an origin of the test environment devices. On another hand, we explain the second metric value by the influence of the participant's other biological signals on the EEG acquisition system. Indeed, the participant's cheeks and neck movements (muscle movements in general) generate electrical potentials which are intercepted by the EEG acquisition system.

Table 2 Metrics of the eye movement detection algorithm for controlling the robot

	Sensitivity (%)	Specificity (%)
Representative participant	86	69

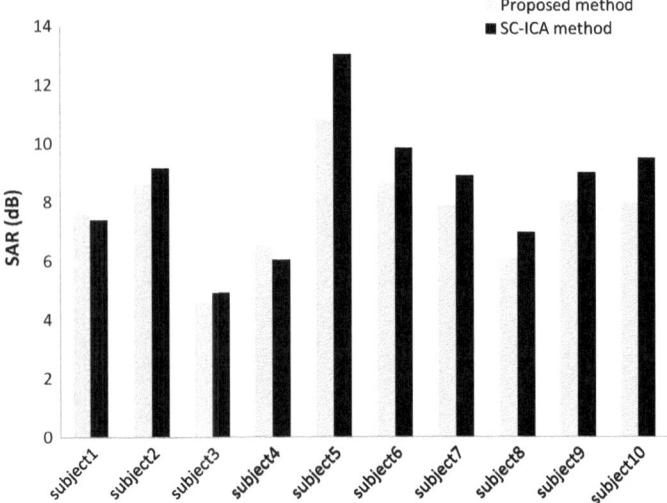

Fig. 6 Comparison of the SARof the proposed method and SC-ICA. Comparison made on the existing dataset

4 Conclusion

In this paper we presented a new architecture of an autonomous mobile robot intended for motor disability assistance through a future implementation on a wheelchair. The designed architecture offers two modules of use. The first module allows the intramurals mapping in a typical building. Through experimentation, we have demonstrated that it is possible to map based on a combination of distances and angles measured by an ultrasonic sensor and servomotors. However the approach incorporated in this fist module presented some limitations, especially in corners mapping, we can overcome these limitations by using a filter layer applied on the resulting map in order to eliminate unnecessary points. Another improvement could be envisaged by using data fusion based on Bayesian networks to improve the robot position estimation.

References

1. Bell, C. J., Shenoy, P., Chalodhorn, R., & Rao, R. P. N. (2008). Control of a humanoid robot by a noninvasive brain—Computer interface in humans. *Journal of Neural Engineering, 5*(2), 214–220.
2. Brose, S. W., Weber, D. J., Salatin, B. I., Grindle, G. G., Wang, H., Vazquez, J. J., & Cooper, R. A. (2010). The role of assistive robotics in the lives of persons with disability. *American Journal of Physical Medicine & Rehabilitation, 89*(6), 509–521.
3. Cisotto, G., Pupolin, S., Cavinato, M., & Piccione, F. (2014). An EEG-based BCI platform to improve arm reaching ability of chronic stroke patients by means of an operant learning training with a contingent force feedback. *International Journal of E-Health and Medical Communications, 5*(1), 114–134.

4. Huang, Q., Zhang, Z., Yu, T., He, T., & Li, Y. (2019). An EEG-/EOG-based hybrid brain-computer interface: Application on controlling an integrated wheelchair robotic arm system. *Frontiers in Neuroscience, 13.*
5. Jain, S., Farshchiansadegh, A., Broad, A., Abdollahi, F., Mussa-Ivaldi, F., & Argall, B. (2015). Assistive robotic manipulation through shared autonomy and a body machine interface. In *International Conference on Rehabilitation Robotics* (pp. 526 531). IEEE.
6. Kalunga, E. K., Chevallier, S., Rabreau, O., & Monacelli, E. (2014). Hybrid interface: Integrating BCI in multimodal human-machine interfaces. In *Proceedigns of AIM* (pp. 530–535).
7. Kraetzschmar, G. K., Hochgeschwender, N., Nowak, W., Hegger, F., Schneider, S., Dwiputra, R., Berghofer J., & Bisschoff, R. (2014). Robocup@work: Competing for the factoryof the future. In*RoboCup2014: Robot World Cup XVIII* (pp. 171–182). Springer.
8. Mühl, C., Jeunet, C., & Lotte, F. (2014). EEG-based workload estimation across affective contexts. *Frontiers in Neuroscience, 8.*
9. Rebsamen, B., Burdet, E., Guan, C., Teo, C. L., Zeng, Q., Ang, M., & Laugier, C. (2007). Controlling a wheelchair using a BCI with low information transfer rate. In *Proceedings of ICORR* (pp. 1003–1008).
10. Rupp, R., Kleih, S. C., Leeb, R., del R. Millan, J., Kübler, A., & Müller-Putz, G. R. (2014). Brain–computer interfaces and assistive technology. In G. Grübler & E. Hildt (Eds.), *Brain-computer-interfaces in their ethical, social and cultural contexts*, Vol. 12 (pp. 7–38), Dordrecht: Springer Netherlands.
11. Tonin, L., Carlson, T., Leeb, R., & del R Millan, J. (2011). Brain-controlled telepresence robot by motor-disabled people. In *Proceedings of EMBC* (pp. 4227–4230).
12. Trénors, L., Yin, C., Hafsia., M., Monacelli, E., & Benali, A. (2019). Gyrolift, a new way of verticalisation on mobile personal transporter. *Simulation Modelling Practice and Theory, 90,* 98–115.
13. van Dinteren, R., Arns, M., Jongsma, M. L. A., & Kessels, R. P. C. (2014). Combined frontal and parietal P300 amplitudes indicate compensated cognitive processing across the lifespan. *Frontiers in Aging Neuroscience, 6.*
14. Wolpaw, J. R., Birbaumer, N., McFarland, D. J., Pfurtscheller, G., & Vaughan, T. M. (2002). Brain-computer interfaces for communication and control. *Clinical Neurophysiology, 113*(6), 767–791.
15. Zammouri, A., & Ait Moussa., A. (2017). Eye blinks artifacts detection in a single EEG channel. *International Journal of Embedded Systems, 9*(4), 321–337.

Fuzzy Dynamic Airspace Sectorization Problem

Gabli Mohammed and Mermri El Bekkaye

Abstract In recent years, the number of controlled flights has increased steadily. At the same time, air traffic management has slowly been organized. To ensure safe, regular and efficiency air traffic, an airspace must be divided into several sectors, each one is assigned to a team of controllers. The latter have many tasks that create a workload. Generally, there are three kinds of workload, the monitoring workload, the conflict workload, and the coordination workload; the first two workloads occur inside the sector, and the third one between a sector and its adjacent sectors. The problem addressed in this study concerned the dynamic airspace sectorization problem (DASP) under certain constraints. Our objective is to find an optimal sectorization of the airspace in order to balance the workloads of air traffic controllers. For that, we first modeled this problem in the form of a dynamic multi-objective optimization problem. Secondly, we considered that there are not only quantitative factors to be taken into account but also psychological factors which have an important influence that should not be neglected. In this context and in order to collect subjective data such as stress, concentration, etc., we developed a questionnaire to be distributed to airspace controllers. Third, to deal with imprecise, uncertain and subjective information, we have developed a fuzzy model and we used metaheuristics to solve the resulting fuzzy DASP problem. The proposed model is illustrated by a numerical example based on a data simulation.

G. Mohammed (✉)
Department of Computer Science, Faculty of Science (FSO),
University Mohammed Premier, Oujda, Morocco
e-mail: medgabli@yahoo.fr; medgabli@fso.ump.ma

M. El Bekkaye
Department of Mathematics, FSO, University Mohammed Premier,
Oujda, Morocco
e-mail: b.mermri@fso.ump.ma

© The Author(s), under exclusive license to Springer Nature Switzerland AG 2021   229
U. Ghosh et al. (eds.), *Machine Intelligence and Data Analytics for Sustainable Future Smart Cities*, Studies in Computational Intelligence 971,
https://doi.org/10.1007/978-3-030-72065-0_13

1   Introduction

Air traffic has grown considerably over the past several decades and planes are equipped with increasingly sophisticated means of navigation. In recent years, the number of controlled flights has continued to increase significantly. In parallel, air traffic management has slowly been organized and is now divided into three main modules: airspace management, flow management and traffic control.

The official definition of air traffic control is as follows (see [1], for example).

The primary purpose of air traffic control is to ensure traffic safety and therefore avoid collisions between aircraft operating in the system, then to optimize traffic flows.

The primary mission therefore remains security, the notion of capacity only intervening afterwards. In this chapter, we look at the first module by trying to model and optimize the sectorization of airspace.

Sectorization is a fundamental architectural feature of the Air Traffic Control (ATC) system. The duties of ATC are to provide safe, regular and efficiency air traffic in the airspace in consideration. To carry out these duties the airspace is divided into a number of sectors, each of them is assigned to a team of controllers. Each sector has a certain capacity depending on several factors (see [2] for instance): ATC system, controller's experience, traffic characteristics, scenarios (overflights, climbing, descending, military activity, . . .), etc. The sector capacity can be defined in several ways.

- this is the number of take-off and landing operations in a given time period corresponding to the allowable delay level;
- this is the practical capacity for the season declared by the airport operator to plan the time of flight operations (seasonal flight schedule);
- is the maximum number of aircraft which can be served during a certain time period (for example in Europe during one hour [3]).

In this chapter, we will adopt the third definition of the capacity of a sector. Therefore, we consider sector capacity as the maximum number of take-off and landing operations that a sector can serve during a given period.

Generally modern jet aircraft do not enable pilots to solve conflicts because of their high speed and their ability to fly with bad visibility [4, 5]. Therefore pilots must be helped by air traffic controller. Controllers of a given sector have many tasks which induce a workload. The definition of workload is critical, it needs to take into account human factors issues, which include subjective estimations of psychological/physiological state and mental workload. Workload has several origins which can be divided into two categories:

- *Quantitative factors.* For each sector, there are three kinds of workload (see [6–8], for instance): the monitoring workload, the conflict workload, and the coordination workload; the first two workloads occur inside the sector, and the third one between a sector and its adjacent sectors.

- *Psychological/physiological factors.* These include stress, concentration, visual and auditory perception, memory, etc. which have no evident mathematical formulation, see [4, 9–13] for instance.

Related Works

In the literature, research concerning airspace sectorization is structured around several criteria (approach, frequency, dimensionality, etc.).

1. There are at least two approaches: an approach using a graph-based model (see for instance [14, 15]), and an approach using a region-based model (see for instance [16–19]). In the past decades a variety of deterministic and stochastic methods have been developed to solve complex traffic engineering problems: genetic algorithm [7, 14, 20, 21], neural network [4, 19], geometric algorithms [9], fusion-fission [22], etc. In the other hand, a few of researchers have used a fuzzy logic approach (see for instance [2]), or stochastic approach (see for instance [23]).
2. Airspace sectorization can be studied with different frequencies: static frequency where sectorization is strategic or pre-tactical (see for instance [4, 17]), and dynamic frequency where sectorization is tactical (it occurs at pre-determined times). In the latter case we have different configurations of airspace, see for instance [21, 24].
3. An airspace sectorization can be computed in different dimensions: 2D (see for instance [4, 9, 15, 18]), and 3D (see for instance [14, 17, 20, 25]).
4. Airspace sectorization aims to satisfying some constraints: sector convexity, safety constraint, minimum stay time constraint, connectivity constraint, etc. [4, 14]. In the literature, researchers have used some or all of these constraints.

In most of these studies, authors are interested in quantitative factors. However, psychological factors are strongly present and hence they have an influence that should not be overlooked. In [26] for instance, authors proposed a fatigue detection method based on the fractal dimension (FD). To do this, they first built a special voice database for radiotelephone communications. Then, the speech signals implement a wavelet decomposition and an FD calculation. Also in [27, 28], it has been proposed that the amount and type of voice communications between air traffic controllers and pilots is a good indicator of a controllers workload.

Although fuzzy logic turns out to be a very promising mathematical approach to model processes characterized by subjectivity, uncertainty and imprecision, the majority of researchers only introduce this approach to optimize control, manage conflicts and resolve air traffic safety problems (see [29] for instance). Very few of them have used fuzzy logic to solve airspace sectorization problem. In this context, we can cite [2] which used fuzzy logic to determine the number of sectors that should remain open during a given period such as a day or a week, in order to ensure an acceptable workload for the controllers.

In this chapter, we are interested in dynamic airspace sectorization problem (DASP) with constraints. The objective is to minimize the coordination workload between adjacent sectors and to balance the workload across the sectors. The second objective can be transformed in minimizing the standard deviation of the traffic inside sectors. We modeled this problem in the form of multi-objective optimization problem that can be transformed into a dynamic mono-objective one. To deal with the imprecise, uncertain and subjective informations we developed a fuzzy model. To solve the fuzzy DASP problem we used a hybrid genetic algorithms (HGA). In order to collect subjective (psychological) data such as stress, concentration, etc., we developed a questionnaire which was distributed to a group of airspace controllers.

The chapter is outlined as follows. In Sect. 2, we describe the problem and we present its formulation. In Sect. 3 we present the dynamic fuzzy model of this problem taking into account the psychological factors of the workload. In Sect. 4, we propose metaheuristic approach to solve the resulting problem. In Sect. 5, we give an application of our approach to some problems, then we present the obtained numerical results. Finally, in Sect. 6 we give some concluding remarks.

2   Problem Statement and Model Presentation

Consider an airspace A of a country or a region to be divided into n small volume units (elementary sectors) denoted by $x_1, x_2, \ldots, x_n$. The set of these units is denoted by E. A sector S is obtained by joining some of these elementary units. We denote by P a partition of the set E, for example a partition P of cardinality m can be defined as follows.

$$P = \{S_j \mid S_j \text{ is a sector}, 1 \leq j \leq m, \bigcup_{j=1}^{m} S_j = E \text{ and } \bigcap_{j=1}^{m} S_j = \emptyset\}$$

We denote by Π the set of all possible partitions of the set E. For example if $E = \{x_1, x_2, x_3\}$, the set of all partitions is given by: $\Pi = \{P_1, P_2, P_3, P_4, P_5\}$ where

- $P_1 = \{\{x_1\}, \{x_2\}, \{x_3\}\}$;
- $P_2 = \{\{x_1, x_2\}, \{x_3\}\}$;
- $P_3 = \{\{x_1\}, \{x_2, x_3\}\}$;
- $P_4 = \{\{x_1, x_3\}, \{x_2\}\}$;
- $P_5 = \{\{x_1, x_2, x_3\}\}$.

In this chapter, we study the dynamic airspace sectorization problem in 2D. The objective is to find, during a certain time period Δt, the partition P which minimizes the coordination workload between adjacent sectors and minimizes the standard deviation of the traffic inside sectors without exceeding the capacity. The number of sectors is not known in advance. The determination of the optimal partition P

Fig. 1 Connectivity
constraint

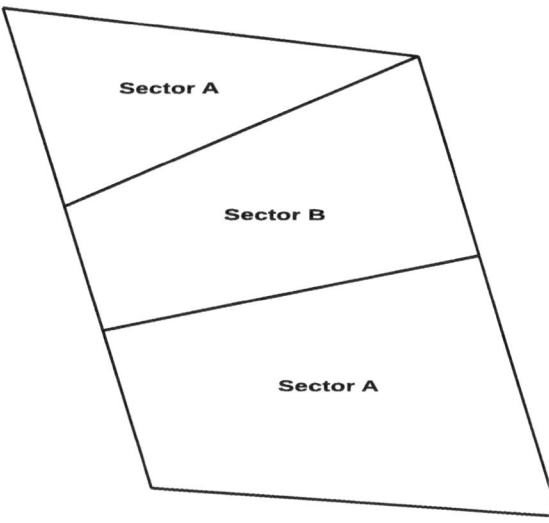

Fig. 2 Minimum stay time
constraint

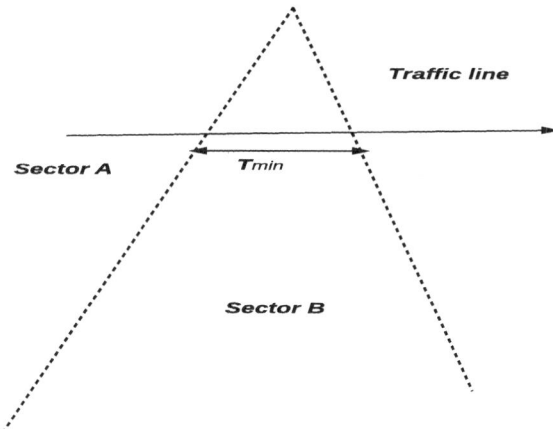

gives us the number of sectors during a time period Δt. Our model consider the two
following constraints:

- *Connectivity constraint.* The sector can not be fragmented. This constraint is hard,
  and must be satisfied in a solution. Figure 1 shows a solution which is not feasible.
- *Minimum stay time constraint.* An aircraft has to stay at least a given amount of
  time in each sector it crosses. This constraint ensures the controller has enough
  time to control the aircraft, see Fig. 2.

We denote by $f(X, \Delta t)$ the number of overflight (traffic) inside the sector X during the time period Δt. This traffic induces the monitoring workload and the conflict workload. We denote by $g(X, Y, \Delta t)$ the function which measures the coordination workload between sectors X and Y during Δt, and we denote by $cap(\Delta t)$ the capacity of a sector during Δt. Since we wish to minimize the coordination workload between adjacent sectors and minimize the standard deviation of the traffic inside sectors without exceeding the capacity, the problem can be expressed as:

$$\begin{cases} \underset{P \in \Pi}{Minimize} \sqrt{\frac{1}{n} \sum_{X \in P} (f(X, \Delta t) - \mu)^2}, \\ \underset{P \in \Pi}{Minimize} \sum_{X \in P} \sum_{\substack{Y \in P \\ Y \neq X}} g(X, Y, \Delta t), \end{cases} \tag{1}$$

subject to:

$$f(X, \Delta t) \leq cap(\Delta t), \quad X \in P, \tag{2}$$

$$Sector\ connectivity\ constraint, \tag{3}$$

$$Minimum\ stay\ time\ constraint. \tag{4}$$

where $\mu = \frac{1}{n} \sum_{X \in P} (f(X, \Delta t)$ and $n = |P|$ is the cardinality of the partition P.

Whenever a solution P violates any of the constraints (2) or (4), a penalty will be awarded to the objective functions. constraint (3) is ensured by a suitable choice of the chromosome of the genetic algorithm solution, see Algorithm 2 in Sect. 4.1.1.

We transform the multi-objective problem (1) into a mono-objective one as follows:

$$\underset{P \in \Pi}{Minimize}\ w_1 \sqrt{\frac{1}{n} \sum_{X \in P} (f(X, \Delta t) - \mu)^2} + w_2 \sum_{X \in P} \sum_{\substack{Y \in P \\ Y \neq X}} g(X, Y, \Delta t), \tag{5}$$

subject to the constraints (2), (3) and (4), where the weights w_1 and w_2 are positive values satisfying

$$0 \leq w_1 \leq 1,\ 0 \leq w_2 \leq 1\ \text{and}\ w_1 + w_2 = 1. \tag{6}$$

To balance the objective function weights, we use dynamic weights, see our papers [30, 31] (more details in Sect. 4.3). And in order to deal with the imprecise, uncertain and subjective informations, we introduce a fuzzy logic model. Therefore, our problem model becomes as follows:

$$\underset{P \in \Pi}{Minimize}\ w_1(t) \sqrt{\frac{1}{n} \sum_{X \in P} (\tilde{f}(X, \Delta t) - \mu)^2} + w_2(t) \sum_{X \in P} \sum_{\substack{Y \in P \\ Y \neq X}} \tilde{g}(X, Y, \Delta t), \tag{7}$$

subject to the constraints (2), (3) and (4), where $w_1(t)$ and $w_2(t)$ are two dynamic weights satisfying condition (6) and t is a time-step. $\tilde{f}$ is the fuzzy traffic inside each sector during Δt and $\tilde{g}$ is the fuzzy workload between adjacent sectors during Δt.

3   Fuzzy Logic Model

In order to investigate more realistic systems, it is necessary to consider the situation that one makes a decision on the basis of data involving fuzziness (see [32]). In this section we develop an approach based on fuzzy logic (see [33]). A fuzzy inference system (FIS) is composed of four components: fuzzification, IF-THEN rules (fuzzy rule base), fuzzy inference engine and defuzzification, see Fig. 3. In general, FIS has multiple inputs and multiple outputs. In this chapter, we apply four inputs to produce two outputs as follows.

- We apply two inputs: the traffic f inside each sector and the psychological factors to produce one output: $\tilde{f}$ (the estimated monitoring workload and conflict workload).
- We apply two inputs: the traffic g between adjacent sectors and the psychological factors to produce one output: $\tilde{g}$ (the estimated coordination workload).

The major question is how to calculate the psychological factors. In the literature there are few attempts to answer this question. For instance, in [27, 28] it has been proposed that the amount and type of voice communications between air traffic controllers and pilots is a good indicator of a controllers workload. From our perspective, interviews and questionnaires are the most appropriate tools to collect subjective information. Therefore, we developed a questionnaire, which aims to collect subjective (psychological) data such as stress, concentration, etc. The questionnaire was distributed to a group of airspace controllers.

We consider that the traffic inside each sector, the traffic between adjacent sectors and the two outputs $\tilde{f}$ and $\tilde{g}$ can be defined as follow: "very small", "small", "medium", "high" and "very high". The membership function values of these fuzzy input and output variables are presented in Fig. 4. Fuzzy data about the psychological factors are collected from the questionnaire. We applied a set of fuzzy IF-THEN rules to obtain fuzzy decision sets. The fuzzy rule base is obtained on the basis of expert knowledge. Table 1 presents the IF-THEN rules for the monitoring and conflict workload, and Table 2 presents the IF-THEN rules for the coordination workload. For example, if the traffic f inside each sector is *high* and the impact of psychological factors is *small*, then the output $\tilde{f}$ is *medium*. In our model, we use the Zadeh operators and the Mamdani-type fuzzy model. For the defuzzification, we use the center of gravity method (see [34], for instance).

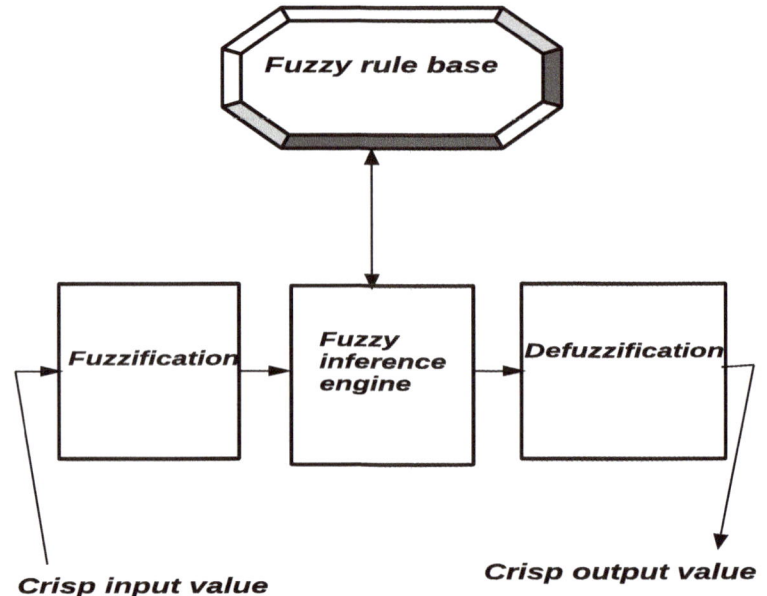

Fig. 3 Fuzzy inference system (FIS)

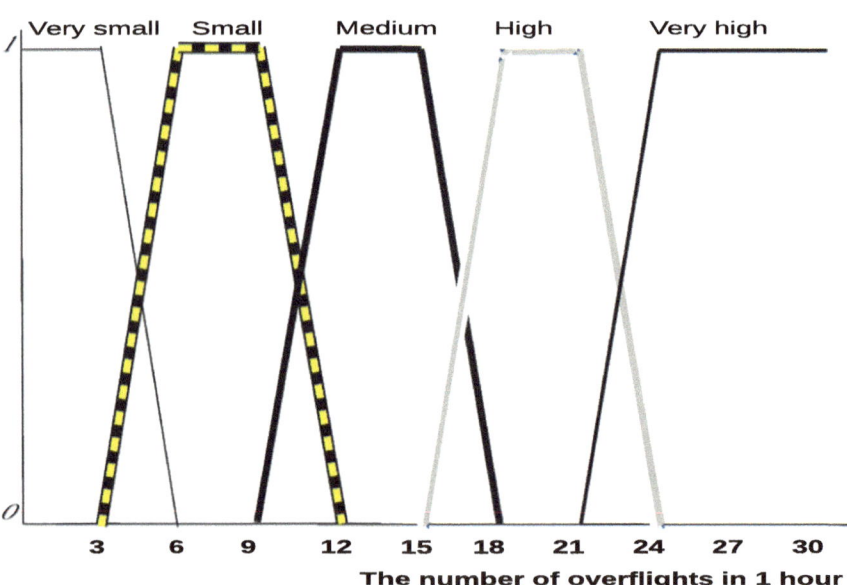

Fig. 4 Membership function values of the fuzzy input and output variables

Table 1 IF-THEN fuzzy rules for the monitoring and conflict workload

		Traffic f inside each sector				
		V. small	Small	Medium	High	V. high
Impact of psycho. factors	V. small	V. small	V. small	Small	Medium	High
	Small	V. small	Small	Medium	Medium	High
	Medium	V. small	Small	Medium	High	High
	High	Small	Small	Medium	High	V. high
	V. high	Small	Small	Medium	High	V. high

Table 2 IF-THEN fuzzy rules for the coordination workload

		Traffic g between adjacent sectors				
		V. small	Small	Medium	High	V. high
Impact of psycho. factors	V. small	V. small	V. small	Small	Medium	High
	Small	V. small	Small	Medium	Medium	High
	Medium	V. small	Small	Medium	High	High
	High	Small	Small	Medium	High	V. high
	V. high	Small	Small	Medium	High	V. high

4 Dynamic Genetic Algorithm Approach and Hybridization

Genetic algorithms (GAs) are optimization algorithms based on techniques derived from genetics and natural evolution. They belong to the family of evolutionary algorithms (a subset of metaheuristics). They already have a relatively old history since the first works of John Holland [35] on adaptive systems date back to 1962. The works of David Goldberg [36] have largely contributed to popularize them.

In genetic algorithm (GA) terminology, a solution vector x is called an individual or a chromosome. Chromosomes are made of discrete units called genes. Each gene controls one or more features of the chromosome. The main components of a GA are: selection, crossover and mutation.

The general functioning of a genetic algorithm is represented by Algorithm 1. Consider the problem (7) presented in Sect. 2. In this section, we have used genetic algorithms (GAs) as they are described in [35, 36].

Algorithm 1. Genetic algorithms

Begin
    Generate the initial population P
    While the stopping condition is not satisfied **do**
        Evaluate individuals of the chromosome P
        Select the new population P_1
        Cross the individuals of P_1 to get P_2
        Mutate the individuals of P_2 to obtain a new population P
    End While
End

4.1 Genetic Algorithm Approach

4.1.1 Chromosome Encoding

Let $n = |E|$ be the number of volume units (VUs). Each VU is characterized by its adjacent VUs and the time that an aircraft has to stay in this VU. We introduce a sequence of n digits, where each digit is an integer taking values between 1 and n. If the digit in a position j, takes a value k, $d_j = k$, that means the VU_j is assigned to the sector k; For example, if $n = 13$, the code *4;3;1;7;1;7;4;7;7;3;4;11;11* means that the first volume unit VU_1 is assigned to the sector number 4, VU_2 to the sector number 3, ..., VU_{13} to the sector number 11. We see that there are 5 sectors: $\{VU_1, VU_7, VU_{11}\}$, $\{VU_2, VU_{10}\}$, $\{VU_3, VU_5\}$, $\{VU_4, VU_6, VU_8, VU_9\}$ and $\{VU_{12}, VU_{13}\}$.

With this encoding method, it is certain that the intersection of sectors is empty, and the union of sectors is the set E.

In the initial population of the genetic algorithm (GA), each chromosome is generated randomly as illustrated in the Algorithm 2. With this construction of the chromosome we guarantee the connectivity constraint. To satisfy the Minimum stay time constraint, we calculate, in each sector, the time that an aircraft remains inside. If this time is insufficient, we penalize the objective function.

Algorithm 2. Sector connectivity constraint algorithm

Step 0. In the first position of chromosome, we generate randomly an integer (gene) g_1 from the set $\{1, \ldots, n\}$. So, the VU_1 is assigned to the sector g_1.

Step 1. In the position i, $1 < i \leq n$, we generate a random gene g_i from the set $\{1, \ldots, n\}$.

Step 2. (*correction*). In the position i, $1 < i \leq n$, for each g_j, $1 \leq j < i$, if $g_i = g_j$ and VU_i hasn't any adjacent VU among the VU in the sector g_j, then we generate a new random gene.

Step 3. We repeat step 2 until a new gene g_i is obtained, or to find an assignment of VU_i in a sector where there are adjacent VU.

Step 4. We repeat steps 1, 2 and 3 until we generate all genes of the chromosome.

4.1.2 Crossover and Mutation Operators

The crossing used by genetic algorithms is the computer transposition of the mechanism which allows, in nature, the production of chromosomes which partially inherit the characteristics of the parents. The aim of the crossing is to enrich the diversity of the population by manipulating the structure of the chromosomes. As for the mutation operator, it acts as a disruptor, it introduces noise in order to maintain the diversity of the population. It leads to explore the search space by preventing the algorithm from converging too quickly towards a local optimum.

Figure 5 presents the crossover and mutation operators. For example, for the first chromosome obtained after crossover, the gene (volume unit) number 10 (VU_{10}) can not belong to the sector containing VU_4, VU_6, VU_8 and VU_9 (connectivity constraint), so we use the previous correction (steps 2 and 3 defined in the previous subsection). Figure 6 illustrates this operation.

4.2 Hybridization Approach

4.2.1 Local Search Method (LS)

Local search (LS) is a metaheuristic method for solving computationally hard optimization problems. LS algorithms are used to explore the nearby region around the current solutions with high intensity. So, by using a LS method, highly accurate solutions can be obtained.

Fig. 5 Example of crossover and mutation operators

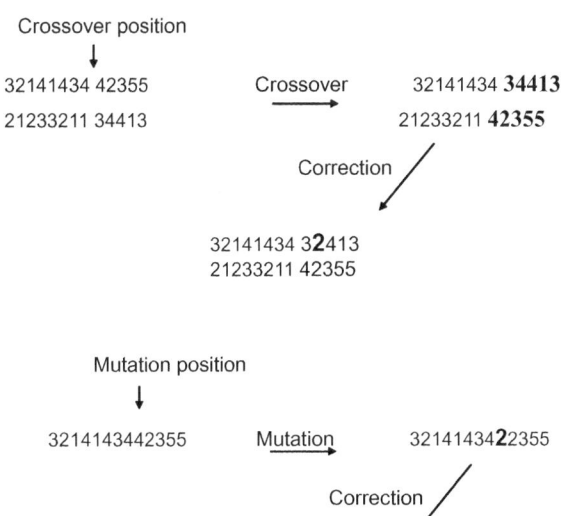

Fig. 6 Illustration of
crossover operator

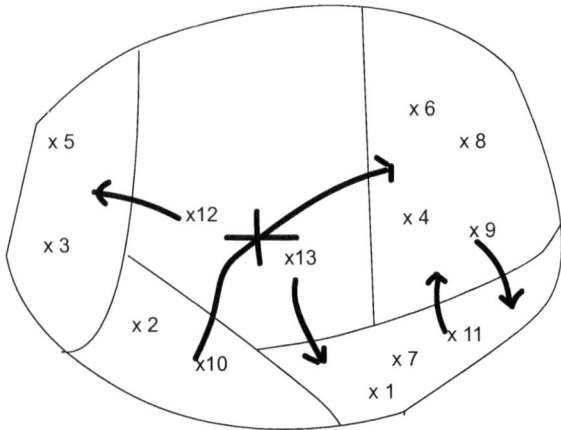

In this chapter we consider the simple local search method (or the *downhill method*) which consists in building a single initial solution and improving it by using a single neighborhood structure until a local optimum is reached, thereby stopping the search. The operation of this method is presented by Algorithm 3.

Often the evolutionary algorithms EAs (including GAs) are hybridized with LS method and these hybridizations (see e.g. [37–39]) are proven to be more efficient than the EAs themselves. The reason behind this is the combination of global exploration and local exploitation [40]. In this chapter we present an algorithm using the hybridization of GA with LS.

Algorithm 3. Local search method

Begin
    Generate initial solution $S := S_0$
    For a predetermined number of times **do**
        Generate a solution S_1 in the neighbourhoud of S
        If $f(S_1) - f(S) < 0$
            $S := S_1$;
        EndIf
    EndFor
    Return the best configuration found
End

4.2.2 Hybridization of GA with LS

For genetic algorithms, local search (LS) is often used for the improvement of the solutions and the intensification of search. In this chapter we exploit this hybridization as follows:

- We take the best solution given by the GA method;
- We set this solution as the initial configuration of LS;
- We apply the LS method on this configuration.

4.3   GA and Dynamic Weights

Consider the optimization model (7), it is of the form of Eq. 8.

$$h = w_1(t)f + w_2(t)g. \tag{8}$$

Assume that f is much greater than g. When applying GA to minimize the objective function h, if we take constant weights w_1 and w_2, then there is a great risk that the procedure of selection of the GA chooses only solutions which improve f by neglecting g, since the function f dominates g. To remedy this problem we should not take fixed weight values, but rather this value must be dynamic and it changes in each iteration of the GA. For this, we consider a GA algorithm using a dynamic weights introduced in our papers [30, 31] as follows (Eq. 9).

$$w_1(t) := \frac{|g(x_{t-1})|}{|f(x_{t-1})| + |g(x_{t-1})|} \qquad \text{and} \qquad w_2(t) = 1 - w_1(t), \tag{9}$$

where x_{t-1} is the best solution of the iteration $(t-1)$ of the GA.

If $f(x_{t-1}) = g(x_{t-1}) = 0$, then take $w_1(t) := w_1(t-1)$. This algorithm (Algorithm 4) has two advantages:

1. By automating the choice of the weights assigned to each objective, it is no longer necessary to define a priori these weights. This task which is very difficult.
2. It ensures an equitable treatment of each objective function, so we have an equitable chance to minimize both functions f and g.

Algorithm 4. Dynamic genetic algorithm

Step 0. To initalize GA, we generate arbitrarily a positive real number $w_1(0)$, so $w_2(0) = 1 - w_1(0)$.
Step 1. Perform an iteration t of the GA, with the fitness function h (Eq. 8).
Step 2. Let x_t be the best solution chosen during the current iteration t.
Step 3. Calculate $f(x_t)$ and $g(x_t)$.
Step 4. Take $w_1(t+1)$ and $w_2(t+1)$ as mentioned in Eq. 9.
Step 5. $t := t + 1$;
Step 6. Repeat steps 1 to 5 until the stop criterion is satisfied.

---

5 Application

5.1 Data Description

In this chapter, we consider the fuzzy DASP problem in an airspace A with 21 volume units $V_i, i = 1, \ldots, 21$. Figure 7 presents this airspace. We have proposed simulation data (quantitative and qualitative data) to apply our approach. By analyzing these data, we decided to take $cap(\Delta t) = 120$ and $\Delta t = 4$ h. So the day is divided into six periods: [6, 10], [10, 14], [14, 18], [18, 22], [22, 2] and [2, 6]. [6, 10] designates the period from the hour 6 a.m to 10 a.m. When analyzing the data, we decided to merge the last two periods, so that the fifth and final period becomes [22, 6] (from 10 p.m to 6 a.m).

The membership function values of the fuzzy input and output variables is presented in Fig. 4. We generated the simulation data in the airspace A for each period, for example simulation data during the period [6, 10] is presented in Table 3. In this

Fig. 7 Airspace with 21 volume units

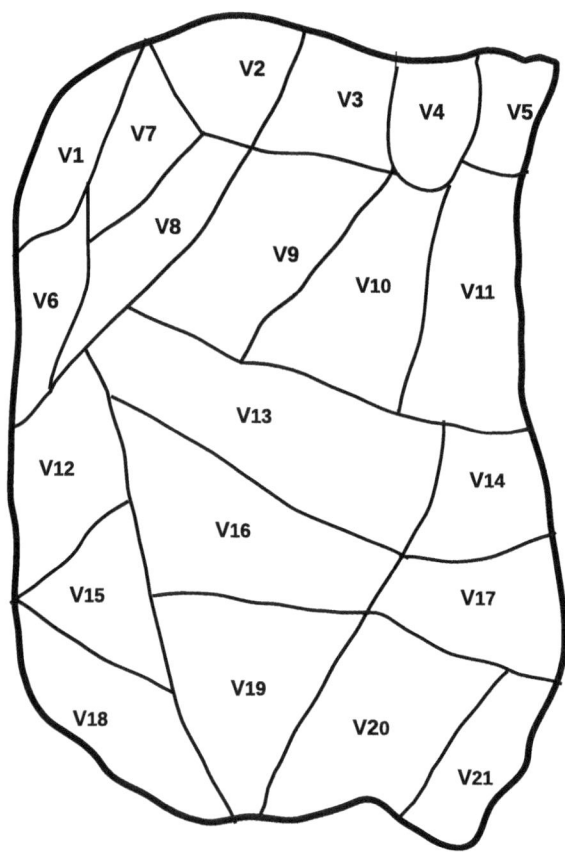

Table 3 Simulation data in an airspace with 21 volume units during the period [6, 10]

	V_1	V_2	V_3	V_4	V_5	V_6	V_7	V_8	V_9	V_{10}	V_{11}	V_{12}	V_{13}	V_{14}	V_{15}	V_{16}	V_{17}	V_{18}	V_{19}	V_{20}	V_{21}
V_1		m	v	v	v	s	h	g	v	s	v	v	s	m	s	v	h	v	m	s	m
V_2			v	s	s	h	v	m	v	v	s	v	v	g	v	s	v	s	v	s	s
V_3				g	m	s	m	s	m	h	h	m	v	s	s	s	v	m	v	v	s
V_4					s	v	s	v	m	v	v	v	s	v	v	v	s	v	s	v	v
V_5						v	v	s	v	s	s	h	v	v	s	s	m	s	v	s	s
V_6							s	h	s	v	s	m	v	h	v	v	m	v	v	m	s
V_7								m	v	s	s	v	s	h	s	v	v	v	s	v	m
V_8									v	s	v	s	s	h	v	v	h	m	v	m	h
V_9										v	s	g	m	g	v	v	v	h	s	m	v
V_{10}											v	h	v	s	h	s	v	m	m	v	s
V_{11}												s	v	v	m	v	v	m	v	v	s
V_{12}													s	s	m	s	s	h	v	m	v
V_{13}														v	v	g	m	v	h	v	h
V_{14}															s	m	v	s	s	h	v
V_{15}																v	s	g	v	v	s
V_{16}																	s	s	g	s	s
V_{17}																		v	s	v	v
V_{18}																			v	s	g
V_{19}																				g	s
V_{20}																					m
V_{21}																					

table we denoted by "v" very small, by "s" small, by "m" medium, by "h" high, by "g" very high and by "V_i", $i = 1, \ldots, 21$ the volume unit number i.

5.2 Computational Results

The algorithms were coded in JAVA programming language and launched on an Intel Core2Duo-2GHz machine with 2 Giga-bytes (GB) in RAM.

In the hybrid GA approach, we have used three selection methods: roulette wheel, scaling and sharing. The parameters of GA are set as follows: crossover probability $p_c = 0.5$, mutation probability $p_m = 0.01$, population size $ps = 25$, and maximum

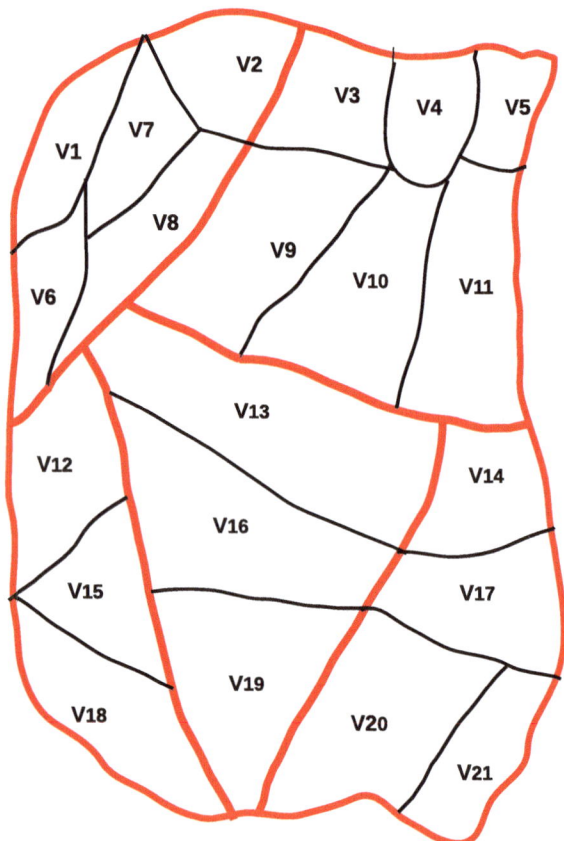

Fig. 8 Airspace sectorization during the period [6, 10]

Fig. 9 Airspace
sectorization during the
period [10, 14]

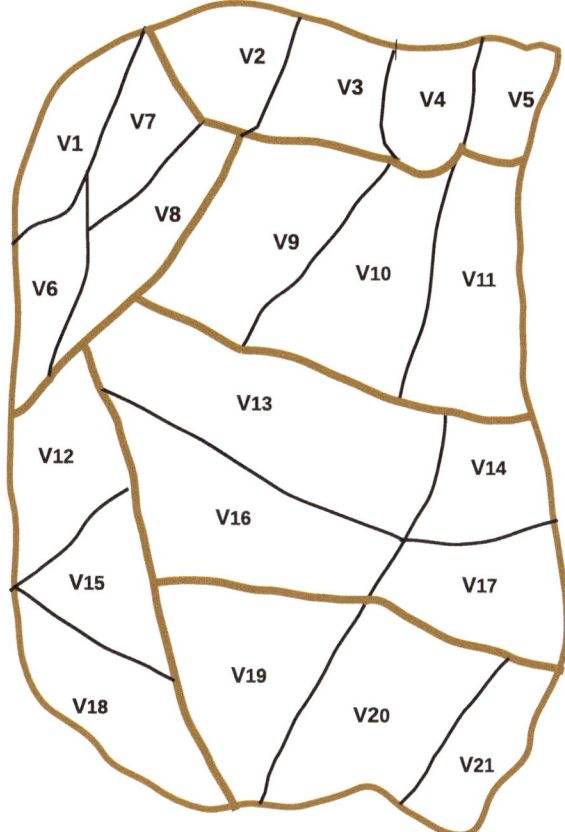

number of generations 5000. In the sharing selection method, the threshold of dissim-
ilarity between two parents is taken as $\sigma_s = ps/2$, and $\alpha = 1$. In the Fuzzy model,
we use the Zadeh operators [33] and the Mamdani-type fuzzy model (see [41], for
instance). For the defuzzification, we use the center of gravity method. Several experiments are done.

Figures 8, 9, 10, 11 and 12 present, respectively, the airspace sectorization during
the five periods mentioned above.

We see that we found 5 sectors in the first period, 6 in the second period, 5 in the
third period, 4 in the fourth period and 3 in the fifth and final period. For the first
period (Fig. 8), the five sectors are formed as follows:

$$P = \{\{V_1, V_2, V_6, V_7, V_8\}, \{V_3, V_4, V_5, V_9, V_{10}, V_{11}\}, \{V_{12}, V_{15}, V_{18}\},$$
$$\{V_{13}, V_{16}, V_{19}\}, \{V_{14}, V_{17}, V_{20}, V_{21}\}\}.$$

Fig. 10 Airspace
sectorization during the
period [14, 18]

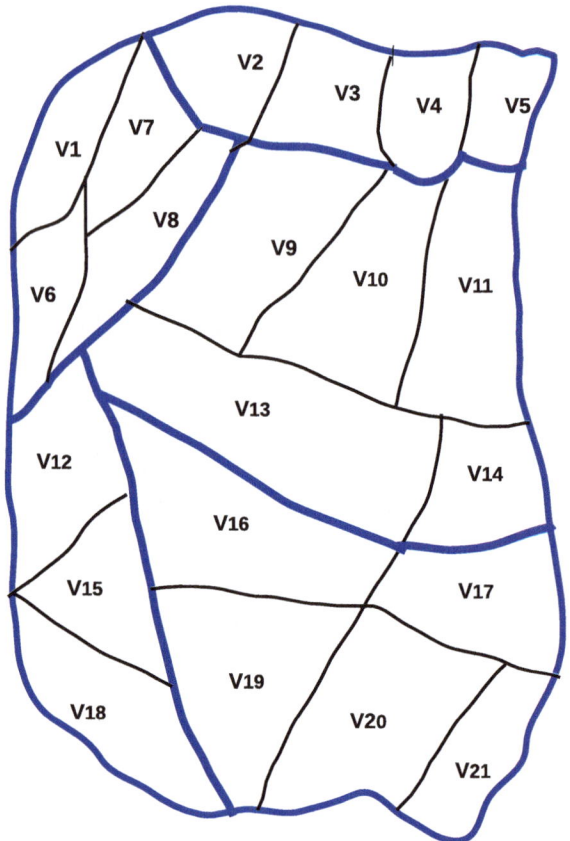

6   Conclusion

In this chapter, we have considered the fuzzy dynamic airspace sectorization problem with constraints. This problem, which is modeled as a multi-objective problem
subjected to some constraints, is transformed into a mono-objective optimization
problem with dynamic weights between the objective functions.

Our model took into account psychological factors such as stress, concentration,
etc. In order to collect subjective data, we developed a questionnaire which was
distributed to a group of airspace controllers.

To make solving the problem more realistic and to deal with imprecise, uncertain
and subjective information, we have developed an approach based on fuzzy logic.

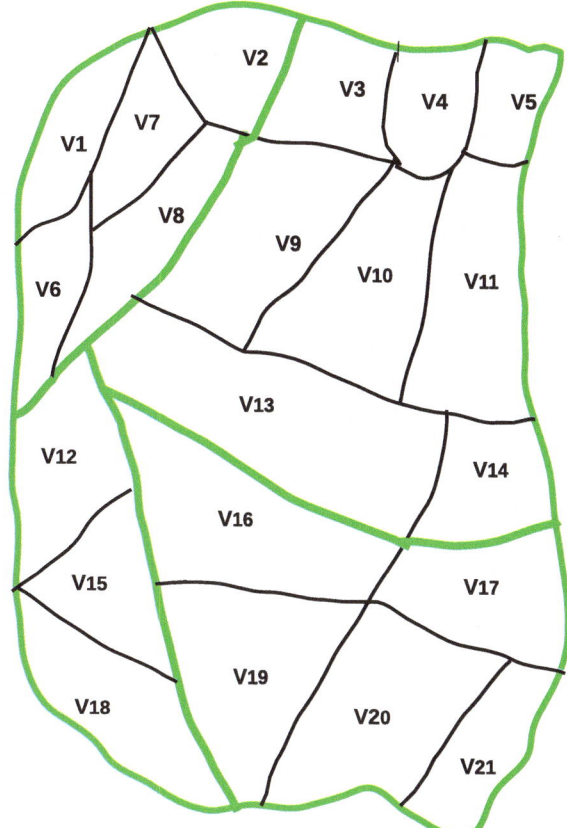

Fig. 11 Airspace sectorization during the period [18, 22]

To solve this problem, we have proposed an approach based on a hybrid genetic algorithm. The coding of the chromosome was done in such a way as to guarantee the connectivity constraint.

As application, we proposed simulation data (quantitative and qualitative data) in an airspace *A* of 21 volume units, then we sought the optimal sectorization.

In future research, we will extend this work in three dimensions and we will take into account other constraints such as safety constraint and convexity constraint.

Fig. 12 Airspace
sectorization during the
period [22, 6]

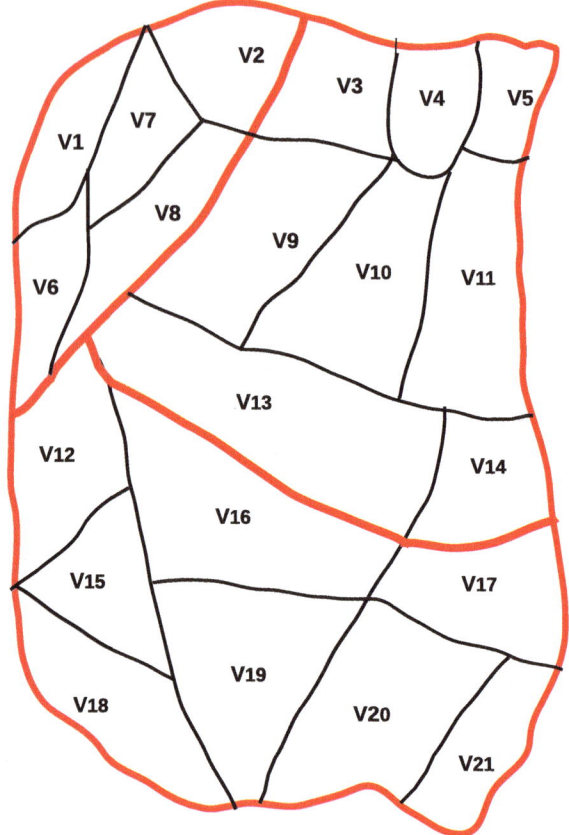

References

1. Riviere, T. (2006). Optimisation de graphes sous contrainte géometrique : création d'un réseau de routes aériennes pour un controle Sector-Less. Thèse de doctorat de l'institut national poly-technique de Toulouse.
2. Babic, O., & Kristic, T. (2000). Airspace daily operational sectorization by fuzzy logic. *Fuzzy Sets and Systems, 116*, 49–64.
3. Allignol, C. (2011). *Planification de trajectoires pour l'optimisation du trafic aerien*. Ph.D. thesis, INP Toulouse.
4. Kumar, K. (2012). ART1 neural networks for air space sectoring. *International Journal of Computer Applications, 37*, 20–24.
5. Riley, V., Chatterji, G., Johnson, W., Mogford, R., Kopardekar, P., Sieira, E. et al. (2004). Pilot Perceptions of Airspace Complexity. In *Part 2, Digital Avionics Systems Conference*, DASC 04. IEEE.
6. Trandac, H., Baptiste, P., & Duong, V. (2003). Optimized sectorization of airspace with con-straints. In *The 5th USA/Europe ATM R&D Seminar* (pp. 23–27). Budapest.
7. Delahaye, D., Schoenauer, M., & Alliot, J. M. (1998). Airspace sectoring by evolutionary computation. In *IEEE International Congress on Evolutionary Computation*.

8. Gabli, M., Jaara, E. M., & Mermri, E. B. (2018). Dynamic airspace sectorization problem using hybrid genetic algorithm. In *Information Systems and Technologies to Support Learning*, EMENA-ISTL 2018, Smart Innovation, Systems and Technologies, *111*, (pp. 282–289). Springer.

9. Basu, A., Mitchell, J. S. B., & Sabhnani, G. (2009). Geometric algorithms for optimal airspace design and air traffic controller workload balancing. *Journal of Experimental Algorithmics 14*(3).

10. de Frutos, P. L., et al. (2019). COMETA: An air traffic controller's mental workload model for calculating and predicting demand and capacity balancing. In *Book: Human Mental Workload: Models and Applications Third International Symposium*, H-WORKLOAD, Rome, Italy (pp. 14–15). Springer.

11. Chen, M. L., Lu, S.-Y., & Mao, I.-F. (2019). Subjective symptoms and physiological measures of fatigue in air traffic controllers. *International Journal of Industrial Ergonomics, 70*, 1–8.

12. Chang, Y.-H., Yang, H.-H., & Hsu, W.-J. (2019). Effects of work shifts on fatigue levels of air traffic controllers. *Journal of Air Transport Management, 76*, 1–9.

13. Kabashkin, I., & Philippov, V. (2020). Distributed ecosystem of voice communications for air traffic control system. *Procedia Computer Science, 177*, 32–39.

14. Delahaye, D., & Puechmorel, S. (2006). 3D airspace sectoring by evolutionary computation, real world applications. GECCO 06. In *Proceedings of the 8th Annual Conference on Genetic and Evolutionary Computation* (pp. 1637–1644).

15. Trandac, H., Baptiste, P., & Duong, V. (2005). Airspace sectorization with constraints. *RAIRO Operations Research, 39*(2), 105–122.

16. Yousefi, A., Donohue, G. L., & Qureshi, K. M. (2003). Investigation of en route metrics for model validation and airspace design using the total airport and airspace modeler. In *The 5th EUROCONTROL/FAA ATM R&D Conference*, Budapest.

17. Bichot, C. E., & Durand, N. (2007). A tool to design functional airspace blocks. In *Proceedings of ATM, the 7th USA/Europe R&D Seminar on Air Traffic Management* (pp. 169–177).

18. Brinton, C. R., Leiden, K., & Hinkey, J. (2009). Airspace sectorization by dynamic density. In *Proceedings of the 9th AIAA Aviation Technology, Integration and Operations (ATIO) Forum*. American Institute of Aeronautics and Astronautics.

19. Gianazza, D. (2010). Forecasting workload and airspace configuration with neural networks and tree search methods. *Artificial Intelligence, 174*, 530–549.

20. Kicinger, R., & Yousefi, A. (2009). Heuristic method for 3D airspace partitioning: Genetic algorithm and agent-based approach. In *9th AIAA Aviation Technology, Integration, and Operations Conference (ATIO) and Aircraft Noise and Emissions Reduction Symposium (ANERS)*.

21. Xue, M. (2008). Airspace sector redesign based on Voronoi diagrams. In *Proceedings of AIAA Guidance, Navigation, and Control Conference*. Honolulu, HI.

22. Bichot, C. E. (2007) Elaboration d'une nouvelle metaheuristique pour le partitionnement de graphe: la methode de fusion-fission. *Application au decoupage de l'espace aerien*. Ph.D. thesis, INP Toulouse.

23. Shone, R., Glazebrook, K., & Zografos, K. G. (2020). Applications of stochastic modeling in air traffic management: Methods, challenges and opportunities for solving air traffic problems under uncertainty. *European Journal of Operational Research*. https://doi.org/10.1016/j.ejor.2020.10.039.

24. Tang, J. (2012). *Large scale multi-Objective optimization for dynamic airspace sectorization*. Ph.D. thesis, University of New South Wales.

25. Li, J., Xue, K., Wang, X., Chang, C., Peng, Z., & Wu, J. (2019). Design and implementation of 3D air traffic control integrated information platform. In *IEEE 2nd International Conference on Electronics Technology (ICET)* (pp. 241–247). Chengdu, China.

26. Shen, Z., Pan, G., & Yan, Y. (2020). A high-precision fatigue detecting method for air traffic controllers based on revised fractal dimension feature. *Mathematical Problems in Engineering, 2020*, (4563962), 13.

27. Morrow, D. G., & Rodvold, M. (1998). Communication issues in air traffic control. In M. Smolensky & E. Stein (Eds.), *Human factors in air traffic control* (pp. 421–456). New York: Academic Press.

28. Porterfield, D. H. (1997) Evaluating controller communication time as a measure of workload. *International Journal of Aviation Psychology, 7*, 171–182.
29. Sarkar, A., Sahoo, G., & Sahoo, U. C. (2012). Application of fuzzy logic in transport planning. *International Journal on Soft Computing (IJSC), 3*, 1–21.
30. Gabli, M., Jaara, E. M., & Mermri, E. B. (2013). Planning UMTS base station location using genetic algorithm with a dynamic trade-off parameter. *Lecture Notes in Computer Science, 7853*, 120–134.
31. Gabli, M., Jaara, E. M., & Mermri, E. B. (2014). A genetic algorithm approach for an equitable treatment of objective functions in multi-objective optimization problems. *IAENG International Journal of Computer Science, 41*(2), 102–111.
32. Katagiri, H., Mermri, E. B., Sakawa, M., Kato, K., & Nishizaki, I. (2005). A possibilistic and stockastic programming approach to fuzzy random MST problem. *IEICE Transactions on Information and Systems, E88–D8*, 1912–1919.
33. Zadeh, L. A. (1965). Fuzzy sets. *Information and Control, 8*, 338–353.
34. Dubois, D., & Prade, H. M. (2000). *Fundamentals of fuzzy sets*. Boston: Kluwer Academic Publishers.
35. Holland, J. (1962). Outline for a logical theory of adaptive systems. *Journal of the Association of Computing Machinery, 9*, 297–314.
36. Goldberg, D. E. (1989). *Genetic algorithms in search, optimization, and machine learning.* (1st ed.). Addison-Wesley.
37. Hoos, H. H., & Stutzle, T. (2004). *Stochastic local search: Foundations and applications.* Morgan Kaufmann, Elsevier.
38. Molina, D., Lozano, M., Garcia-Martinez, C., & Herrera, F. (2010). Memetic algorithms for continuous optimisation based on local search chains. *Evolutionary Computation*, 27–63.
39. Moscato, P. (1989). On evolution, search, optimization, genetic algorithms and martial arts: Towards memetic algorithms. Caltech Concurrent Computation Program (report 826).
40. Mandal, A., Das, A. K., Mukherjee, P., & Das, S. (2011). Modified differential evolution with local search algorithm for real world optimization. In *IEEE Congress on Evolutionary Computation (CEC)* (pp. 1565–1572).
41. Kaur, A. R., & Kaur, A. M. (2012). Comparison of mamdani-type and sugeno-type fuzzy inference systems for air conditioning system. *International Journal of Soft Computing and Engineering (IJSCE), 2*(2). ISSN: 2231-2307.

Virtualization Technology for LoRaWAN Roaming Simulation in Smart Cities

Francesco Flammini, Andrea Gaglione, Dániel Tokody, and Dalibor Dobrilović

Abstract Internet of Things (IoT) low power and long range wireless technologies play a key role as an enabling technology for the development of the communication backbone for future smart cities, which will be increasingly based on multi-sensor intelligent data analytics. At the current state-of-the-art, a number of technologies collectively known as Low Power Wide Area Networks (LPWANs) can provide connectivity for IoT applications. Among LPWAN technologies, Long Range Wide Area Network (LoRaWAN) has been effectively used in a wide range of application domains. An important factor that accelerates the process of LoRaWAN implementation in IoT systems is the use of LoRaWAN simulation tools. Most simulation tools are widely used to simulate various networking technologies. As for LoRaWAN, their focus is on simulation of PHY and MAC layers. In this chapter, we address the possible use of virtualization technologies to simulate LoRaWAN at the application layer. Virtual network laboratories based on virtual technologies have existed for over a decade. These laboratories have been used for the purpose of educational and test environments. This chapter presents a novel model based on virtualization technology for the design, development and testing of the application for roaming in LoRaWAN networks in the context of future smart cities to enable continuous intelligent infrastructure monitoring and surveillance using moving devices such as those wearables or installed on-board vehicles and drones. The virtual network and

F. Flammini
Division of Product Realisation, School of Innovation, Design and Engineering, Mälardalen University, Eskilstuna, Sweden
e-mail: francesco.flammini@mdh.se

A. Gaglione
Brit Insurance, London, UK
e-mail: andrea.gaglione@britinsurance.com

D. Tokody
Next Technologies Ltd. Complex Systems Research Institute, Maglód, Hungary
e-mail: tokody.daniel@nexttechnologies.hu

D. Dobrilović (✉)
Technical Faculty "Mihajlo Pupin" Zrenjanin, University of Novi Sad, Zrenjanin, Serbia
e-mail: dalibor.dobrilovic@uns.ac.rs

its architecture for the simulation of LoRaWAN applications are presented in this chapter together with example use case scenarios.

Keywords LoRaWAN simulation · Virtualization technology · Virtual laboratories · Smart city platform

1 Introduction

In the era of growing Internet of Things (IoT) industry low power and long range wireless technologies, especially Low Power Wide Area Networks (LPWANs), play a key role in IoT systems deployment. In the variety of LPWAN technologies such as SigFox, NB-IoT, Weightless etc., Long Range Wide Area Network (LoRaWAN) has been effectively used in of application domains [1]. According to [2] LoRa and NB-IoT are the current market leaders. LoRaWAN covers 100 countries worldwide, comparing to 36 countries covered SigFox [3]. LoRaWAN also has more than 350 ongoing trials and city developments. Currently, non-cellular networks (LoRaWAN, SigFox, etc.) contributed to almost three-quarters of all LPWA network connections. The projection is that by 2023 non-cellular LPWANs will cede its market share dominance to cellular LPWANs (NB-IoT and LTE-M), since non-cellular will capture over 55% of LPWANs connections [4]. Possible loss of the market in the future does not decrease the LoRaWAN importance for the IoT applications.

Since LoRaWAN still represent one of the key enabling technologies that provide connectivity to IoT systems and that IoT becomes an enabling technology for big data generation, analytics and machine learning, an important factor that can burst the process of LoRaWAN implementation in IoT systems relies on the use of LoRaWAN simulation tools. After presenting existing network simulation tools that can be used for simulating LoRaWAN, such as NS-2, NS-3, OMNET++ and Riverbed, this chapter considers a novel model which includes virtualization technology in an open platform and cloud environment for the engineering, development and testing of the application for roaming in LoRaWAN networks in the context of future smart cities in order to enable continuous intelligent infrastructure monitoring and surveillance using moving devices such as those wearables or installed on-board vehicles and drones. The chapter also provides considerations on how virtualization technology and cloud based systems can support LoRaWAN simulations.

Virtual network laboratories based on virtual technologies exists over decade and with this research is made an effort towards expansion of virtual labs applicability to LoRaWAN as well. Majority of standard simulation tools are focused on simulation of LoRaWAN PHY and MAC layers. With the virtualization technology, the LoRaWAN application layer can be simulated as well. This chapter is structured as follows. After the introductory section, the LoRa and LoRaWAN technologies are briefly described in Sect. 2. In Sect. 3 is presented the review of simulation tools applicable for LoRaWAN simulation. Section 4 presents virtual laboratories and their applicability as well. The description of the proposed virtual laboratory for the LoRaWAN targeting application layer is given in the Sect. 5 with related experience. Finally, Sect. 6 offers brief concluding remarks.

2 LoRa and LoRaWAN

2.1 LoRa and LoRaWAN Basics

LPWANs are wireless wide area network technologies designed to allow long-range communications, at a low bit rate and power consumption. They are widely used in IoT, Machine to machine (M2M) and Smart systems applications.

LoRa (Long Range) is a proprietary standard of Semtech company. It uses license-free sub-gigahertz radio frequency bands (169, 433, 868 MHz in Europe, and 915 MHz in North America) presented in Fig. 1. The range of LoRa is very long and it goes depending of the environmental conditions more than 10 km in rural areas and according to different source much more.

LoRa covers the physical layer and the communication protocol built upon the LoRa physical layer. The MAC layer is covered with LoRaWAN (Long Range Wide Area Network), an open source communication protocol defined by the LoRa Alliance consortium. LoRaWAN defines the communication protocol and system architecture for the network, while the LoRa physical layer enables the communication.

LoRa Alliance is a non-profit technology alliance with more than 500 member companies. The first LoRaWAN standard was announced by the LoRa Alliance in June 2015 and LoRaWAN specification 1.1 was released in October, 2017 [5].

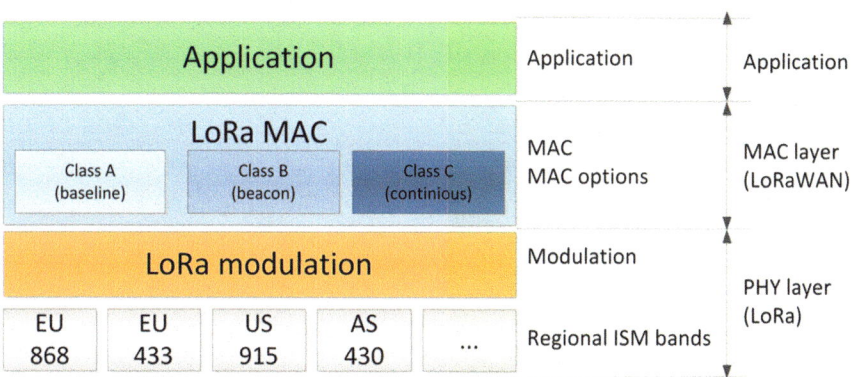

Fig. 1 LoRaWAN protocol stack

As it is stated in the Sect. 1, LoRa and LoRaWAN are currently most popular LPWAN technologies, with the projection that it will keep if not leading, than one of the leading positions in the future.

LoRaWAN protocol stack is presented in Fig. 1.

2.2 LoRaWAN Network Elements

LoRaWAN networks typically use star and star-of-stars topology depending of the type of their appliance and environments. The elements of LoRAWAN network include: end-devices, gateways, networks, application and join server (Fig. 2). LoRaWAN gateways relay messages between end-devices and a central Network Server. The Network Server routes the packets from devices to the associated Application Server. To secure radio transmissions the LoRaWAN protocol relies on symmetric cryptography using session keys derived from the device's root keys. In order to secure the communication between the elements Join Server deals with the storage of the device's root keys and the associated key derivation operations [5].

LoRaWAN End Device is composed of a sensor or an actuator, and it has wireless connection to a LoRaWAN network with LoRa RF modulation. End devices are usually autonomous, battery or solar powered unit. LoRaWAN Gateways receive LoRa modulated RF messages from end devices within the range and forwards it to the LoRaWAN network server. The connection between LoRa Gateways and other servers is based on IP backbone. The same end device can be served by multiple gateways in the area. LoRaWAN Gateways operate at the physical layer and their role is LoRa message forwarding.

The management of network is the task of LoRaWAN Network Server. It dynamically controls the network parameters in order to adapt to the changing conditions of the system. It also establishes secure 128-bit AES connections for the data transfer

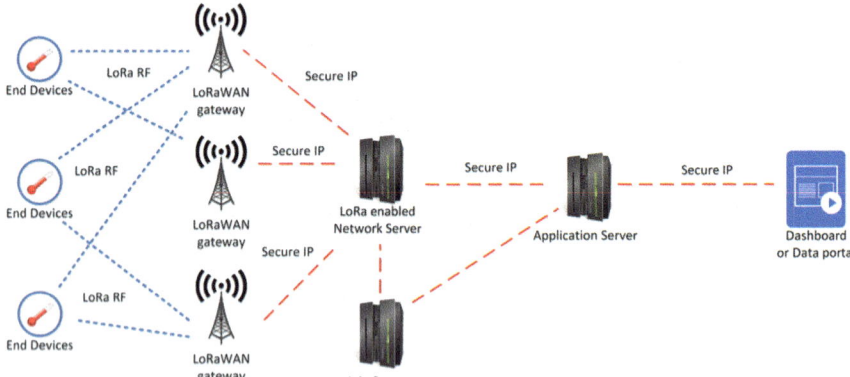

Fig. 2 Basic architecture of LoRaWAN network

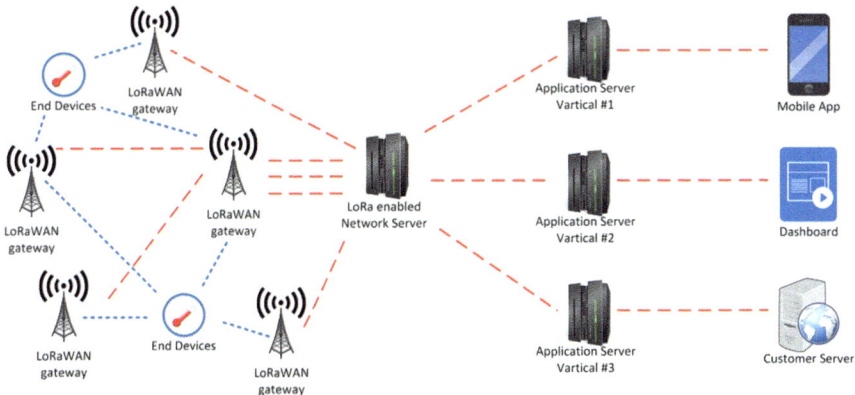

Fig. 3 Complex architecture of LoRaWAN network

between end devices and user applications and to control traffic between end device and Network server. It ensures the authenticity of sensor node on the network and the integrity of messages [6].

Two remaining elements of the LoRaWAN network are the Application Server and the Join Servers. The Application Server is responsible for secure handling, managing and interpreting sensor data. The Join Server manage the process of adding end devices to the network. It contains the information required to process join-request frames and generate join-accept frames. It is responsible for signaling to the network server which application server should be connected to the end-device [6]. The complex architecture of LoRaWAN network is presented in Fig. 3.

3 Network Simulation Tools

There is a large quantity of simulation tools designed for computer and communication networks and related technologies. In this section some of the most popular network simulation tools will be presented. Very interesting overview of network simulators suitable for LoRaWAN can be found in [7].

3.1 Network Simulator NS-2

NS-2 (Network Simulator) is a discrete event simulator targeted at networking research and provides support for simulation of TCP, routing, and multicast protocols over wired and wireless networks. The development of this tool began in 1989 as a variant of the REAL network simulator. Currently ns development is supported through DARPA with SAMAN and through NSF with CONSER, both in collaboration with other researchers including ACIRI. Ns has contributions from other researchers as well, including wireless code from the UCB Daedelus and CMU Monarch projects and Sun Microsystems [8].

3.2 Network Simulator NS-3

NS-3 is another discrete-event network simulator designed for Internet systems, targeted primarily for research and educational use. It is free software, licensed under the GNU GPLv2 license, and is publicly available for research and development [9]. Reynders et al. [10] give example of using ns-3 for LoRAWAN. Authors propose a new long range, low power module for ns-3. Proposed model is compliant with the LoRaWAN v1.0 class A specification. It is designed to study the effects of different parameters on the network performance and therefore is highly configurable. Its flexible architecture allows easy integration of new protocols and simulations—e.g., to study what will happen if a backbone server is congested. One of main contributions in the model is that it supports distributed gateways.

In [11] another example of simulation of LoRa network with a new ns-3 module is presented. This example is designed to study the performance of a LoRa-based IoT network in a typical urban scenario.

3.3 OMNET++

OMNeT++ is an extensible, modular, component-based C++ simulation library and framework designed for building network simulations. Simulator supports wired and wireless communication networks, on-chip networks, queuing networks, etc. By model frameworks domain-specific functionalities are provided. Those functionalities are support for sensor networks, wireless ad hoc networks, Internet protocols, performance modeling, photonic networks, etc. OMNeT++ has an Eclipse-based IDE and a graphical runtime environment. The tool is distributed under the Academic Public License.

FLoRa (Framework for LoRa) is an open-source simulation framework for end-to-end LoRa simulations. It is based on the OMNeT ++ network simulator and uses components from the INET framework as well. FLoRa allows the creation of LoRa networks with modules for LoRa nodes, gateways and network servers [12].

The LOCATE platform is presented in [13]. It allows users to broadcast geo-tagged help requests on the LoRa link. Within this solution a wearable LOCATE prototype has been developed, including the mobile application and the IoT device, composed of LoRa and BLE modules. The performance of the LOCATE platform is evaluated with OMNeT++ tool, by using the FLoRA framework 3 to model the

LoRa operations at the PHY layer. To this purpose, authors used the PDR and Path Loss model derived from the measurements. Another example of using OMNeT++ and FLoRA framework for LoRaWAN simulation is given in [13].

3.4 Riverbed Simulator

Riverbed Modeler [14] represents a packet level, event-based simulation model. It provides a virtual environment for modeling, analyzing, and performance estimation of IT infrastructures, applications, servers, and networking technologies. It is successor of OPNET IT Guru and Riverbed Modeler product, There is Academic Edition of this software designed to enable lab exercises for teaching fundamental networking concepts. Riverbed software is used in commercial and government organizations worldwide and by more than 500 universities.

Authors of [15] are motivated with the fast increase of the number of LoRaWAN end devices deployed in field and therefore for the need of understanding the limits of LoRaWAN scalability. In their research they approach this problem by proposing a new, powerful LoRaWAN network simulation tool compatible with version 1.1 specification, which can be applied to verify technology improvements and to discover possible solutions to optimize the LoRaWAN network performance. Riverbed Modeler was selected to be applied since the authors have successfully used it during their previous research and it appeared to be an extensive and accurate simulation tool.

3.5 SimPy/Python

SimPy [16] is a process-based discrete-event simulation framework based on standard Python. Processes in SimPy are defined by Python generator functions and can be used to model active components (customers, vehicles or agents). SimPy also provides various types of shared resources for modeling limited capacity congestion points (servers, checkout counters and tunnels).

LoRaSim is a discrete-event simulator based on SimPy for simulating collisions in LoRa networks and to analyze scalability. Bor et al. [17] use their practical experimentation results to build the simulator LoRaSim. This simulation tool captures specific LoRa link behavior and enables evaluation of large-scale LoRa networks. In the paper [18] authors investigate the use of directional antennae and the use of multiple base stations as methods of dealing with internetwork interference. The effectiveness of both approaches is compared with LoRaSim and the findings show that both methods are able to improve LoRa network performance in interference settings. Another example of simulating LoRaWAN v1.0 in Python is [19] and it's currently being tested with a RFM95 LoRa modules attached to a Raspberry Pi single board computer.

4 Virtual Network Laboratories

Virtual networks can be described as a virtual computer networking environments based on the virtualization technology [20, 21]. The virtualization technology enables emulation of virtual computers on a single physical one. Virtual computers or virtual machines (VM) emulate the complete hardware of the physical machines including network interface cards. The emulation of several network cards on multiple virtual machines enables virtual networks creation. Nowadays, the hardware and the newest virtualization software are powerful enough to enable the emulation of large-scale computer networks on a single computer. The complex virtual networks can be established in a distributed environment too. The virtual networks provide the platform for network experiments, research, software development and testing. In a higher education virtual laboratories can be effective, adaptive and scalable alternative for the much more expensive physical network laboratories. In this Section some of the virtual networking labs will be presented.

4.1 *VNX and VNUML*

Virtual Network User Mode Linux (VNUML) is an open-source tool for virtualization designed to simulate computer networks [22, 23]. It was initially developed as a part of the Euro6IX1 research project by the Telematics Engineering Department (DIT) of the Technical University of Madrid (UPM) in Spain, in order to simulate the next generation of IPv6-based scenarios. VNUML is based on User- Mode Linux (UML), a modification of the standard Linux kernel, and Zebra routing daemon.

VNX replaces VNUML [24, 25] as its successor. VNX is a general purpose open-source virtualization for building virtual network testbeds. It allows the definition and automatic deployment of network scenarios made of virtual machines of different types of Linux, Windows, FreeBSD, Olive or Dynamips routers, etc. The tool can be used for testing network applications/services for complex network laboratories and to allow students to interact with realistic network scenarios. VNX is composed of two main parts: an XML language that allows description of the virtual network scenario (VNX specification language) and the VNX program. VNX supports two different virtualization techniques and uses an XML-style scripting language to define the virtual network.

4.2 *Kathará and Netkit*

Netkit is an emulator that enables experiments in virtual computer networking, including emulation of hardware devices and virtual links that interconnects these devices. The Netkit was developed as open source tool in 1999, in Rome Tre University, Italy. The tool was used to provide practical experience to students on a single computer. Virtual machine with Debian GNU/Linux OS and networking software enabled those virtual machines to act as network devices (routers and switches). A total of 11 virtual laboratories were developed to provide practical experimentation at computer network disciplines in ICMC/USP (Institute of Mathematics and Computers Sciences, University of São Paulo) [26].

Kathará is successor of Netkit and it is container-based framework to deploy virtual networks featuring NFV (Network Functions Virtualization), SDN (Software-defined networking) and traditional routing protocols such as BGP (Border Gateway Protocol) and OSPF (Open Shortest Path First). It is implementation of Netkit with usage Python and Docker technologies. With these novel technologies this tool becomes faster and lighter solution. Kathará comes with P4, OpenVSwitch, Quagga, Bind, FRRouting and other tools [27].

4.3 IMUNES

IMUNES (Integrated Multiprotocol Network Emulator/Simulator) is a lightweight kernel level network emulator with three main tools of the standard FreeBSD kernel. Those three tools are: FreeBSD jails—a lightweight virtualization solution that enables different jails to share system resources; Netgraph kernel modules—used for emulating network interfaces and linking virtual nodes in the simulation and implementation of lower layer (data link layer) and network equipment such as hubs and switches and ZFS file system—transactional file system developed by Sun Microsystems [28, 29]. IMUNES is used in networked and distributed systems research.

4.4 Remotely Accessible Networking Laboratories

Combination of virtual machines and remotely accessible platforms or cloud systems gives the new perspective for the virtual laboratories. One of the examples is VNLab, a remotely accessible virtual network laboratory based on Microsoft Virtual Server 2005 and designed for teaching [30, 31]. In these related researches is presented the model of establishing the computer networking and computer security classes and of organization of student work with application of these systems.

Gercek et al. [32] presents experiences gained from transforming a traditional computer lab into an online virtual domain at the at University of Houston-Clear Lake (UHCL) MIS Department. The research highlighted technical and operational challenges in building and operating an online virtual lab. The network labs described in this paper are utilized mostly for graduate computer networking, network security and wireless networking courses. The paper explores how advances in virtualization and cloud computing technologies affect transitioning from traditional computing labs to online format.

5   Virtual Laboratory for LoRaWAN

In this section will be presented the virtual laboratory based on Microsoft Azure cloud system and virtual machines designed for the development and testing of LoRaWAN applications. The laboratory is built on virtual machines running Ubuntu 16.04 installation and the Java 10 SDK.

The classical LoRaWAN architecture scenario, based on the LoRaWAN architecture presented in Sect. 2, can be simulated in virtual laboratory with the topology illustrated in Fig. 4. The figure shows scenario with one Network Server, one Join server, one Application Server, three LoRa Gateways and six end devices. The Network Server and the Join Server are installed on the same virtual machine. To simulate each the devices, we developed simulation scripts implemented in the Java programming language. Specifically, two scripts are used for simulating the end device and the messages sent by them, one script is used for simulating the Network Server, and one script for simulating the LoRa Gateway.

The creation of a virtual laboratory was driven by the need for developing and testing novel approaches to LoRaWAN roaming scenarios. To this aim, we introduce the concept of Distribution Server (DS). The experimental scenario for development of roaming procedures with this novel approach is presented in Fig. 5. This scenario introduces three new nodes in the network, being all of them Distribution Servers. The script is developed for the purpose of simulating the Distribution Server as well. For the simulation of this scenario 15 virtual machines are needed.

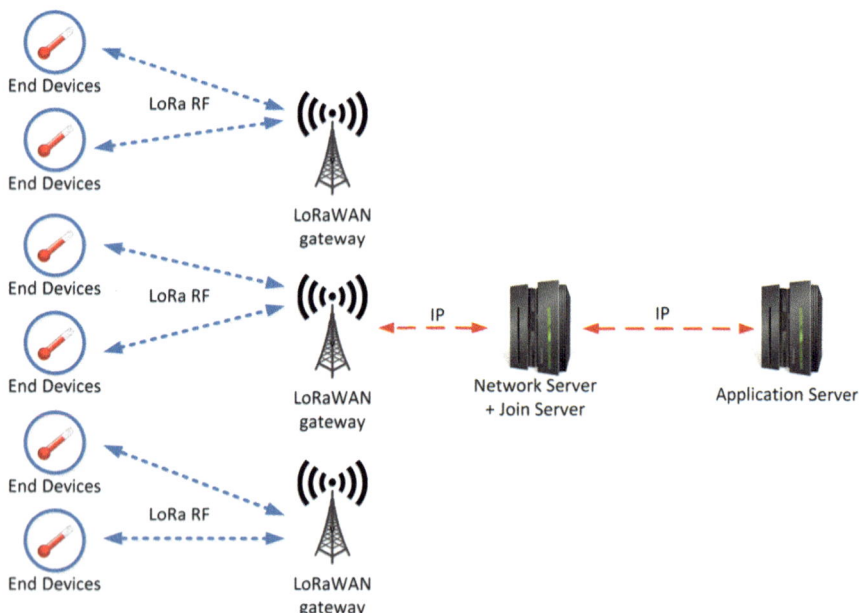

Fig. 4 LoRaWAN virtual laboratory with classical scenario

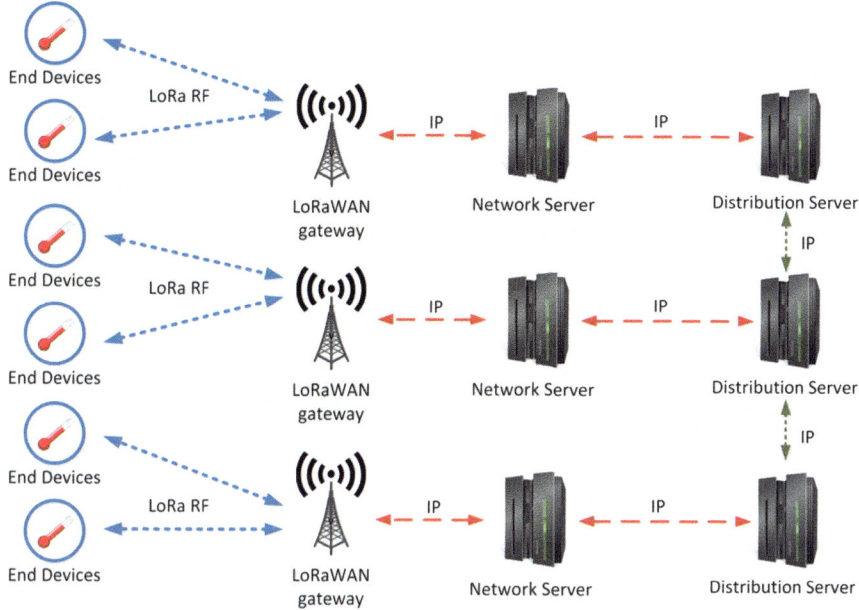

Fig. 5 LoRaWAN virtual laboratory components for experimental scenario

The assumption of flexibility and effectiveness of using virtual machines in testing roaming methodology in LoRaWAN networks was justified after running simulation scenarios. The virtual environment and the developed software showed their high potential in simulating the LoRaWAN networks at application layer. One of the most visible advantages of using virtual network and virtual network machines is easy network scaling and change of profiles of network elements, by multiplying, copying and altering virtual machines easily.

The roaming scenario is designed for situations when two different shipment companies allow roaming between their proprietary networks to enable shipment tracking. Scenario is depicted in Fig. 6, where a national level roaming process is described. As shown in Fig. 5, the scenario have six End Devices, three LoRaWAN Gateways, three Network Servers (with integrated Join and Application server) and three Distribution servers. End Devices, Gateways, Network Servers, and Distribution Servers are Java programs running on separate virtual machines. For a similar deployment, a total of 15 virtual machines should be used. If the number of virtual machines needs to be decreased, some network elements such as gateways, network servers and distribution servers, can be grouped into a single virtual machine.

The proposed platform can be additionally used for LoRaWAN software development and testing in scenarios such as smart railways [33] and intelligent shipment tracking [34].

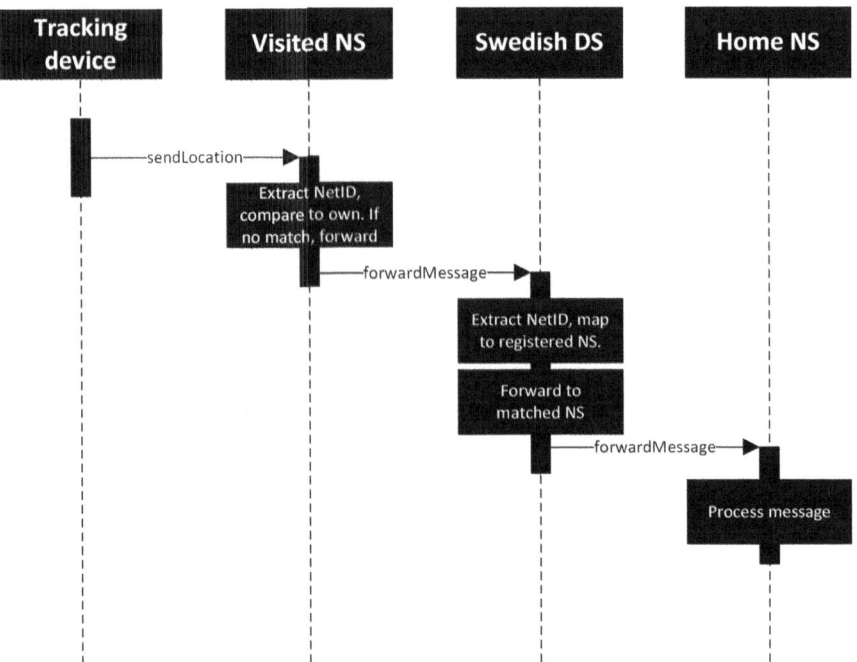

Fig. 6 Sequence diagram of national roaming

6   Conclusion

LoRa and LoRaWAN technologies are widely accepted in the industry. Besides their high performances, their applicability in many fields increases their perspective for the future. The projection for the technology perspective gives researches the motivation for experimentation and development in the field. Considering this wide acceptance in the research community, the simulation tools applicable for LoRa/LoRaWAN simulations become very important. Besides the well known tools, the virtualization technology, and virtual network laboratories are explored in this chapter as possible tool for simulating LoRaWAN networks.

Considering the fact that IoT becomes an enabling technology for big data generation, analytics and machine learning, this chapter gives the example of utilization of virtualization technology, open platforms, and cloud computing systems in testing and development roaming software and scenarios in LoRaWAN networks. The goal of the platform is to support, optimize and speed up LoRaWAN application development. The technological details about the environment, composition of the lab scenarios and the related experience are given in this chapter. In general, the experience with usage of the lab is positive, giving the strong basis for the future utilization of the laboratory and further experience in developing the roaming software and methods.

Future work can continue in several directions. To establish more effective and easy-to-use virtual environment for simulating LoRaWAN networks some of the research efforts should be focused on the right choice of the operating system platform for the virtual machines. Other efforts should be directed to the improvement of integration of multiple virtual machines to one scenario. The integration can be improved with the selection of efficient, configurable and user-friendly integrated platforms for managing and set-ups of LoRaWAN simulation scenarios. The integrated platform can be designed from scratch or can be used as one of existing open-source and freeware software solutions. After the establishment of the integrated platform, the extensive testing and validation of the proposed platform should be conducted.

This virtual platform is primarily designed for the development and testing of the application for roaming in LoRaWAN networks. Its primary purpose is not limited to developers and engineers. On the contrary, the platform can be equally used in the education processes not only for learning LoRAWAN application development, but also for the development and testing of LoRaWAN software in industry. At this stage of research, the application of the platform to educational processes is not extensively addressed; however, it will be considered in future work.

The biggest limitation we are aware of is the integration of the LoRaWAN physical layer in the proposed platform, as the platform is currently focused on LoRaWAN upper layers; one possible future research direction is to address such an issue.

References

1. Markkula, J., Mikhaylov, K., Haapola, J. (2019). Simulating LoRaWAN: On importance of inter spreading factor interference and collision effect. In *Proceedings of. ICC 2019-2019 IEEE International Conference on Communications (ICC)*, Shanghai, China (pp. 1–7). https://doi.org/10.1109/icc.2019.8761055.
2. IoT Analytics GmbH. (2018). LPWAN emerging as fastest growing IoT communication technology–1.1 billion IoT connections expected by 2023, LoRa and NB-IoT the current market leaders, September 27, 2018, Retrieved May 2020, from https://iot-analytics.com/lpwan-market-report-2018-2023-new-report/.
3. Weissberger, A. (2017). LoRaWAN and Sigfox lead LPWANs; interoperability via compression. Retrieved May 2020, from https://techblog.comsoc.org/2017/10/25/lora-wan-and-sigfox-lead-lpwans-interoperability-via-compression/.
4. About ABI Research. (2018). NB-IoT, CAT-M, SIGFOX and LoRa battle for dominance drives global LPWA network connections to Pass 1 Billion By 2023, Oyster Bay, New York-11 Jun 2018. Retrieved May 2020, from, https://www.abiresearch.com/press/nb-iot-cat-m-sigfox-and-lora-battle-dominance-drives-global-lpwa-network-connections-pass-1-billion-2023/.
5. LoRaWAN® Specification v1.1 (2017). LoRa Alliance Inc. Retrieved May 2020, from, https://lora-alliance.org/resource-hub/lorawanr-specification-v11.
6. LoRa® and LoRaWAN®: A Technical Overview, December 2019, Semtech Corporation. Retrieved May 2020, from https://lora-developers.semtech.com/library/tech-papers-and-guides/lora-and-lorawan/.
7. Marais, J. M., Abu-Mahfouz, A. M., & Hancke, G. P. (2019). A review of LoRaWAN simulators: design requirements and limitations. In *Proceedings of International Multidisciplinary*

Information Technology and Engineering Conference (IMITEC). Vanderbijlpark, South Africa, (pp. 1–6). https://doi.org/10.1109/imitec45504.2019.9015882.

8. The Network Simulator-ns-2. Retrieved May 2020, from https://www.isi.edu/nsnam/ns/.

9. The Network Simulator-ns-3. Retrieved May 2020, from https://www.nsnam.org/.

10. Reynders, B., Wang, Q., & Pollin, S. (2018). A LoRaWAN module for ns-3: implementation and evaluation. In *Proceedings of the 2018Workshop on ns-3 (WNS3 2018)*, (June 2018), Surathkal, India. ACM, New York, NY, USA (pp. 61–68). https://doi.org/10.1145/3199902.3199913.

11. Magrin, D., Centenaro, M., & Vangelista, L. (2017). Performance evaluation of LoRa networks in a smart city scenario. In *Proceedings of 2017 IEEE International Conference on Communications (ICC), Paris* (pp. 1–7). https://doi.org/10.1109/icc.2017.7996384.

12. Slabicki, M., Premsankar, G., & Di Francesco, M. (2018). Adaptive configuration of lora networks for dense IoT deployments, In *Proceedings of NOMS 2018-2018 IEEE/IFIP Network Operations and Management Symposium, Taipei* (pp. 1–9). https://doi.org/10.1109/noms.2018.8406255.

13. Sciullo, L., Trotta, A., & Di Felice, M. (2020). Design and performance evaluation of a LoRa-based mobile emergency management system (LOCATE). *Ad Hoc Networks, 96*(2020), https://doi.org/10.1016/j.adhoc.2019.101993.

14. Riverbed Technology, Riverbed Modeler. Retrieved Oct 2020, from https://www.riverbed.com/gb/products/steelcentral/steelcentral-riverbed-modeler.html.

15. Markkula, J., Mikhaylov, K., & Haapola, J. (2019). Simulating LoRaWAN: On importance of inter spreading factor interference and collision effect. In *Proceedings of ICC 2019-2019 IEEE International Conference on Communications (ICC)*, Shanghai, China (pp. 1–7). https://doi.org/10.1109/icc.2019.8761055.

16. SimPy, Team SimPy. Retrieved May 2020, from https://simpy.readthedocs.io/en/latest/.

17. Bor, M., Utz, R., Voigt, T., Alonso, J. (2016). Do LoRa low-power wide-area networks scale? In *Proceedings of the 19th ACM International Conference on Modeling, Analysis and Simulation of Wireless and Mobile Systems*, Nov 59–67. https://doi.org/10.1145/2988287.2989163.

18. Voigt, T., Bor, M., Roedig, U., & Alonso, J. (2017). Mitigating inter-network interference in LoRa networks. In *Proceedings of the International Conference on Embedded Wireless Systems and Networks, EWSN'17*, (February 2017), (pp. 323–328).

19. Jeroen Nijhof. Retrieved May 2020, from https://github.com/jeroennijhof/LoRaWAN.

20. Gordon, A. D. (2016). V for Virtual. *Electronic Notes in Theoretical Computer Science, 162*(2006), 177–181. https://doi.org/10.1016/j.entcs.2006.01.030.

21. Dobrilovic, D., Stojanov, Z., Odadzic, B., & Markoski, B. (2012). Using network node description language for modeling networking scenarios. *Advances in Engineering Software, 43*(1), 53–64. https://doi.org/10.1016/j.advengsoft.2011.08.004.

22. Virtual Network User-Mode-Linux (VNUML) web server. Retrieved Oct 2020, from https://www.dit.upm.es/vnumlwiki/index.php/Main_Page.

23. Virtual Networks over linuX (VNX) web site. Retrieved Oct 2020, from https://www.dit.upm.es/~vnx/.

24. Fernández, D., Ruiz, F. J., Bellido, L., Pastor, E., Walid, O., & Mateos, V. (2016). Enhancing learning experience in computer networking through a virtualization-based laboratory model. *International Journal of Engineering Education, 32*(6), 2569–2584.

25. Barona, L. I., Valdivieso, A. L., & García, L. J. (2015). Extending openflow in virtual networks. In *Proceedings of the 7th International Conference on Information Technology ICIT 2015* (pp. 252–258). https://doi.org/10.15849/icit.2015.0037.

26. Gurgel, P. H. M., Branco, Luiz H. C., Barbosa, E. F., & Branco, Kalinka R. L. J. C. (2013). Development of a practical computer network course through Netkit virtualization tool. *Procedia Computer Science, 18*, 2583–2586. https://doi.org/10.1016/j.procs.2013.05.445.

27. Kathara Framework. Retrieved May 2020, from https://github.com/KatharaFramework/Kathara/wiki.

28. Salopek, D., Vasic, V., Zec, M. Mikuc, M., Vasarevic, M., & Koncar, V. (2014). A network testbed for commercial telecommunications product testing. In *Proceedings of the 22th International Conference on Software, Telecommunications and Computer Networks Softcom 2014*, Split, (September 2014) (pp. 372–377). https://doi.org/10.1109/softcom.2014.7039061.

29. Vasic, V., Suznjevic, M., Mikuc, M., & Matijasevic, M. (2012). Improving distributed traffic generation performance by using IMUNES network emulator. In *Proceedings of the Softcom 2012 20th International Conference on Software, Telecommunications and Computer Networks*, Split, (September 2012) (pp. 1–5).
30. Dobrilovic, D., Stojanov, Z., Odadzic, B., & Sinik, V. (2015). Platform for teaching communication systems based on open-source hardware. In *Proceedings of IEEE Global Engineering Education Conference (EDUCON 2015)*, Tallin, Estonia (pp. 737–741). https://doi.org/10.1109/educon.2015.7096051.
31. Dobrilovic, D., Jevtic, V., & Odadzic, B. (2013). Expanding usability of Virtual Network Laboratory in IT Engineering Education. *International Journal of Online and Biomedical Engineering (iJOE), 9*(1), 26–32. http://dx.doi.org/10.3991/ijoe.v9i1.2388.
32. Gercek, G., Saleem, N., & Steel, D. (2016). Implementing cloud based virtual computer network labs for online education: Experiences from a phased approach. *International Journal of Online and Biomedical Engineering (iJOE), 12*(03), 70–76. https://doi.org/10.3991/ijoe.v12i03.5564.
33. Dirnfeld, R., Flammini, F., Marrone, S., Nardone, R., & Vittorini, V. (2020). Low-power wide-area networks in intelligent transportation: Review and opportunities for smart-railways. In *Proceedings of 23rd IEEE International Conference on Intelligent Transportation Systems (ITSC)*, 20–23 Sept 2020.
34. Flammini, F., Gaglione, A., Tokody, D., Dobrilovic, D. (2020) LoRaWAN roaming for intelligent shipment tracking. In *Proceeding of 4th IEEE Global Conference on Artificial Intelligence and Internet of Things-GCAIoT 2020 (in press)*, 12–16 Dec, Virtual Conference, 2020.

Computational Intelligence for Automatic Object Recognition for Vision Systems

Belhedi Wiem, Hireche Chabha, and Kammoun Ahmed

Abstract Computer vision combines cameras, edge or cloud computing, software and artificial intelligence (AI) to enable systems to automatically recognise and identify objects. In this context, this paper details machine intelligence and data analysis algorithms that can be used for real-time object recognition. This task is essential to design an automatic vision system that can be integrated in multiple applications in future sustainable smart cities, for example, self-driving cars, robots for home help/assistance or security. Therefore, this paper is firstly a bibliographical study of algorithms for real time object recognition, and secondly a comparison of these different approaches. The comparison will consist in highlighting the specificities of each algorithm and the common points or similarities that may exist between them. Also, the results of a concrete comparison in terms of time and recognition rate were also reported. The algorithms that were included in this study are: Convolutional Neural Network (CNN), Region-based CNN (R-CNN), Fast R-CNN, Faster R-CNN for region-based algorithms, and YOLO, Tiny-YOLO, Nano-YOLO, Mini-YOLO, Slim-YOLO, MobileNet, Single Shot Multibox Detector (SSD), and RetinaNet for non-region-based algorithms. Furthermore, this review addresses in its different sections the issue of implementing AI algorithms, whether in software or hardware architectures, or in a co-design approach which is considered an appropriate method capable of simultaneously taking advantage of software and hardware advantages.

B. Wiem (✉) · H. Chabha · K. Ahmed
Department of research and innovation, Altran Technologies, Rennes, France
e-mail: wiem.belhedi@altran.com

H. Chabha
e-mail: chabha.hireche@altran.com

K. Ahmed
e-mail: ahmed.kammoun@altran.com

1   Computer Vision Applications for Future Smart Cities

Image recognition, which is a sub-category of Computer Vision and Artificial Intelligence, is a set of methods used to detect and analyze images in order to automate a specific task. It is a technology that is capable of identifying places, people, objects and several other types of elements within an image and drawing conclusions from them through analysis. Technically, image recognition consists of analyzing each pixel of an image in order to extract information, the same way the human eye does [1].

In future smart cities, several applications require computer vision such as object detection and recognition for self driving cars. In fact these cars can be driven on their own. One of the main capabilities of autonomous cars is to detect pedestrians, cars, trucks, traffic signs, etc. These detections are essential for their correct functioning of autonomous cars [2]. Another application that requires computer vision algorithms is automation of robotic processes that can be used for domestic assistance or even in smart industry. In this latter, one of the most common applications is object picking and sorting. Through object detection techniques, the robot can understand the location of objects, and then select the object and sort it using that information [3].

Computer vision algorithms are also necessary for the security of smart cities, for example for fraud detection with regard to banking security. In fact, in smart cities, banking terminals should be equipped with computer vision capabilities allowing face recognition. This, in fact, provides an additional layer of authentication and could identify when an unauthorized user attempts to access an account [4].

One of the main goals in creating a smart city is to improve the quality of life. Since health is an important factor in this regard, a smart health system must be an integral part of the smart city. With increased connectivity and the availability of sophisticated tools and sensors, smart cities have the potential to provide health care services that can truly meet the needs of citizens [5].

Smart health extends the concept of health care to cover not only hospitals and homes, but the entire city. For example, the patient, when leaving home, gets information about temperature, weather, traffic jams, safe routes, etc. This will enable effective disease prevention and management. The system will also provide advice to patients and notifications to health practitioners. Patients can be guided to the nearest health care provider in case of minor problems or, in case of emergency, ambulances can be guided to the patient's location. Sensors can continuously monitor factors such as blood glucose, blood pressure and heart rate, as well as data such as location and activity that would allow the system to identify critical events.

A recent research has been done in this context using computer vision, a facial expression recognition system that can be used in intelligent health care. With this system, doctors and caregivers can continuously monitor patients' feelings and facial expressions from distance and take appropriate action if necessary. The system can also give caregivers automatic feedback from patients without having to ask for their input [6].

This paper is interested in reviewing the state of the art on Computer vision algorithms for object recognition and detection. It presents an overview of these different and consistent algorithms in order to facilitate their evaluation and compare their characteristics. The algorithms studied in this paper are divided in mainly two categories: Region-based and non-region-based algorithms. The first category consists of convolutional neural network (CNN) and its respective extensions such as RCNN, Fast RCNN and Faster RCNN while the second category unites YOLO, MobileNet, Single Shot Multibox Detector (SSD), and RetinaNet. In addition, several YOLO extensions are also included in this study that are Tiny YOLO, Nano YOLO, Mini YOLO, and Slim YOLO. The sections below give some details on each algorithm regarding methodology, characteristics and the added value it brings to the state of the art. Almost all algorithms in each category are cited in chronological order to help the lecturer recognize the new features of each algorithm compared to its predecessors. This makes comparison easier and can help when selecting the one that meets your requirements. Several application fields such as smart cities, smart vehicles and security are subject to such algorithms and this is exactly the context of this review. Indeed, it gives an assessment of the state of the art and provides some comparisons to help making the appropriate choice for the desired application. For example, for our future work on smart vehicles such as ADAS and UAVs, hardware-friendly algorithms (such as CNN and YOLO) are more suitable for co-design applications where hardware resources and software flexibility as well as HW-SW distributions are all required. Within this context, Sect. 4 also presents a review of existing work on FPGA-based architectures fo AI algorithms.

2 Region-Based Artificial Intelligence Algorithms for Object Recognition/Detection

This section is dedicated to report and compare the Region-based artificial intelligence algorithms for object recognition and detection algorithms. These algorithms are mainly: convolutional neural network (CNN), RCNN, Fast RCNN, Faster RCNN.

2.1 Convolutional Neural Network

The main steps of a CNN are described as follows: the convolutional layer is responsible for extracting the local features in the image, a pooling layer is used to significantly reduce the parameter magnitude (dimension reduction), a fully connected (FC) layer is used to output the desired result. The architecture of the convolutional neural network (CNN) is illustrated in Fig. 1 [7, 8].

The first step of CNN is the convolution layer, also known as feature extraction layer [9].

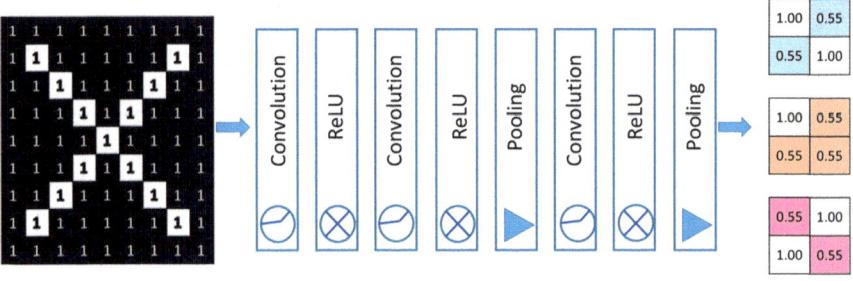

Fig. 1 CNN architecture

After processing an image through a convolutional layer, the output normally goes through an activation function. Common activation functions include the sigmoid function (Eq. 1) and the ReLU function (Eq. 2).

$$f_{Sigmoid}(x) = \frac{1}{1 + e^{-x}} \tag{1}$$

$$f_{ReLU}(x) = max(0, x) \tag{2}$$

Once the activation function is applied, the pooling layer is set up. Pooling consists of downsampling, which consists of reducing data dimensionality in order to avoid overfitting. This step can significantly reduce data dimensions. The reason is that even after convolution, the image remains large (because the convolution kernel is small), so to reduce data dimensions, downsampling is performed. The pooling layer can reduce data dimension more efficiently than the convolutional layer. This can not only greatly reduce the amount of computation, but also effectively avoid overfitting [10].

The fully connected layer is the last step before the loss layer. The data processed by the convolutional layer and the pooling layer is passed to the fully connected layer to achieve the desired end result. After having the data reduced by the convolutional layer and the pooling layer, the fully connected layer can take place, otherwise the data volume is too large, resulting in a high cost calculation and therefore low efficiency. The flattened output is transmitted to a feed-forward neural network and back-propagation applied at each iteration of the training. Over a series of epochs, the model is able to distinguish dominant and certain low-level features in images and classify them using the Softmax classification Technique [11].

Finally, the loss layer specifies how the training penalizes the gap between the predicted (output) and actual labels and is normally the last layer of a neural network. Various loss functions might be used, such as cross-entropy, which is a measure of network accuracy. When the network is initialized with random values, the loss function will be high, and the aim of network training is to reduce the loss function as low as possible [12].

2.2   Region-Based CNN

The Region-based CNN (RCNN) was first proposed in [13]. As illustrated in Fig. 2, the RCNN-based object detection process includes three models which are: CNN for feature extraction, the Support Vector Machine (SVM) classifier for object identification, and the regression model for narrowing the enclosing boxes.

The RCNN-based object detection process is summarized as follows:

- First, a pre-formed CNN is envisaged.
- Then the last network layer is trained according to the number of classes to be detected.
- The third step consists in obtaining the regions of interest (ROI) for each image. Then, all these regions are reshaped to match the CNN input size.
- After obtaining the regions, a Support Vector Machine (SVM) is trained in order to classify the objects and the background. A binary SVM is formed for each class.
- Finally, a linear regression model is formed in order to generate tighter bounding boxes for each identified object.

Despite the efficiency of this architecture, the combination of these models makes the RCNN very expensive in terms of computation. It takes about 40–50 s to make predictions for each new image, making the model impossible to use to build against for large datasets.

2.3   Fast RCNN

In order to reduce the computation time of a RCNN algorithm, the main idea is to run one RCNN model per image and get all RoI, i.e. regions containing an object,

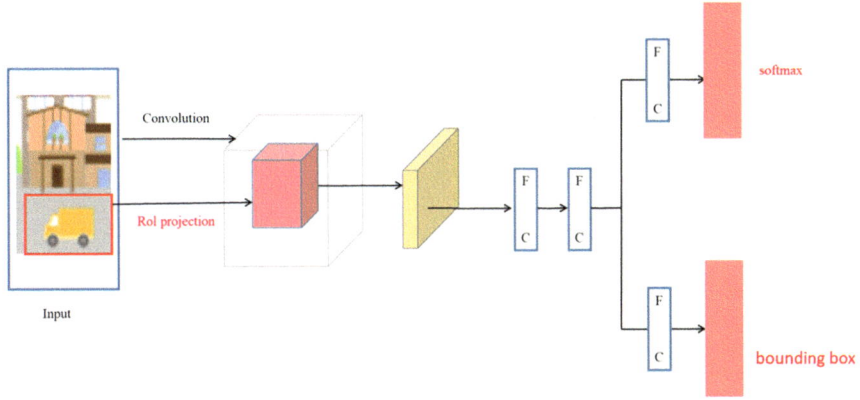

Fig. 2 RCNN architecture [13]

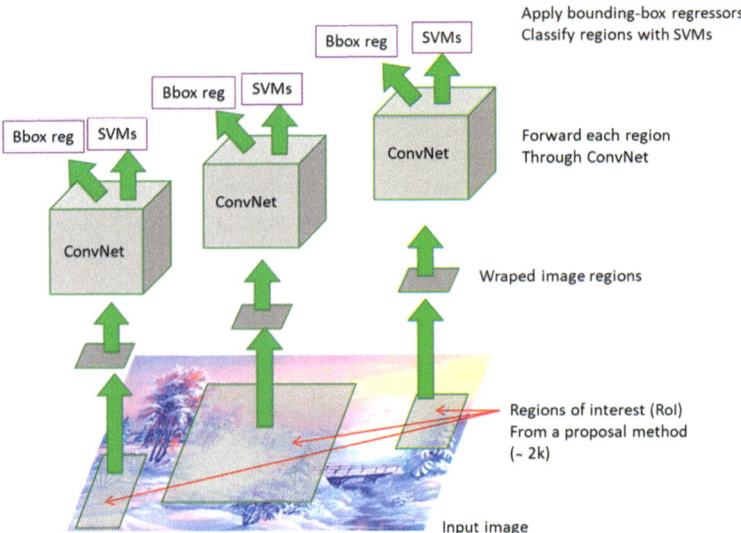

Fig. 3 Fast RCNN architecture [14]

instead of running a CNN 2,000 times per image. Hence, in a fast RCNN, the input image is fed to a CNN network which, in turn, generates the convolutional feature maps. These maps are used to extract regions from the proposals. We then used a RoI pooling layer to reshape all proposed regions to a fixed size so that they can be fed into a fully connected network [14].

The architecture of the Fast RCNN is shown in Fig. 3:

- ConvNet generates the ROI for each image.
- An RoI grouping layer is applied on all these regions to reshape them according to the input of the ConvNet. Each region is then transmitted to a FC network.
- A softmax layer is used above the FC network to output the classes. With the softmax layer [11], a linear regression layer is also used in parallel to output the bounding box coordinates for the predicted classes.

Thus, instead of using three different models (as in RCNN), Fast RCNN uses a single model that extracts the entities from the regions, divides them into different classes and simultaneously returns the bounding boxes of the identified classes [15]. As a result, ROIs on the same image share memory and computational effort in both forward and backward passes. This results in a reduction of the computation time of Fast RCNN compared to its predecessor (the RCNN).

2.4   Faster RCNN

Faster RCNN enhances Fast RCNN by using a Region Proposal Network (RPN) with the CNN model [16]. In fact, Faster RCNN takes image feature maps as input and generates a set of object proposals, each with an objectivity score as output. The optimizations performed in Faster RCNN make it efficient for real-time recognition tasks [16].

The main steps of Faster RCNN are as follows:

- ConvNet returns the feature map of the image.
- The region proposal network is applied to the feature maps which returns the object proposals together with their objectivity score.
- A RoI grouping layer is applied on these proposals for reshaping task.
- Finally, the proposals are passed to a FC layer, softmax layer, and linear regression layer in order to classify and output the bounding boxes of the objects.

2.5   Comparison

A comparison of the CNN versions is given in Table 1. For each algorithm, the specifications, computational time and limitations are reported.

CNN requires a lot of RoI to predict accurately and makes the algorithm computationally expensive. RCNN, however, uses selective search to generate regions of interest from each image and transmits each region to the CNN separately. In addition, RCNN uses three different models to make predictions. This results in high computation time for the algorithm. In order to solve this problem, Fast RCNN transmits each image only once to the CNN and then extracts the feature maps. Selective search is used on these maps to generate predictions. Hence, Fast RCNN combines

Table 1 Different CNN versions together with their respective Computational times, and Limitations

Algorithm	Computational time (s)	Limitations
CNN	300	Requires a lot of regions to predict accurately and therefore a high computationally expensive
RCNN	50	Since each region is transmitted to the CNN separately, it also uses three different models to make predictions
Fast RCNN	2	The selective search is computationally expensive hence the computing time is still high
Faster RCNN	0.2	The different systems work one after the other. Hence the performance of each system depends on the performance of the previous one

the three models used in RCNN. Despite the improvements made to reduce computation time, the latter is still highly consuming due to the selective search. As a solution, Faster RCNN replaces the selective search method with region proposals network, which makes the algorithm much faster. Thus, Faster RCNN can be employed for computer vision tasks under real-time constraints [17].

3 Non Region-Based Artificial Intelligence Algorithm for Object Recognition/Detection

This section is devoted to the presentation of Non-Region-based artificial intelligence algorithms for real-time object recognition, namely You Only Look Once (YOLO), MobileNet, Single Shot Multibox Detector (SSD) and RetinaNet.

3.1 YOLO

There are different methods to perform object and human detection using a neural network. In this part of the section, we present the YOLO (You Only Look Once) method.

As mentioned previously in Sect. 2, the R-CNN method uses region proposal methods to generate potential bounding boxes of an image prior to initiating the classification process. Then, post-processing is used to refine the bounding boxes, to rescore the boxes based on other objects in the image, and to suppress duplicate detection. In the case of YOLO, it formulates the object detection as a regression problem where the detector runs from the image pixels to the coordinates of the bounding boxes and the class probabilities of the objects. Thus, with YOLO, the system looks once at an image to predict which objects and where they are present in the image. In addition, YOLO has the ability to learn only global representations of objects. Thus, it can detect the object regardless of its position in the image and regardless of its total or partial visibility.

The main difference between YOLO and RCNN (or RCNN variants), which is the best state-of-the-art detection method in terms of accuracy, is that YOLO can encode contextual information because it processes the complete image during the training phase compared to RCNN or fast RCNN. As a result, YOLO makes fewer background errors than RCNN or Fast RCNN.

For greater efficiency in the detection process, YOLO is combined with CNN and takes advantage of CNN's convolution layer to simultaneously predict several bounding boxes and class probabilities for these boxes. This combination allows YOLO to perform the training step for the entire image and then optimize the detection performance. Thus, YOLO algorithm aims at predicting a class of an object and the bounding box specifying the location of the object. To do so, each bounding box can be described by the following parameters:

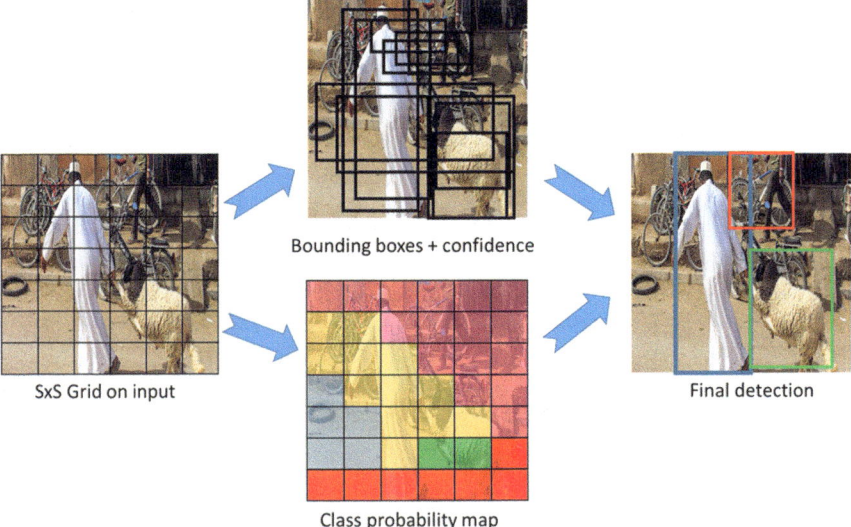

Fig. 4 Illustration of YOLO process

- The center of the bounding box (bxby).
- The width (bw).
- The height (bh).
- The class of the object denoted by c.

In addition to these parameters, YOLO computes the probability Pc, of an object being present in the bounding box. Figure 4 illustrates the process of the YOLO algorithm. It consists, first, in dividing the input image into SxS grid cell, then calculating both the bounding boxes and the confidence factor indicating the presence of an object in these boxes. Second, it computes the class probability map which returns the probability value of belonging to a object class. Finally, the object detection is given with the accuracy rate.

However, YOLO also has some limitations because it imposes strong constraints on bounding box predictions. Indeed, each box can predict only one object class. In addition, the algorithm cannot predict and detect small size objects in the image. To overcome these limitations, a new version has been released, called YOLOv3. This new version uses more convolution layers than the traditional YOLO version. This new structure allows to predict and detect multiple object classes in a grid cell of an image. The famous YOLOv3 algorithm is based on Darknet-53 network [18], it contains 53 layers and offers a balance between accuracy and detection speed. Despite this, YOLOv3 still requires high computation cost, FLOP and large memory footprint at runtime to maintain good detection performance, it brings a high level of computing overhead and power consumption of embedded devices [18, 19].

3.1.1 Mini and Tiny YOLOv3

A new version of YOLOv3 has been released in order to reduce resource consumption and memory footprint. The first optimized version is Mini-YOLOv3.The mini version of YOLOv3 is obtained by performing channel pruning on the convoluted layers of classic YOLOv3. Pruning is a kind of optimization method that is incremental and iterative, this optimization method follows the steps below:

- Evaluate the importance of each component in a pre-formed deep model.
- Remove components that are less important for inference phase.
- Fine tuning of the pruning model to compensate for a potentially temporary performance degradation.
- Evaluate the refined model to determine if the pruning model is suitable for deployment. Note that an incremental pruning strategy is preferable to avoid oversizing.

Figure 5 shows the structure of Mini-YOLOv3. Based on YOLOv3, a lightweight network structure without reduced detection accuracy is redesigned. A feature extraction backbone with a pruning parameter of only 16% darknet-53 is built. To compensate the degradation in accuracy, a multi-scale feature pyramid network based on a simple U-shaped structure is added to improve the performance of multi-scale object detection. The model is smaller and has fewer training parameters and FLOP

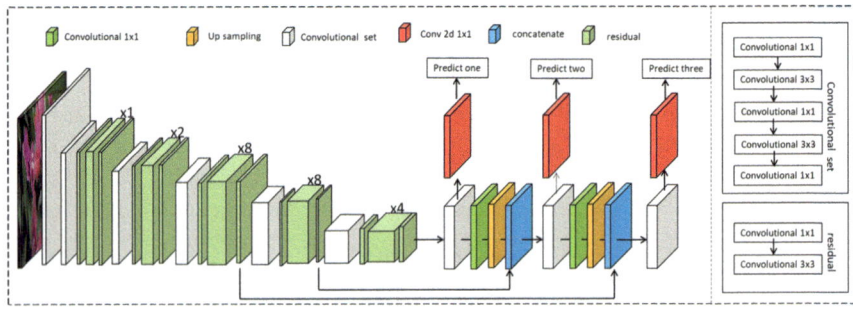

(a) Structure of YOLOv3

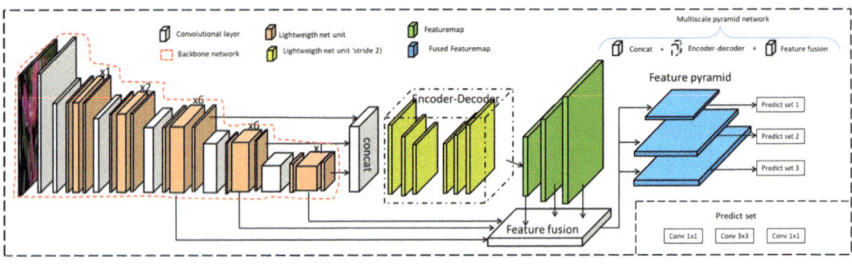

(b) Structure of Mini-YOLOv3

Fig. 5 Pruning YOLOv3 to Mini-YOLOV3 [20]

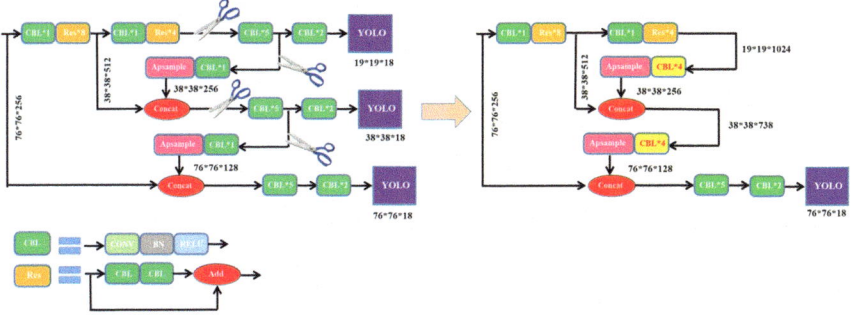

Fig. 6 Zoom on how the pyramid network feature, used in the Mini-YOLOv3 structure, is built [21]

operations than the original YOLOv3. Mini-YOLOv3 is evaluated on a MSCOCO reference dataset [20]. Mini-YOLOv3 achieves a 77.6% decrease in FLOP, 76.7% decrease in parameter size and detection accuracy compared to YOLOv3. Figure 6 gives further details on the construction of the pyramid feature network part. We can see that the basic structure is pruned by applying channel pruning, some connections are removed and a new one is built in order to reduce the cost of communications and computation.

Another extension of YOLOv3, called Tiny-YOLOv3, is obtained by performing pruning parameters. Tiny-YOLOv3 has fewer parameters than Mini-YOLOv3, the netwrok structure is softer but the issue is that Tiny YOLOv3 gives sub-optimal detection results with less accuracy than previous versions.

3.1.2 Slim-YOLO

Like the Mini-YOLOv3 version, the Slim-YOLOv3 is a pruned YOLOv3. It is not guaranteed that each component plays an important role in direct inference. Channel pruning on the convolution layers is performed, a search for a more compact and efficient channel configuration, to further reduce the learning parameters and FLOPs. Channel pruning is used on YOLOv3 in order to design Slim-YOLOv3, following the procedure shown in Fig. 7.

Slim-YOLOv3, with a smaller structure and less trainable parameters than YOLOv3, is able to achieve faster detection accuracy with less FLOP than YOLOv3. As a result, Slim-YOLOv3 is more suitable for real-time UAV applications [22]. In [22], Slim-YOLOv3 is trained on VisDrone2018-Det dataset which contains images captured by an UAV. Slim-YOLOv3 learns multiple classes of objects from an image, the point is that the object category with the highest number of instances can dominate the detectors optimization. Therefore, the mAP score of the dominant category (ex: car) is significantly higher than that categories with a smaller number of instances (ex: bicycle) as shown in Table 2.

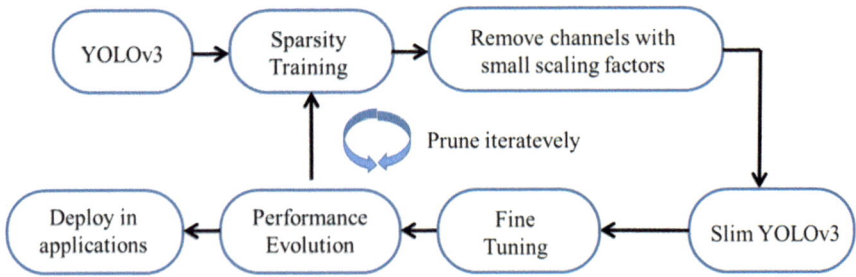

Fig. 7 An iterative procedure for channel pruning for Slim-YOLOv3 [22]

Table 2 Results obtained by applying Slim-YOLOv3 on VisDrone2018-Det dataset [22]

Class	Images	Instances	Precision	mAP
Pedestrian	548	8.840	33.0	25.8
People	548	5.120	31.4	17.0
Bicycle	548	1.290	14.4	2.7
Car	548	1.4100	60.3	76.0
Van	548	1.980	43.8	27.1
Truck	548	750	26.8	16.4
Tricycle	548	1.040	26.9	6.8
Awning-tricycle	548	532	33.0	3.0
Bus	548	251	55.9	22.8
Motor	548	4.890	35.6	23.0
Overall	548	3.8800	36.1	21.2

3.1.3 Nano-YOLO

Nano-YOLO is a compact convolutional neural network (CNN) with macro and micro architecture designs at the module level suitable for the task of embedded object detection. The Nano-YOLO has a model size of 4.0 MB (15.1 times and 8.3 times smaller than Tiny YOLOv2 and Tiny YOLOv3 respectively.). It requires 4.57B operations for inference (34% and 17% less than Tiny YOLOv2 and Tiny YOLOv3 respectively.). While reaching mAP of 69.1% on the 2007 VOC dataset (12% and 10.7% higher than Tiny YOLOv2 and Tiny YOLOv3 respectively.). In [23], authors experiment on inference speed and energy efficiency on a Jetson AGX Xavier integrated module demonstrate the efficiency of Nano-YOLO for embedded scenarios. The Nano-YOLO network provides a robust balance of accuracy, size and computational complexity that makes it well suited for embedded object detection in peripheral and mobile scenarios. Table 3 shows the object detection accuracy results of the compact arrays tested on the VOC 2007 test set. The input size is 416 * 416 for all arrays tested.

Table 3 Results obtained by applying Nano-YOLO on VOC dataset [23]

Model name	Model size (MB)	mAP (VOC 2007) (%)	Computational cost (ops)
Tiny-YOLOv2	60.5	57.1	6.97 B
Tiny-YOLOv3	33.4	58.4	2 B
Nano-YOLO	4.0	69.1	4.57 B

To summarize, Sect. 3 explains the process of object detection using YOLO algorithm, as well as the advantages and limitations of YOLO. Several state of the art distributions of YOLO are proposed such as YOLOv3, Mini-YOLOv3, Tiny-YOLOv3, Slim-YOLOv3, Nano-YOLOv3, and so on. All these distributions take into account the limitations of standard YOLO and the context of applications. For example, in our context of human detection in a SaR (Search and Rescue) UAV mission, it is necessary to integrate the AI detection application on board the UAV and accelerate the computation using FPGA platform. Thus, we have to respect certain constraints, such as execution time, resource utilization, etc. Therefore, YOLO is suitable for this case study because it is possible to adapt to the context and constraints by applying Slim-YOLOv3 for instance.

3.2 Single Shot Multibox Detector

The Single Shot Multibox Detector (SSD) works similarly to YOLO. The only difference is that YOLO uses fully connected layers instead of convolutional layers at the top of the network. In fact, the SSD process consists of two main parts: extraction of feature maps and applying a convolution filter to detect objects [24].

More specifically, there are two SSD models which are SSD300 and SSD512. In the SSD300 model, the input size is fixed at 300*300. It is used in lower resolution images, has a faster processing speed and is less accurate than the SSD512. However, in the SSD512 model, the input size is fixed at 500*500. It is used at higher resolution images and is more accurate than other models [24].

3.3 MobileNet

MobileNet is in fact a CNN-based algorithm, whose convolution phase is divided into two simpler and therefore faster operations [24].

MobileNet has a simple architecture consisting of a 3*3 depth convolution followed by a 1*1 point convolution.

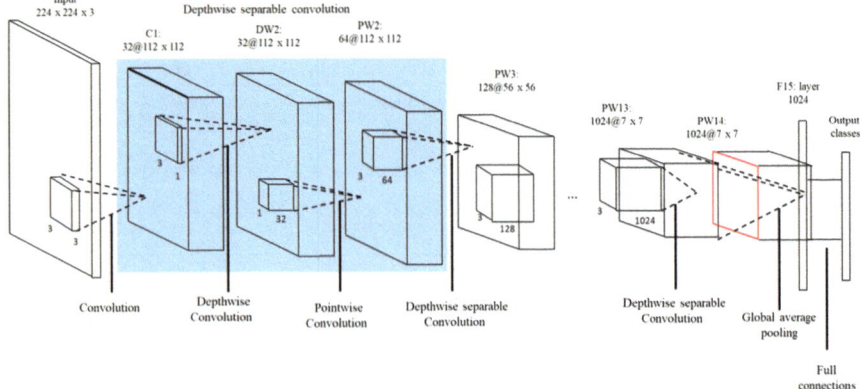

Fig. 8 MobileNet architecture

As shown in Fig. 8, the MobileNets architecture is composed of 28 layers including 13 Depthwise Convolutions and 13 Pointwise Convolutions.

MobileNet applies filters to each layer and then recombines the results so that the model requires much less computation than the CNN model. This incredible speed is nevertheless paid for by a few percentage errors. For example, in road recognition task, MobileNets' recognition rate was 95.5% compared to 95.9% for CNN. In this task, thanks to the lightness of the algorithm in terms of operations, it can reach up to 95% accuracy in only 4 min.

3.4 RetinaNet

RetinaNet is a single-stage detector using focal loss, lower loss is provided by negative samples so that the loss is concentrated on samples, which improves the accuracy of prediction [25]. As illustrated in Fig. 9, the architecture of RetinaNet is composed of subnets to solve the imbalance problem encountered in two- and multi-stage architectures.

3.5 Comparison

In Table 4, a comparison in terms of accuracy and computing time is presented. The algorithms studied are YOLOv1, YOLOv2, YOLOv3, Faster RCNN, SSD and RetinaNet.

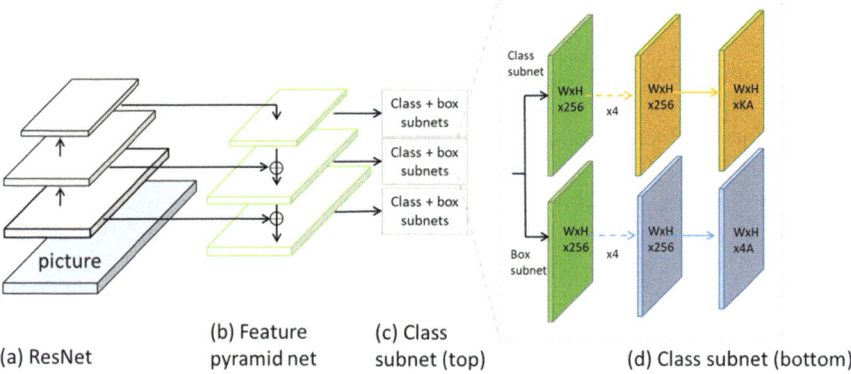

(a) ResNet   (b) Feature pyramid net   (c) Class subnet (top)   (d) Class subnet (bottom)

Fig. 9 RetinaNet architecture

Table 4 Comparison in terms of accuracy and computational time is reported for Faster RCNN, RetinaNet, SSD, YOLOv1, YOLOv2, YOLOv2-Tiny, YOLOv3 and, YOLOv3-Tiny

Algorithm	Accuracy (%)	Computational time (s)
Faster RCNN	81.15	327.48
RetinaNet	93.12	314.43
SSD	85.15	48.86
YOLOv3	**93.62**	22.30
YOLOv3-Tiny	77.58	10.25
YOLOv2	84.92	19.01
YOLOv2-Tiny	73.73	**9.43**
YOLOv1	84.53	21.95

4 Hardware Artificial Intelligence Algorithms for Object Recognition/Detection

In this section, the hardware implementation of a neural network architecture using Field Programmable Gate Arrays (FPGA) is studied. In fact, most existing FPGA-based AI designs are focused on classification and neglect the training stage. For example, several implementations have focused on designing a single neuron on FPGA. In [26], a neural network hardware implementation using FPGA has been proposed. To acheive a multi-layer feedforward neural network, a digital system architecture is conceived. A 351 backpropagation network (three neurons in the input layer, five neurons in the hidden layer and one output layer) has been implemented. It is important to note that the suggested architecture is modular and allows to increase or decrease the number of neurons as well as the layers. The results of the evaluation show, using internal layer parallelism, that the approach takes 10 clock cycles to calculate the outputs. This approach provides good ideas and results on the implementation of hardware IA, but they are still unable to give practical solutions for the training of CNN.

In the same context, the paper [27] deals with image recognition by suggesting a hardware architecture system based on TNNs (tiny neural networks). The TNN architecture is known for its scalability by mapping layers and dynamic configuration, and for its learning capability. TNNs follow a strategy of collaborative work to solve complex recognition issues. In [27], an implementation of TNNS on FPGA is used in advanced driver assistance systems.

For training purposes, the authors of [28] proposed an approach to a new reconfigurable framework that has been used in CNN training. The idea is to involve streaming reconfiguration via F-CNN and to generate a data path to cover the training cycle for the CNN layers. These F-CNN data paths are designed to be customizable and reusable. This approach has proven effective in maximizing performance despite bandwidth and hardware resource constraints. It is important to note that both FPGA (calculation accelerator) and the CPU (the controller) are included in the proposed architecture.

The evaluation results show that the FPGA-based design is about 4 times faster than CPU implementation, and 7.5 times more energy efficient than the GPU implementation [28].

Other work focuses on the acceleration of AI algorithms via FPGA. For example, in [29], some improvements to accelerate neural network computations have been proposed for use in mobile devices. More precisely, it consists in the design and implementation of an FPGA-based CNN with fixed-point calculations. Thus, the proposed solution is based on displacing computation from software to hardware using fixed point calculations. It is important to note that, thanks to the reduced number of parameters, the problem of memory bandwidth is avoided. The results of the evaluation show that this approach requires a total number of 236746 clock cycles to run an image classification operation.

Moreover, with respect to the implementation and acceleration of IA algorithms, in [30], the authors evaluated three convolution variants which are Direct convolution, Convolutional Fast Fourier Transform and Convolutional Fast Fourier Transform Overlap (FFT-OVA-Conv).

However, most AI algorithms are computationally expensive in terms of resources when used to recognize and identify objects in real time [31, 32]. Consequently, their use poses constraints related to their embeddability on reconfigurable targets. Due to their limited computational performance, AI algorithms should not be implemented on embedded CPUs. Moreover, as we have seen in this section, the implementation of AI algorithms on FPGA-based architectures makes these algorithms more suitable for real-time use on embedded systems.

Nevertheless, this approach has a major drawback [33]. Indeed, some parts of the AI algorithms are much lighter than others in terms of computation, and would therefore benefit from being executed on a CPU (or GPU) rather than on an FPGA [34]. Therefore, in future work, a hybrid architecture based on both the FPGA and CPU (and/or GPU) will be developed. The dynamic aspect at the FPGA level can be ensured by exploiting the partial dynamic reconfiguration (PDR) technique. Indeed, PDR will allow the abstraction of hardware computing units into tasks that can be freed or loaded from the FPGA during execution.

5 Conclusions and Future Work

A basic approach for object recognition/identification that takes a different RoI from the image is CNN. However, the limitation of this approach is that RoI may have different spatial locations in the image and different ratios. Therefore, a large number of regions must be selected, which is computationally expensive. As a solution, algorithms such as RCNN, fast/faster RCNN, YOLO, etc. have been developed for real-time use.

The main difference between CNN and R-CNN extensions is that convolution is performed only once per image and a feature map is therefore generated. Unlike RCNN and Fast RCNN, which use selective search, Faster RCNN uses a separate network to predict and find RoI.

As a result, RCNN is much faster than other CNN extensions. This makes this algorithm useful in real-time object detection tasks. However, the main limitation of Faster RCNN is that it consists of different systems that work one after the other. Therefore, the performance of each system depends on the performance of the previous one, which affects the overall performance of Faster RCNN.

Unlike region-based algorithms, YOLO identifies the parts of the image with a high probability of containing the object. Thus, YOLO uses a single CNN to predict the bounding boxes and the class probabilities for these boxes. This makes it much faster, allowing it to be used in real-time tasks. However, due to spatial constraints, the main limitation of YOLO is its inability to detect small objects in the image. SSD is another non region-based algorithm that works the same way as YOLO, with the difference that it uses fully connected layers instead of convolutional layers at the top of the network. Finally, MobileNet is also a non-region based algorithm, whose convolution phase is divided into two simpler and therefore faster operations.

This study has identified the artificial intelligence algorithms used for a computer vision system, to compare them in terms of algorithm complexity, recognition accuracy, computation time and limitations. Based on this study, the first step will be to improve the state-of-the-art of Faster RCNN. This can be achieved by optimizing a number of strategies, including model pre-training and correct calibration of several key parameters. As Faster RCNN is mainly based on CNN, then the latter will be implemented on FPGA. In fact, we chose a combination of CNN-YOLO architecture because it presents a compromise between response time and detection accuracy in the context of embedded systems. Thus, in future work, CNN-YOLO will be implemented in a hybrid architecture based on both FPGA and CPU while exploiting the technique of partial dynamic reconfiguration. This in order to operate an optimized real-time object detection/recognition on embedded systems (FPGA-based) using Faster RCNN-YOLO.

References

1. Viola, P., & Jones, M. (2001). Robust real time object detection. *International Journal of Computer Vision, 4*(34–47), 4.
2. Baroffio, L., Bondi, L., Cesana, M., Redondi, A. E., & Tagliasacchi, M. (2015). A visual sensor network for parking lot occupancy detection in smart cities. In *2015 IEEE 2nd World Forum on Internet of Things (WF-IoT)* (pp. 745–750).
3. Garcia, C. G., Meana-Llorian, D., G-Bustelo, B. C. P., Lovelle, J. M. C., & Garcia-Fernandez, N. (2017). Midgar: Detection of people through computer vision in the Internet of Things scenarios to improve the security in Smart Cities, Smart Towns, and Smart Homes. *Future Generation Computer Systems, 76*, 301–313.
4. Gade, R., Moeslund, T. B., Nielsen, S. Z., Skov-Petersen, H., Andersen, H. J., Basselbjerg, K., et al. (2016). Thermal imaging systems for real-time applications in smart cities. *International Journal of Computer Applications in Technology, 53*(4), 291–308.
5. Pramanik, Md. I., et al. (2017). Smart health: Big data enabled health paradigm within smart cities. *Expert Systems with Applications, 87*, 370–383.
6. Sundaravadivel, P., et al. (2017). Everything you wanted to know about smart health care: Evaluating the different technologies and components of the Internet of Things for better health. *IEEE Consumer Electronics Magazine, 7*(1), 18–28.
7. Lawrence, S., Giles, C. L., Tsoi, A. C., & Back, A. D. (1997). Face recognition: A convolutional neural network approach. *IEEE Transactions on Neural Networks, 8*(1), 98–113.
8. Liao, X., Li, K., Zhu, X., & Liu, K. R. (2020). Robust detection of image operator chain with twostream convolutional neural network. *IEEE Journal of Selected Topics in Signal Processing, 14*(5), 955–968.
9. Lee, S., Kim, H., Lieu, Q. X., & Lee, J. (2020). CNN-based image recognition for topology optimization. *Knowledge-Based Systems, 105887*.
10. Dheir, I. M., Mettleq, A. S. A., Elsharif, A. A., & Abu Naser, S. S. (2020). Classifying Nuts Types Using Convolutional Neural Network.
11. Wiem, B., Mowlaee, P., & Aicha, B. (2018). Unsupervised single channel speech separation based on optimized subspace separation. *Speech Communication, 96*, 93–101.
12. Dong, C., Loy, C. C., & Tang, X. (2016). Accelerating the super resolution convolutional neural network. In *European Conference on Computer Vision* (pp. 391–407). Springer, Cham.
13. Girshick, R., et al. (2014). Rich feature hierarchies for accurate object detection and semantic segmentation. In *Proceedings of the IEEE Conference on Computer Vision and Pattern Recognition*.
14. Girshick, R. (2015). Fast RCNN. In *Proceedings of the IEEE International Conference on Computer Vision* (pp. 1440–1448).
15. Li, J., Liang, X., Shen, S., Xu, T., Feng, J., & Yan, S. (2017). Scale aware fast RCNN for pedestrian detection. *IEEE Transactions on Multimedia, 20*(4), 985–996.
16. Ren, S., He, K., Girshick, R., & Sun, J. (2015). Faster RCNN: Towards real time object detection with region proposal networks. In *Advances in Neural Information Processing Systems* (pp. 91–99).
17. Sun, X., Wu, P., & Hoi, S. C. H. (2018). Face detection using deep learning: An improved faster RCNN approach. *Neurocomputing, 299*, 42–50.
18. Redmon, J., & Farhadi, A. (2018). Yolov3: An incremental improvement. arXiv:1804.02767.
19. Liu, J., &Wang, X. (2020). Tomato diseases and pests detection based on improved YoloV3 convolutional neural network. *Frontiers in Plant Science, 11*, 898.
20. Mao, Q. C., Sun, H. M., Liu, Y. B., & Jia, R.-S. (2019). Mini-YOLOv3: Real-time object detector for embedded applications. *IEEE Access, 7*, 133529–133538.
21. Wang, B., et al. (2020). Icing-EdgeNet: A pruning lightweight edge intelligent method of discriminative driving channel for ice thickness of transmission lines. *IEEE Transactions on Instrumentation and Measurement*.

22. Zhang, P., Zhong, Y., & Li, X. (2019). SlimYOLOv3: Narrower, faster and better for real-time UAV applications. In *Proceedings of the IEEE International Conference on Computer Vision Workshops*.

23. Wong, A., Famuori, M., Shafiee, M. J., Li, F., Chwyl, B., & Chung, J. (2019). YOLO nano: A highly compact you only look once convolutional neural network for object detection.

24. Li, Y., Huang, H., Xie, Q., Yao, L., & Chen, Q. (2018). Research on a surface defect detection algorithm based on MobileNet-SSD. *Applied Sciences, 8*(9), 1678.

25. Ale, L., Zhang, N., & Li, L. (2018). Road damage detection using RetinaNet. In *2018 IEEE International Conference on Big Data (Big Data)* (pp. 5197–5200). IEEE.

26. Savran, A., & Ünsal, S. (2003). Hardware implementation of a feed forward neural network using fpgas. In *The Third International Conference on Electrical and Electronics Engineering (ELECO 2003)*.

27. Moreno, F., et al. (2009). Reconfigurable hardware architecture of a shape recognition system based on specialized tiny neural networks with online training. *IEEE Transactions on Industrial Electronics, 56*(8), 3253–3263.

28. Zhao, W. (2016). F-CNN: An FPGA-based framework for training convolutional neural networks. In *IEEE 27th International Conference on Application-specific Systems* (p. 2016). IEEE: Architectures and Processors (ASAP).

29. Solovyev, R. A., et al. (2018). FPGA implementation of convolutional neural networks with fixed-point calculations. arXiv:1808.09945.

30. Abtahi, T., et al. (2018). Accelerating convolutional neural network with fft on embedded hardware. *IEEE Transactions on Very Large Scale Integration (VLSI) Systems, 26*(9), 1737–1749.

31. Wiem, B., Ahmed, K., & Chabha, H. (2021). Incremental learning for real-time hardware software partitioning for FPGA Applications, In *Accepted for Publication in 13th International Conference on Agents and Artificial Intelligence (ICAART 2021)*.

32. Wiem, B., & Marwa, H. (2020). Supervised hardware-software partitioning algorithms for FPGA-based applications. In *Proceedings of the 12th International Conference on Agents and Artificial Intelligence* (Vol. 2, pp. 860–864), (ICAART 2020).

33. Kammoun, A., et al. (2019). Hardware acceleration of approximate transform module for the versatile video coding standard. In *27th European Signal Processing Conference (EUSIPCO)*. IEEE.

34. Kammoun, A., et al. (2018). Hardware design and implementation of adaptive multiple transforms for the versatile video coding standard. *IEEE Transactions on Consumer Electronics, 64*(4), 424–432.

Development of an IoT Based Systems to Mitigate the Impact of COVID-19 Pandemic in Smart Cities

H. M. K. K. M. B. Herath, G. M. K. B. Karunasena, and H. M. W. T. Herath

Abstract The dramatic growth of urbanization in modern cities calls for smart strategies to resolve crucial problems such as transportation, healthcare, energy, and civil infrastructure. The Internet of Things (IoT) is one of the most exciting enabling technologies to address smart city problems by creating a large global network of interconnected physical objects embedded with electronics, software, sensors, and network connectivity. Since the end of 2019, the world has been confronted with the challenge of the COVID-19 virus originating in Wuhan, China. Precautionary measures are expected to be needed in the world to combat the pandemic of COVID-19 until an effective vaccine is developed. The disease has already proven the value of modern smart healthcare which plays a significant role in preventing COVID-19. The aim of this research is to develop the COVID-19 cluster tracking system, and face mask detection system for public safety. The proposed cluster tracking system was validated for 26 test subjects in an experimental scenario of a potential COVID-19 cluster. The accuracy of the proposed face mask detection system was 86.96% observed at 0.9756 precision. According to the testing results, the proposed system was showed promising outcomes for the prevention of the COVID-19 pandemic in a smart city.

Keywords COVID-19 · Global pandemic · Internet of Things (IoT) · Smart city · Smart healthcare · Smart tracking · Urban IoT

1 Introduction

Over the past few decades, three major global outbreaks of pandemics have spread, including Severe Acute Respiratory Syndrome (SARS), Middle East Respiratory Syndrome (MERS), and Ebola Virus (EVD) [1]. Liver, neurologic, gastrointestinal,

H. M. K. K. M. B. Herath (✉) · G. M. K. B. Karunasena
Faculty of Engineering Technology, The Open University of Sri Lanka, Nugegoda, Sri Lanka

H. M. W. T. Herath
Faculty of Engineering, University of Moratuwa, Moratuwa, Sri Lanka

U. Ghosh et al. (eds.), *Machine Intelligence and Data Analytics for Sustainable Future Smart Cities*, Studies in Computational Intelligence 971, https://doi.org/10.1007/978-3-030-72065-0_16

and respiratory diseases may be caused by this group of crown-like viruses [2]. SARS, caused by the SARS-associated coronavirus (SARS-CoV) with ready transmission through droplets, first appeared in Southern China in November 2002. The time of incubation was usually 2–7 days. By late July 2003, the epidemic was declared to be over; this global pandemic resulted in a total of 8098 likely cases and 774 associated deaths [3]. In Saudi Arabia, new MERS cases have been identified, in contrast to SARS. Since it was first identified in 2012, a total of 2519 laboratory-confirmed MERS cases and 866 related deaths were reported globally by the end of January 2020 [4]. According to the World Health Organization (WHO), the novel coronavirus (SARS-CoV-2) outbreak first started in Wuhan, China on 12 December 2019 with 79.6% similarity in sequence identity as the SARS-CoV [5]. Table 1 depicts the comparison between the different viruses and their characteristics.

The SARS-CoV-2 outbreak was declared as a Public Health Emergency of International Concern (PHEIC) by the WHO Director-General on 30, January 2020. This incident was characterized as a pandemic on 11, March 2020. On February 11, 2020, SARS-CoV-2 was named 'COVID-19' by the World Health Organization.

As of November 2020, the WHO was announced that 217 countries have recorded 52.7 million confirmed cases and over 1.29 million deaths. Figure 1 demonstrates confirmed global COVID-19 cases as of November 2020. A study carried out by Jiang et al. [6] showed that the death rate of COVID-19 is 4.5% globally. The mortality rate for patients is 8.0% in the age range of 70-79 years, while 14.8% for patients above 80 years. Patients above the age of 50 with chronic illnesses are at the greatest risk and finding a way to diagnose illness before heading into severe conditions is increasingly necessary.

The outbreak of COVID-19 has emphasized the importance of technology in daily living. The need for rapid digital urban planning and greater communication

Table 1 Comparison between Coronavirus disease 2019 (COVID-19), Severe Acute Respiratory Syndrome (SARS), Middle East Respiratory Syndrome (MERS), and Ebola Virus Disease (EVD)

Description	COVID-19	SARS	MERS	EVD
Incubation period (days)	2-14	2-10	2-14	2-21
Contagious during incubation	Yes	Yes	Yes	No
Vaccine as of 2020	No	No	No	Yes
Infected body part	Blood vessels	Respiratory system		All muscles
Main transmission	Respiratory droplet secretions			Body fluids
Shape and size in nm	Spherical 80–120	Spherical 80–90	Spherical 90–125	Filament 14,000 × 80 nm
Mortality rate %	2.65	14–15	34	Up to 90
Pronounced R-naught	2–2.5	3.1–4.2	<1	1.5–1.9

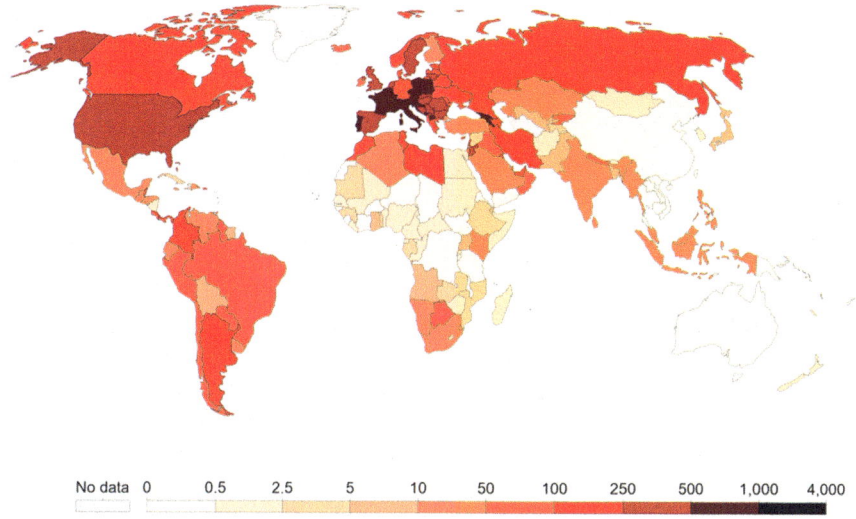

Fig. 1 Global COVID-19 confirmed cases (per Million People) as of 11, Nov 2020 [7]

with citizens was highlighted by the pandemic and some governments are taking advantage of the opportunity to increase their investments in technology, specifically in smart cities.

Current cities are dynamic structures marked by a large number of interconnected people, businesses, facilities and utilities, communication networks, and diverse modes of transportation. Increased urbanization and population growth in the technological, economical, organizational, and social sectors pose a different form of problems that threatens to jeopardize urban economic and environmental sustainability [8]. The rapid growth of the population faced by several towns has caused traffic congestion, health problems, environmental pollution, and growing social inequality (Kim and Khan 2012). In recent years, due to the development of hardware and software designs, there has been an exponential increase in information and communication technologies (ICTs). There is, however, a consensus that smart cities are distinguished by the extensive use of information and communication technology (ICT).

A smart city is a notion and there is still no simple and concise meaning between academia and practitioners of the term. In a simplified explanation, a smart city is a place where, with the use of information, digital and telecommunication technology, conventional networks and services are made more versatile, effective, and sustainable to enhance its operations for the benefit of its inhabitants [9]. A number of major cities around the world are studying the implementation of the smart city model to increase the quality of life of their inhabitants and maximize the use of urban infrastructure and resources [10]. In order to enhance efficiency and operations in healthcare, transport, energy, education and many other sectors are include

Fig. 2 Components of a
modern smart city

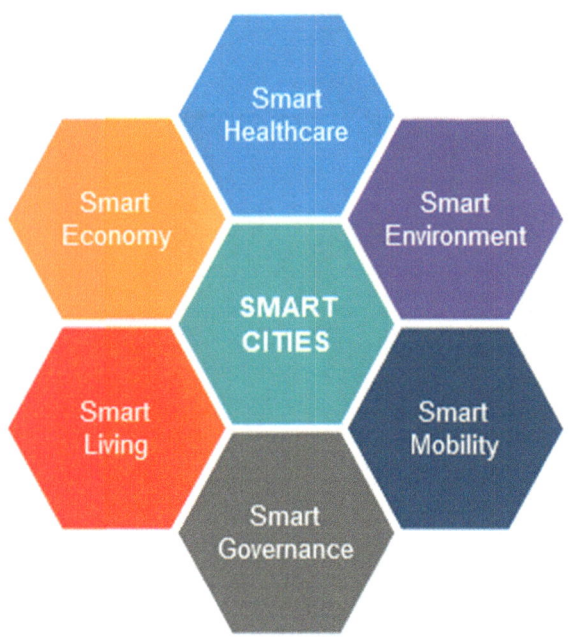

various advanced technologies and techniques supporting models in smart services. It is known that cities can be defined as "Smart" if they have met the components as shown in Fig. 2.

The smart city environment, from a technical point of view, is a dynamic one containing several areas of technology. In several fields, major players operate, providing solutions that complement other players. Those capable businesses are working towards a convergence point where they can provide end-to-end solutions for the needs of urban technology.

The United Nations Population Fund (UNFPA) indicated that around 3.3 billion (54%) of the world's population lived in urban areas in 2014. Approximately 5 billion people (66%) will live in urban areas by the end of 2050. The recent U.S. Census Bureau data revealed that all but one of the twenty major cities in the U.S. experienced population growth last year. If this migration persists, cities will need to become more innovative to keep up with the rising population. Smart cities will then continue to become the norm in the big metropolitan areas of the country. According to the Juniper Research study, Singapore is the world's leading smart city in the fields of mobility, health, safety, and productivity. The study found that when it comes to innovative ways to target the elderly, Singapore and Seoul are leading the way, emphasizing their embrace of digital service systems as well as remote surveillance devices to provide their citizens with better access to health care and health-related information. The government acknowledged the increased burden on their healthcare system by the aging population in Seoul, and so they introduced the 'U-Health' (Ubiquitous Health) policy. The plan aimed to enhance the remote surveillance of elderly residents.

Table 2 Impact of the COVID-19 pandemic on top-ranked smart cities of the world

Rank	Country	Smart City	City Score	Population	COVID-19 Cases	Impact (%)
1	UK	London	33.5	8.982 M	0.118 M	1.31
2	Singapore	Singapore	32.3	5.639 M	0.053 M	0.94
3	South Korea	Seoul	31.4	9.776 M	6505	0.06
4	USA	New York	31.3	18.59 M	0.281 M	1.51
5	Finland	Helsinki	31.2	1.180 M	5642	0.48
6	Canada	Montreal	30.1	3.981 M	0.044 M	1.10
7	USA	Boston	29.6	4.249 M	0.023 M	0.54
8	Australia	Melbourne	29.5	4.203 M	0.198 M	0.47
9	Spain	Barcelona	29.4	5.258 M	0.064 M	1.21
10	China	Shanghai	29.2	23.74 M	1254	0.005

The Institute for Management Development (IMD) and the Singapore University for Technology and Design (SUTD) published the Smart City Index for the year 2020. A simple comparison of COVID-19 cases from WHO and data obtained from their records is shown in Table 2. According to the analysis, the spread of COVID-19 is astonishingly low (~2%) in smart cities. Smart city enable designs could play an important role in alleviating the pandemic has been observed.

Modern Internet of Things (IoT) technology has recently been widely used in smart city applications. The term "Internet of Things (IoT)" was first mentioned in 1999 following the advent of Internet-based techniques in the 1990s [11]. IoT can be described as a global infrastructure that facilitates advanced services by connecting physical and virtual objects using interconnected information and communication technologies (ICTs) [12]. The Internet of Things (IoT) uses networking and advanced sensing technologies, as well as ubiquitous and widespread computing, to update physical objects to smart objects [13]. IoT offers the ability to monitor, manage, and control smart devices remotely, and to generate new insights and actionable information from massive streams of real-time data. The uses of sensors have grown significantly over the past few decades due to the rapid development of cloud technology, enhanced storage space, processing speed, decreased manufacturing, and installation costs. Physical objects in the IoT are equipped with sensors to collect heterogeneous data. These sensors have the low computing power and limited life. The more details they obtained the more valuable decisions that can be made [14].

IoT designs allow the implementation of smart services to consumers to enhance the quality of their lives [15] and by the end of 2020, between 50 and 100 billion users will be connected to the internet using different ICTs [16]. This research is mainly focused on healthcare designs in smart cities, where the enhanced quality of patient life is guaranteed by IoT-based medical services.

The IoT architecture is comprised of a system of multiple components such as cloud services, protocols, sensors, actuators, layers, etc. The IoT layer is primarily composed of physical, network, and application layers [23]. Alternative protocols for

Table 3 Comparison of various network protocols used in smart city applications

Protocol	Frequency	Data rate	Transmission range	Power usage
2G/3G	Cellular bands	10 Mbps	Several miles	High
Bluetooth/BLE	2.4 GHz	1, 2, 3 Mbps	~300 feet	Low
IEEE 802.15.4	Sub GHz, and 2.4 GHz	40, 250 kbps	>100 square miles	Low
LoRa	Sub GHz	<50 kbps	1–3 miles	Low
LTE Cat 0/1	Cellular bands	1–10 Mbps	Several miles	Medium
NB-IoT	Cellular bands	0.1–1 Mbps	Several miles	Medium
SigFox	Sub GHz	<1 kbps	Several miles	Low
Weightless	Sub GHz	0.1–24 Mbps	Several miles	Low
Wi-Fi	Sub GHz, 2.4, and 5 GHz	0.1–54 Mbps	<300 feet	Medium
WirelessHart	2.4 GHz	250 kbps	~300 feet	Medium
ZigBee	2.4 GHz	250 kbps	~300 feet	Low
Z-Wave	Sub GHz	40 kbps	~100 feet	Low

communication between IoT devices using IP have been developed by the Internet Engineering Task Force (IETF) due to the flexible and reliable standard of IP [24, 25]. Table 3 depicts the various IoT networks used for the smart city application.

IoT guarantees that health-care facilities are customized by creating a digital profile for each patient. Due to the lack of easy access to treatment services, numerous health conditions in traditional healthcare facilities have gone undetected [17, 35, 36]. Through IoT based healthcare, various distributed devices collect, process, and transfer medical information to the cloud through real-time, allowing vast data streams to be captured, processed, and analyzed in many new ways and context-dependent alarms to be enabled. This groundbreaking data collection approach allows continuous and invisible control to medical equipment through the internet from every connected device [18–20]. Mobile computing provides IoT services through the use of mobile phone devices, software, or m-health care [21]. The m-healthcare contributes to the IoT by offering different services such as compactness, IP connectivity, low power consumption, and protection [22].

This debate explains that the implementation of an IoT-based system for the healthcare sector in smart cities is a highly effective way of preventing the spread of the COVID-19 virus. In the development of the proposed system, the IoT described here plays a major role. The main objectives of this research are the development of a COVID-19 patient tracking and cluster recognition system, and the development of a face mask detection system for public safety. In Sect. 2, the literature survey of the research has been carried out. The design and implementation of the proposed system have been explained in Sect. 3. The test results of the proposed system have described in Sect. 4. Section 5 has discussed the advantage and disadvantages of the proposed system and, finally, Sect. 6 has concluded the chapter and proposed some future works.

2 Related Works

The novel coronavirus is a new strain not previously identified in the human body. Many nations have now left the clusters undetected and this causes the immense problem of spreading the virus. The aim of this research is to develop a system for tracking, identifying clusters, and monitoring safety precautions in the COVID-19 pandemic. To achieve the research objectives, previous studies of GPS acquisition for healthcare applications, IoT frameworks for cloud-based applications, reliable network protocols used in smart cities, and computer vision techniques have been focused on primarily. Previous studies led by researchers in the field of healthcare and IoT have been discussed briefly in this session.

Milon and his research team [26] have suggested a smart IoT healthcare system to track the patient's individual health signs as well as the condition in the room where patients are actually in real-time. Five sensor data have been collected from the hospital environment, such as heart rate, body temperature, room temperature, CO, and CO2. It has noticed as, for each case, the error percentage of the developed scheme was within a certain limit (<5%). The status of the patients has transmitted via a portal to the medical staff, where they processed and evaluated the current situation of the patients. A health monitoring system has developed by Tamilselvi et al. [27] that can track basic patient symptoms such as heart rate, body temperature, eye movement, and the percentage of oxygen saturation in the IoT network. Their system used heartbeat, temperature, eyeblink, and SpO2 sensors as sensing methods for the experiment, while Arduino-UNO used as a processing unit.

Sareen et al. [28] have engaged in a study to prevent the spread of the disease at the early stage of the outbreak. In order to describe and monitor the real-time state of the outbreak using CPI data, temporal network analysis method was used. Using synthetic data from two million users, the performance and accuracy of the proposed model have assessed on the Amazon EC2 cloud. The proposed model from Sareen provided 94% accuracy for the classification and 92% of the utilization of resources. Mohan [29] has proposed a data privacy scheme for IoT personal medical devices (PMDs) to improve patient mobility. It has made for better monitoring of the patient's wellbeing when moving. It posed the security challenges and vulnerabilities of PMD IoT that pose a challenge in resolving these threats. They have suggested some initial solutions to counter these security challenges.

Mallik et al. [30] have proposed the development of a tracking app based on the real-time global positioning system for ambulances that carrying COVID-19 patients. The goal of their research was to help the traffic police and ensure that the public is distanced from patients. A mobile's in-built global positioning system has used for their study. With the help of Google API and its services, they have acquired the location of the device and upload the coordinates to a real-time database (Firebase). It has also been observed that the proposed system provided promising results for a real-time scenario. A study on sharing multimedia healthcare data through a cloud-based body area network approach has been followed by Hassan et al. [31]. The proposed network architecture has composed of four layers, such as perception, network, cloud computing, and application. For the network layer, the incorporation of Zigbee and TCP/IP into the coordinator devices has used. Various simulations have conducted using the OPNET simulator to show the feasibility of the proposed architecture. Susi et al. [32] has engaged in the development of motion mode detection algorithm and step detection for smartphone users. Their algorithm, using smart devices, was able to characterize a pedestrian's gait cycle. To detect motion modes called class 1, class 2, class 3, and class 4, a classifier has developed. To implement their system, Micro-electromechanical Systems (MEMS) technology has been used. With 97% accuracy, they found that their method was more accurate. The mobile sensor technologies proposed by Susi [32] was promisingly helpful for the development of our system.

For the face mask detection system, the Haar classification-based research and digital image processing methods were studied. The knowledge of the previous research studies discussed here was very useful for the development of the system. The purpose of our research was to detect and alert the public monitoring center to whether a mask is worn by the user or not. Meenpal et al. [33] have proposed a method for generating precise face segmentation masks from any size of arbitrary image input. Their method has used the predefined training weights of VGG-16 architecture to extract features, starting with RGB images of any size. The Training process has carried out through a fully convolutional neural network in order to semantically segment the human faces presented in the image. An appearance-based approach to facial expression recognition based on the Haar characteristics has investigated by Whitehill [34] from the University of the Western Cape, South Africa. Due to the high accuracy of recognition and rapid performance, he has used the Adaboost boosting algorithm along with the Haar classification technique. His study suggested that the classification of Haar features had the same performance as the Gabor + SVM approach.

With a careful review, it has been discovered that not many potential systems have been developed for the COVID-19 patient tracking and in particular, IoT solutions for cluster identification. The next section of this chapter describes the proposed designs of the COVID-19 patients tracking, cluster identification system and, face mask detection systems for safety precautions.

3 Proposed Design

The design section of this chapter consists of two subsections. Section 1 describes the development of the COVID-19 patients tracking and cluster identification system and Sect. 2 describes the development of a face mask detection system for safety precautions. Both designs proposed here were linked with the IoT network and functioned as a single operation. IoT gateway was connected with smart devices and surveillance cameras. GPS data and Accelerometer data were extracted using the inbuilt sensors of smart mobile phones.

Google Firebase is a Backend-as-a-Service also known as Baas that developed by the Google. It offers a number of tools and resources for developers to design a

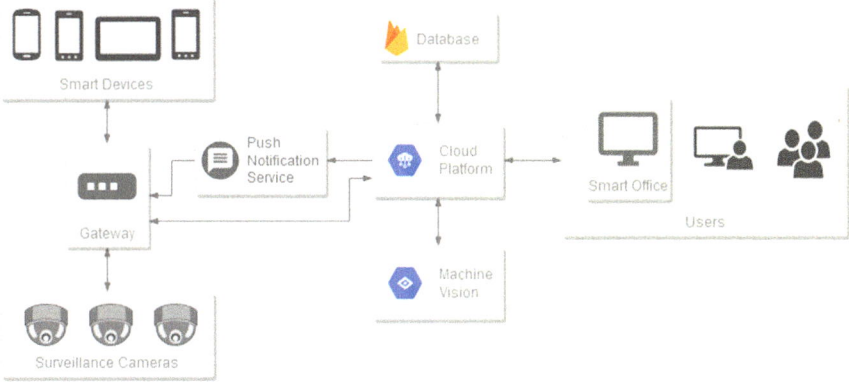

Fig. 3 IoT architecture of the proposed system

variety of software, expand their user base, and earn profits. Firebase is categorized as a database program called NoSQL, which stores data in records identical to JSON. In this research, the Google Firebase platform was used to store data on the cloud database.

The smart office has access to the IoT framework and should initially define recognized COVID-19 infected patients on the system. The image processing framework was attached to the IoT platform and used to communicate with IP cameras to recognize individuals who have face masks on their faces. Figure 3 illustrates the IoT architecture of the proposed design with gateways, cloud databases, and machine vision management systems.

3.1 COVID-19 Patient Tracking and Cluster Recognition System

This section describes the development procedure of the COVID-19 patient tracking and cluster recognition system. The study was used 26 test subjects at 11 different locations in the Western Province of Sri Lanka. Figure 4 illustrates the experimental scenario of a potential COVID-19 cluster. The cluster scenario was based on past COVID-19 incidents in the Western Province of Sri Lanka. As in Figure, P0 is the recognized patient zero of the cluster. Once the user was identified by clinical trials, the smart office marked the patient's profile as a COVID-19 positive case. The cluster starts to be recognized according to the GPS history of the user that was stored on the smartphone. The cluster sequence was based on the date/time that they interacted with each other and the data were sorted according to the date/time.

GPS historical data of P0 contains location 0 (6.83121, 79.98123), therefore location 0 was marked and stored in the cloud database. After P0 was marked as a

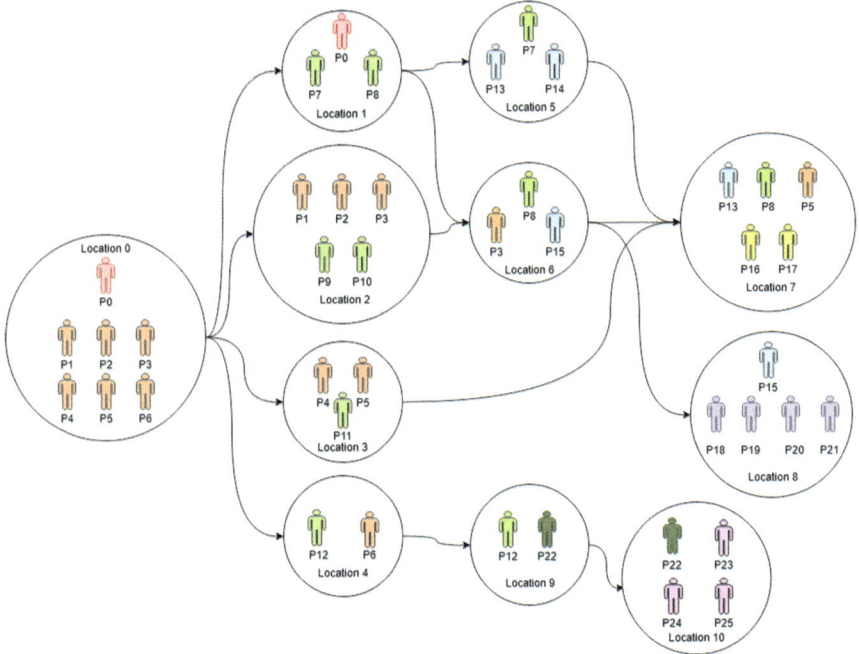

Fig. 4 An experimental scenario of a possible COVID-19 cluster

COVID-19 patient, the test subjects at location 0 (6.83121, 79.98123) were recognized as the initial suspects of the COVID-19 by the system, then it is necessary to perform the PCR test to determine the critical paths of the cluster. In this case, we assume that P1, P2, P3, P4 were tested positive for COVID-19, and P5, P6 tested negative for COVID-19. Based on this information, the critical paths of the cluster were determined by the system and continues to draw a critical path based on the GPS historical data. Table 4 depicts the GPS data of eleven different locations in the Western Province of Sri Lanka.

The DISTANCE-CALC algorithm was developed to calculate the distance between two different GPS coordinates. The 'Haversine' formula was used to calculate the great-circle distance between two points. Obtaining the distance was performed using the Eqs. 1, 2, and 3.

$$a = sin^2\left(\frac{\Delta\varphi}{2}\right) + \left(cos\varphi_1 \times cos\varphi_2 \times sin^2\left(\frac{\Delta\lambda}{2}\right)\right) \tag{1}$$

$$c = 2 \times \text{atan2}\left(\sqrt{a}, \sqrt{1-a}\right) \tag{2}$$

$$d = R \times c \tag{3}$$

Table 4 GPS coordinates of the locations used for the experiment

Location ID	GPS coordinate	Location ID	GPS coordinate
0	6.83121, 79.98123	6	6.88852, 79.88084
1	6.88268, 79.88388	7	6.97416, 79.95076
2	6.91358, 79.90951	8	6.96901, 79.94758
3	6.87223, 79.89108	9	6.94736, 79.98897
4	6.88445, 79.88394	10	6.98562, 80.01251
5	6.88509, 79.88335		

Where,

φ is latitude, λ is longitude, R is radius of earth (mean radius = 6,371 km)

The 3D acceleration of smart devices was measured using the built-in accelerometer sensor and determined the user's activity. The X, Y, and Z values represented the acceleration of the unit where the acceleration arose in each direction. Figure 5 shows the GPS acquisition subsystem of the proposed design. Prior to the GPS acquisition, it is very important to identify the status of the user. It is assumed that users are not running or driving while performing the experiments in this research. Users who carrying out normal activities were taken into account.

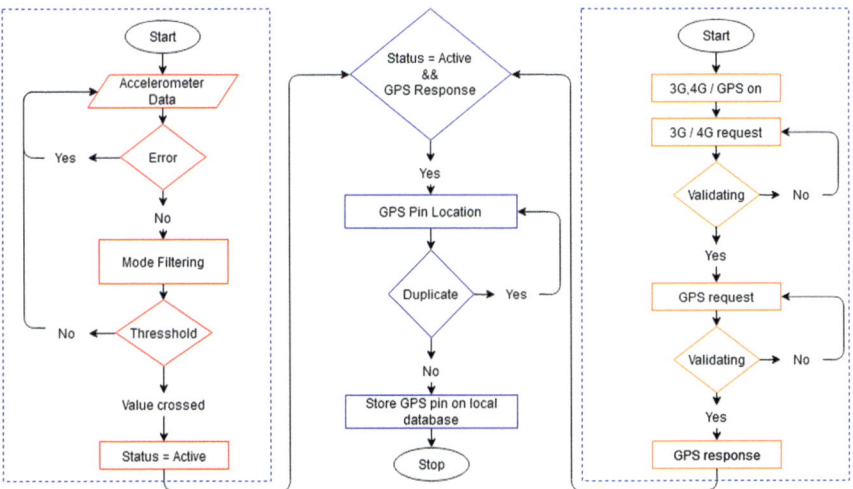

Fig. 5 Program flow chart of the GPS acquisition algorithm

The accelerometer data was originally obtained by the system when the program was running, and there is no error in the verified data. Through the step of mode filtering, the data was processed. When the user is not running or driving, the status marked as 'active'. Initially, the GPS acquisition sub-system checked the availability of the 3G/4G access. Then request the current GPS location and confirmed that whether the request had a response. Once the validation is passed by both sub-systems, the system checks itself for duplicated GPS pins generated within one hour. The system does not store the same pin that was generated for a period of one hour and this action was carried out for the purposes of database management. It is also benefits from eliminating unwanted information from the system. The acquired data was finally stored in the local database. Algorithm 1 shows the pseudocode of distance calculation process.

Algorithm 1 GPS Distance Calculation

1: **procedure** DISTANCE-CALC
2:     $\varphi_0 \leftarrow$ initial point latitude
3:     $\lambda_0 \leftarrow$ initial point longitude
4:     $R \leftarrow 6.371$
5: *top*:
6:     **if** $\varphi_i, \lambda_i = exist$, **then return** true
7:     $\Delta\varphi_i \leftarrow \varphi_i - \varphi_0$
8:     $\Delta\lambda_i \leftarrow \lambda_i - \lambda_0$
9:     $a_i \leftarrow f(\Delta\varphi_i, \Delta\lambda_i, \varphi_0, \lambda_0)$
10:     $C_i \leftarrow f(a_i)$
11:     $d_i \leftarrow f(C_i, R)$
12: *loop*:
13:     $\sum_{i=1}^{N} d_i$
14:     **if** $d_i <= 15$ **then**
15:         $location \leftarrow (\varphi_i, \lambda_i)$
16:         **goto** *loop*.
17:         close;
18:     $i = i+1$
19:     **goto** *top*.

When the GPS data are storing in the local database, the GPS locations were recorded on a time-based approach by storing locations every minute. This action had a disadvantage when uploading data to the cloud database because there were approximately 21,600 records for 15 days. Therefore we proposed a method to store data using a location and time-based approach. When the user stays at the same GPS position for an hour, the system does not duplicate the location for a 5-meter radius distance. This replication is only valid for a one-hour time span. After one hour, the same place will be added but on a different time slot. When the user is in stationary mode, the accelerometer senses this action. After implementing this new method, the database stored approximately 6,000 records of nearly 1 MB of data.

The IoT framework consists of user handling, database management, map services such as map and cluster visualization, and monitoring services. Cloud database manager and Local database manager were used to manipulate data on the cloud

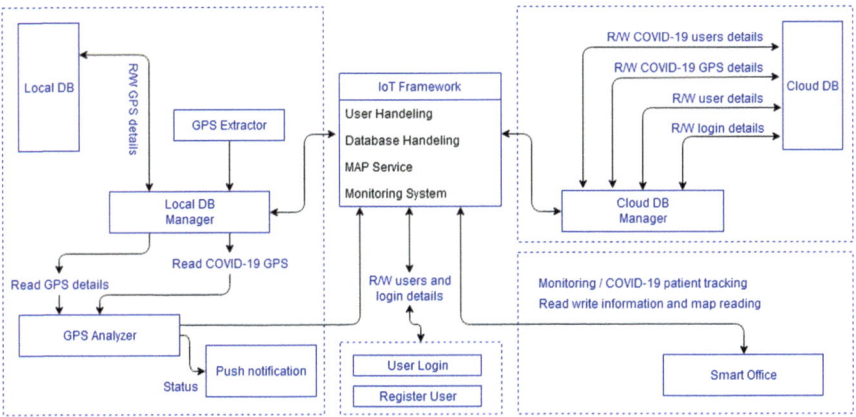

Fig. 6 Flow diagram of the database management system

and local databases. Once a new user is registered with the system, the data will be stored in the cloud database with the help of the Cloud DB manager. The cloud database consists of user information, login details, COVID-19 positive users, and the GPS location history of the COVID-19 patients. The cloud database stored the GPS data of the user if the user is suspected of having COVID-19. Otherwise, all the user's GPS data was stored in the local database. The local database contains GPS data acquired by the GPS extractor for the past 15 days. The GPS extractor is composed of a location extraction algorithm. The proposed design model of GPS extraction and database management is shown in Fig. 6.

After the acquisition of the GPS location, the users were sorted according to the several sorting processes such as date-base, time-base, and distance-base. The cluster was rearranged according to the sorting algorithm of the system. The simulation was developed using the JavaScript language. Plotting of the cluster was especially used the layer ID and Track ID attribute values of the cloud database.

The system detects the area within a radius of 100 m and security measures are needed when COVID-19 suspects are located in the region (details that based on the location ID and the track ID variable of the cloud database). The handling of safety measurements is very critical without an area lockdown. Wearing a face mask is a new practice for people these days and most of them sometimes forget to wear a mask. The next subsection of this chapter describes the development of a face mask detection system.

3.2 Development of a Face Mask Detection System

The objective of this section is to development of a face mask detection system for safety precaution against the COVID-19 virus. The system was developed with the aid of an IP camera (Hikvision DS-2CD1230-I 3mp IP CCTV network camera) which is remotely accessible for the research studies. The video feed from the remote IP camera was acquired by using the machine vision process that depicts on the IoT architecture. Once data was acquired, Python and image processing techniques were used to detect the features on the video feeds. Following pseudocode (Algorithm 2) represents the flow of the video streaming and data acquisition process.

Algorithm 2 Video Stream Acquisition Process

1: **procedure** VIDSTREAM
2: *Initialization;*
3:     *host* ← 192.168.100.23
4:     *output* ← img (true/false)
5:     *k:0xFF (hold key)*
6:     **while** host = true **do**
7:         **if** $k = true$ **then**
8:             *img (false)*
9:             **if** $k <> false$ **then**
10:                *img (true)*
11:            *localhost (stream)* ← img

The Support Vector Machine (SVM) training process was carried out with a perspective on the training of the features extracted. The Haar Cascades algorithm is the most popular algorithm in the field of computer vision that used to detect faces on an image. The Haar Cascade algorithm is not only used to detect faces on images, but we can also train the model to detect different artifacts. Figure 7 shows the working principle of the proposed face mask detection system.

The original dataset consists of 220 masked face images and 220 mask-free images. The dataset was taken from the Kaggle, which is publicly available for researchers that have been published by D. Makwana. Initially, 100 images were

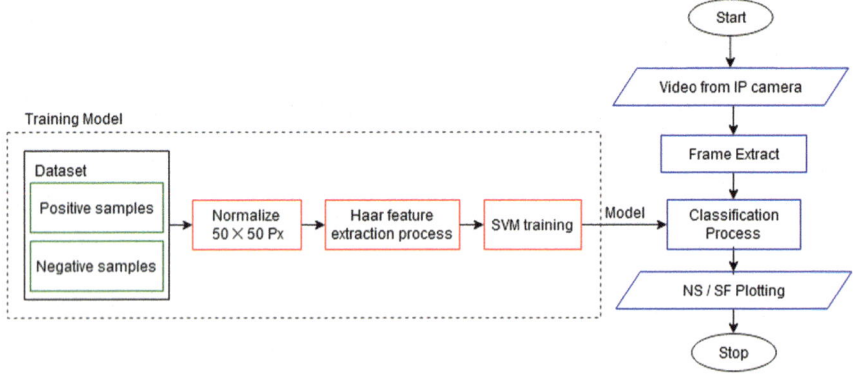

Fig. 7 Flow diagram of the proposed image processing algorithm

extracted from masked faces and 200 images without masks. All images were pre-processed by cropping and re-sizing to 50×50 pixels. This normalization process was carried out to increase the performance of the training. After the dataset was pre-processed, mask and face features were extracted. Once the video feed was acquired by the IP camera, the extraction of the frames was performed. In the classification process, all frames were compared to the training model. Once the mask has been detected, the SF (Safe Face) tag was marked on each face of the frame.

The following pseudocode (Algorithm 3) represents the processing steps of the Support Vector Machine (SVM) classifier.

Algorithm 3 Support Vector Machine Algorithm

1: **procedure** SVMPROCESS
2:     **input** *dataset D*
3:     **output** *confusion matrix, validation.*
4: *Initialization;*
5:     *train* $D_{j=0}^{N} \leftarrow$ split [D, size=300]
6:     *test* $D_{j=0}^{N} \leftarrow$ split [D, size=92]
7:     *SVMClassifire(Haar features extraction)*
8:     $X_{k=0}^{N} \leftarrow$ no of samples
9:     $Y_{k=0}^{N} \leftarrow$ labels
10: *loop:*
11:     $X_i \leftarrow$ f(Di,Yi)
12:     *return Yi*
13:     **goto** *loop.*
14:     *calculate score*
15:     *calculate confusion matrix*

---

4 Results

This section discusses the results and observations of each objective. Section 1 describes the results and observations of the COVID-19 patient tracking and cluster recognition system. Section 2 discusses the experimental results and accuracy of the face mask detection system.

4.1 COVID-19 Patient Tracking and Cluster Recognition System

The proposed design was subjected to different levels of experimentation. In order to obtain GPS (Global Positioning System) data, the measurements of the accelerometer were taken. Users' actions were determined by the data acquisition of the built-in smartphone accelerometer sensor. Figure 8 demonstrates preliminary findings when

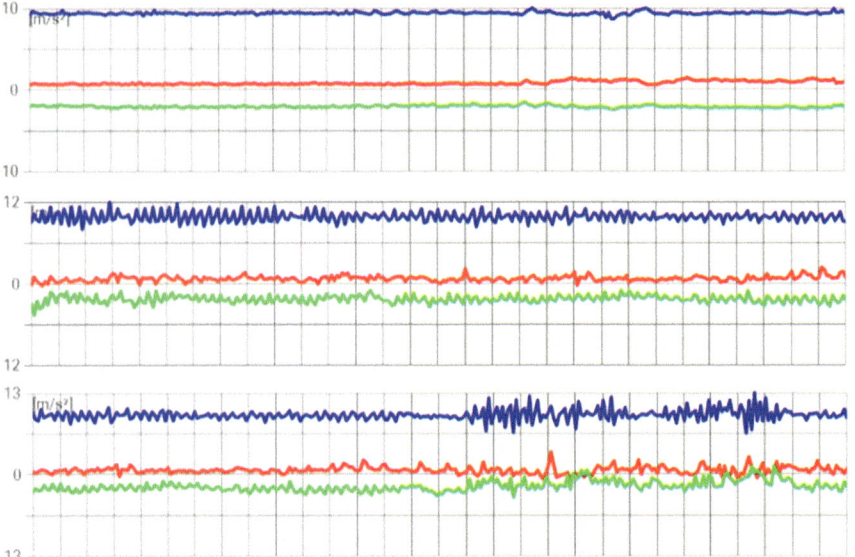

Fig. 8 Accelerometer sensor reading of the smart phone while performing activities, from top to bottom: resting, walking, and running

undertaking a variety of tasks, such as resting, walking, and running. The acquisition of the GPS was done only if the user performed a walking or resting activity.

Tables 5, 6, and 7 depict the observation results of the possible COVID-19 patients and their locations acquired by the GPS acquisition process.

Table 5 GPS extraction results of the COVID-19 cluster (Location 0-3)

Subject	Location 0	Location 1	Location 2	Location 3
P0	6.83117, 79.98122	6.88279, 79.88371	–	–
P1	6.83126, 79.98118	–	6.91358, 79.90954	–
P2	6.83127, 79.98114	–	6.91357, 79.90953	–
P3	6.83121, 79.98115	–	6.91357, 79.90954	–
P4	6.83125, 79.98122	–	–	6.8722, 79.89102
P5	6.83117, 79.98116	–	–	6.87217, 79.8911
P6	6.83121, 79.9812	–	–	–
P7	–	6.88269, 79.88373	–	–
P8	–	6.88278, 79.88388	–	–
P9	–	–	6.91357, 79.90954	–
P10	–	–	6.91357, 79.90953	–
P11	–	–	–	6.87216, 79.89097

Table 6 GPS extraction results of the COVID-19 cluster (Location 4–7)

Subject	Location 4	Location 5	Location 6	Location 7
P3	–	–	6.8885, 79.88085	–
P5	–	–	–	6.97417, 79.9508
P6	6.88438, 79.88376	–	–	–
P7	–	6.88507, 79.88337	–	–
P8	–	–	6.88851, 79.88087	6.97414, 79.95079
P12	6.94744, 79.98888	–	–	–
P13	–	6.88504, 79.88334	–	6.97416, 79.95077
P14	–	6.88506, 79.88339	–	–
P15	–	–	6.88853, 79.88087	–
P16	–	–	–	6.97413, 79.95083
P17	–	–	–	6.97415, 79.95082

Table 7 GPS extraction results of the COVID-19 cluster (Location 8-10)

Subject	Location 8	Location 9	Location 10
P12	–	6.94744, 79.98888	–
P15	6.96889, 79.94751	–	–
P18	6.96887, 79.94745	–	–
P19	6.96885, 79.9475	–	–
P20	6.96889, 79.94756	–	–
P21	6.96891, 79.94742	–	–
P22	–	6.94739, 79.98891	6.98561, 80.01236
P23	–	–	6.98561, 80.01243
P24	–	–	6.98554, 80.0124
P25	–	–	6.98566, 80.01239

Figure 9 shows the COVID-19 patient monitoring system seen from the admin side. Data from the Cloud database was acquired for visualization and mapping purposes. The Google MAP API (Application Programming Interface) was used to identify the GPS location on the map. The data shown in the data table were the COVID-19 suspects and once the patient was tested positive, the Virus Status variable

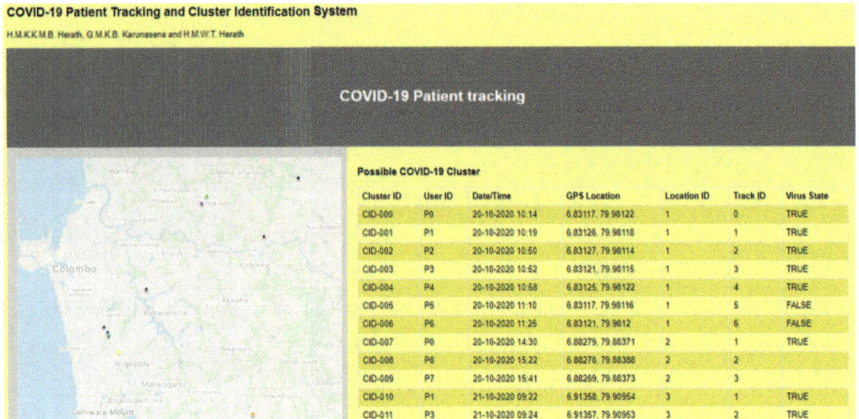

Fig. 9 Admin access window of the COVID-19 suspected patients mapping and data acquisition system

on the data-table update as "True". This action needs to be taken from the admin side, and no individual has the ability to access the status of the virus. According to the data acquired by the system, when a location is found to have more than 50 track IDs, the location represented on the location ID shall be required to monitor the safety precaution immediately or even to follow the lock-down measures.

JavaScript (JS) is a dynamic computer programming language developed in 1995. It is lightweight and is most widely used as part of web pages whose implementations allow the user to connect and build interactive web pages with the client-side script. The simulation of the possible COVID-19 cluster was developed on the basis of GPS data, location ID and patient ID using JavaScript language. We assumed that COVID-19 was negative for P5, P6 in this cluster scenario and labelled them as negative by the smart office. According to the simulation, due to the P5 and P6 from the COVID-19 virus, the green-colored path was marked as completely safe from COVID-19. The simulation result of a possible COVID-19 cluster is shown in Fig. 10.

4.2 Development of a Face Mask Detection System

This section discusses the experimental results of the face mask detection system. With multiple skin types, the face masks detection system was tested for real-time feed and the results are very promising. For the verification of the detection of face masks on different skin types, the result shown in Fig. 11 was obtained while performing the experiment. Figures 12 and 13 demonstrate the test results of the experiments on different images taken from internet sources.

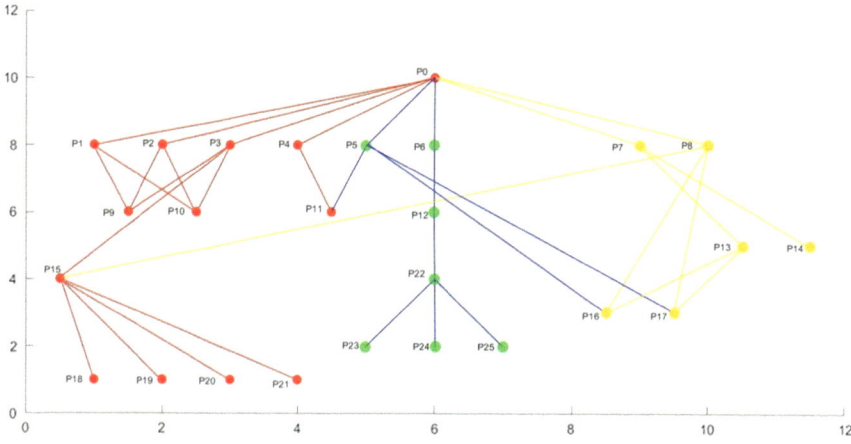

Fig. 10 Simulation of the COVID-19 cluster identification system

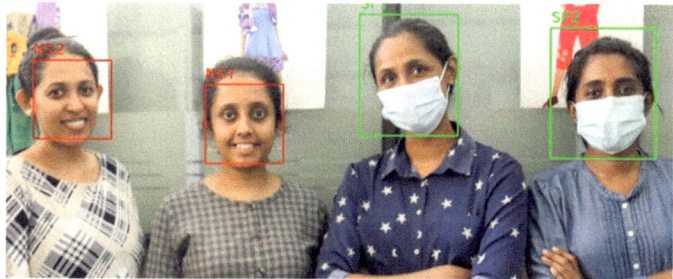

Fig. 11 Real-time testing of the face mask detection algorithm

Fig. 12 Testing of the face mask detection algorithm on internet resource 1

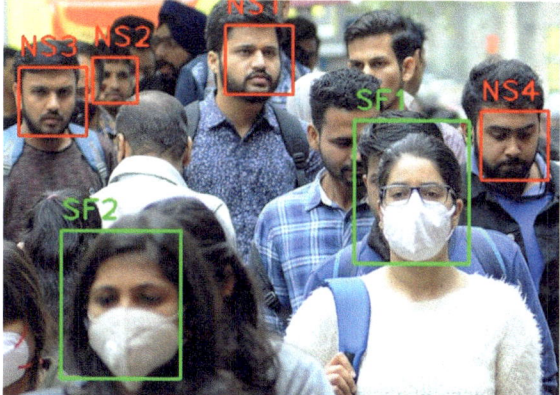

Fig. 13 Testing of the face
mask detection algorithm on
internet resource 2

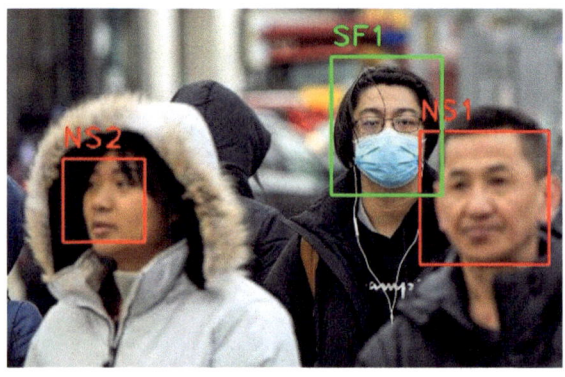

Table 8 Confusion matrix of
the SVM results

	True positive	**True negative**	
Predicted positive	TP = 40	FP = 01	41
Predicted negative	FN = 11	TN = 40	51
	51	41	92

The experiment was tested with 10 online image sources and each test were showed an error percentage of 0%, 2.105%, 1.05%, 0%, 2.106%, 1.053%, 3.158%, 1.054%, and 3.159%. The proposed design was tested with 92 face models with and without wearing face masks. Table 8 depicts the confusion matrix of the SVM algorithm. The precision of the proposed system was 0.9756 with an accuracy of 86.96%.

5 Discussions

Using IoT designs and smart healthcare concepts have been proposed as the next level of revolution in the healthcare sector which brings about a social transition that can't compete with anything in the intervening decades. All these concepts have begun to be acknowledged by the community around the world, and the governments remain reluctant to implement them due to the many constraints attached to these ideologies. An accurate implementation of the concepts of smart healthcare and IoT designs in smart cities requires a vast change in management strategy. The first aspect of this change of management is the recognition of the immediate needs and the creation of a vision and a mission for how these needs can be solved. Many challenges have to be faced when implementing the concepts of smart healthcare and IoT designs to prevent the COVID-19 pandemic in smart cities, so it needs to build the framework to face the challenges as the next step. There might be a few common difficulties that governments could have to endure, along with ways to face and overcome those difficulties when implementing the concepts of smart healthcare and IoT designs to prevent the COVID-19 pandemic in smart cities.

A significant investment will be needed to complete the entire project of integrating smart healthcare and IoT designs in smart cities to avoid COVID-19 pandemic concepts. It should then be considered in one phase at a time, but to meet all the conditions as quickly as possible, because this is a global emergency. Although the final vision and goal may have implausible benefits, their costs may preclude the authorities from approving the initiative. The proposed designs of this writing have been suggested to be applied with a view to risk control and cost reduction, using multiple consecutive small-scale IoT project implementations full of short and strong milestones with realistic costs. Starting at a minimal level with prototype technology and only investing in foundational parts carried out in stages would certainly reduce the expense of the initial investment. Furthermore, it can be introduced as a public infrastructure with the aid of more investment from private organizations.

It is possible to suggest that placing data online, especially relating to sensitive materials such as those used in smart healthcare and IoT designs discussed in this writing seems unsafe. Many IoT applications consider protection to be a crucial aspect and seek to ensure that all possible threats relating to data leakage are detected. In order to minimize risk, if data were to be intercepted in some manner, certain steps can be taken, such as decoupling information, encryption, secure by one-way or outbound messages, etc.

6 Conclusions

The proposed designs for smart healthcare and IoT designs in this writing display positive outcomes so that they can be successfully used in smart cities to avoid the COVID-19 pandemic. The experiment was more accurate in several trials, as discussed in the results section of this chapter. The COVID-19 Cluster identification system has demonstrated 96.15% accuracy for several adequate trials. The system developed to identify people who are not wearing face masks demonstrated an accuracy of 86.96% at 0.9756 precision.

Therefore as per the results shown, a precise strategy has been proposed during this writing which will help to simplify the challenges of the healthcare system in smart cities facing during the COVID-19 Pandemic. And also addressed how it helps to communicate the value of each phase to the overall project and how to make it work. The overall possibility to develop an IoT-based system to manage and balance the impact of COVID-19 in smart cities has been widely discussed. So, the proposed designs are showing promising results to prevent COVID-19 pandemic.

References

1. Drosten, C., Günther, S., Preiser, W., Van Der Werf, S., Brodt, H. R., Becker, S., et al. (2003). Identification of a novel coronavirus in patients with severe acute respiratory syndrome. *New England Journal of Medicine, 348*(20), 1967–1976.
2. Zaki, A. M., Van Boheemen, S., Bestebroer, T. M., Osterhaus, A. D., & Fouchier, R. A. (2012). Isolation of a novel coronavirus from a man with pneumonia in Saudi Arabia. *New England Journal of Medicine, 367*(19), 1814–1820.
3. Lau, S. K., Woo, P. C., Li, K. S., Huang, Y., Tsoi, H. W., Wong, B. H., et al. (2005). Severe acute respiratory syndrome coronavirus-like virus in Chinese horseshoe bats. *Proceedings of the National Academy of Sciences, 102*(39), 14040–14045.
4. Al Zobbi, M., Alsinglawi, B., Mubin, O., & Alnajjar, F. (2020). Measurement method for evaluating the lockdown policies during the COVID-19 pandemic. *International Journal of Environmental Research and Public Health, 17*(15), 5574.
5. World Health Organization. (2020). Coronavirus disease 2019 (COVID-19): Situation report, 72.
6. Jiang, F., Deng, L., Zhang, L., Cai, Y., Cheung, C. W., & Xia, Z. (2020). Review of the clinical characteristics of coronavirus disease 2019 (COVID-19). *Journal of General Internal Medicine*, 1–5.
7. Roser, M., Ritchie, H., Ortiz-Ospina, E., & Hasell, J. (2020). Coronavirus pandemic (COVID-19). *Our World in Data, 2020,* 1.
8. Neirotti, P., De Marco, A., Cagliano, A. C., Mangano, G., & Scorrano, F. (2014). Current trends in Smart City initiatives: Some stylised facts. *Cities, 38,* 25–36.
9. Mohanty, S. P., Choppali, U., & Kougianos, E. (2016). Everything you wanted to know about smart cities: The Internet of things is the backbone. *IEEE Consumer Electronics Magazine, 5*(3), 60–70.
10. Jawhar, I., Mohamed, N., & Al-Jaroodi, J. (2018). Networking architectures and protocols for smart city systems. *Journal of Internet Services and Applications, 9*(1), 26.
11. Ashton, K. (2009). That 'internet of things' thing. *RFID Journal, 22*(7), 97–114.
12. Guillemin, P., & Friess, P. (2009). Internet of things strategic research roadmap. *The Cluster of European Research Projects*. In Technical Report. River Publishers.
13. Al-Fuqaha, A., Guizani, M., Mohammadi, M., Aledhari, M., & Ayyash, M. (2015). Internet of things: a survey on enabling technologies, protocols, and applications. *IEEE Communiations Surveys Tutorials, 17*(4), 2347–2376.
14. Perera, C., Zaslavsky, A., Christen, P., & Georgakopoulos, D. (2013). Context aware computing for the internet of things: a survey. *IEEE Communications Surveys Tutorials, 16*(1), 414–454.
15. Dastjerdi, A. V., & Buyya, R. (2016). Fog computing: Helping the Internet of Things realize its potential. *Computer, 49*(8), 112–116.
16. Alavi, A. H., Jiao, P., Buttlar, W. G., & Lajnef, N. (2018). Internet of Things-enabled smart cities: State-of-the-art and future trends. *Measurement, 129,* 589–606.
17. Kodali, R., Swamy, G., & Lakshmi, B. (2015). An implementation of IoT for healthcare. *2015 IEEE Recent Advances in Intelligent Computational Systems (RAICS)* (Vol. 10, pp. 411–416).
18. Nazir, S., Ali, Y., Ullah, N., & García-Magariño, I. (2019). Internet of Things for healthcare using effects of mobile computing: A systematic literature review. *Wireless Communications and Mobile Computing, 2019,* 1–20.
19. Almotiri, S., Khan, M., & Alghamdi, M. (2016) Mobile Health (m-Health) system in the context of IoT. In *2016 IEEE 4th International Conference on Future Internet of Things and Cloud Workshops (FiCloudW)* (pp. 39–42).
20. Karaca, Y., Moonis, M., Zhang, Y., & Gezgez, C. (2019). Mobile cloud computing based stroke healthcare system. *International Journal of Information Management, 45,* 250–261.
21. Nisha, N., Iqbal, M., Rifat, A., & Idrish, S. (2015). Mobile health services. *International Journal of Asian Business and Information Management, 6*(1), 1–17.
22. Nandana, W., Mel, W., & Priyankara, H. (2012). Online remote laboratory for open distance learning. In *First International Conference on Open and Distance e-Learning* (pp. 1–10).

23. Zhao, K., & Ge, L. (2013). A survey on the internet of things security. In *9th International Conference on Computational Intelligence and Security (CIS)*, IEEE (pp. 663–667).
24. Sheng, Z., Yang, S., Yu, Y., Vasilakos, A. V., McCann, J. A., & Leung, K. K. (2013). A survey on the IETF protocol suite for the internet of things: Standards, chal-lenges, and opportunities. *IEEE Wireless Communications, 20*(6), 91–98.
25. Vasseur, J. P., & Dunkels, A. (2008). Ip for smart objects. *White Paper, 1,* 1–6.
26. Islam, M. M., Rahaman, A., & Islam, M. R. (2020). Development of smart healthcare monitoring system in IoT environment. *SN Computer Science, 1*(3), 1–11.
27. Tamilselvi, V., Sribalaji, S., Vigneshwaran, P., Vinu, P., GeethaRamani, J. (2020) IoT based health monitoring system. In *2020 6th International Conference on Advanced Computing and Communication Systems (ICACCS)* (pp. 386–389). IEEE. (March).
28. Sareen, S., Sood, S. K., & Gupta, S. K. (2018). IoT-based cloud framework to control Ebola virus outbreak. *Journal of Ambient Intelligence and Humanized Computing, 9*(3), 459–476.
29. Mohan, A. (2014) Cyber security for personal medical devices internet of things. In *2014 IEEE International Conference on Distributed Computing in Sensor Systems* (pp. 372–374). IEEE. (May).
30. Mallik, R., Sing, D., &Bandyopadhyay, R. (2020). GPS tracking app for police to track ambu-lances carrying COVID-19 Patients for ensuring safe distancing. *Transactions of the Indian National Academy of Engineering*, p. 1.
31. Hassan, M. M., Lin, K., Yue, X., & Wan, J. (2017). A multimedia healthcare data sharing approach through cloud-based body area network. *Future Generation Computer Systems, 66,* 48–58.
32. Susi, M., Renaudin, V., & Lachapelle, G. (2013). Motion mode recognition and step detection algorithms for mobile phone users. *Sensors, 13*(2), 1539–1562.
33. Meenpal, T., Balakrishnan, A., & Verma, A. (2019). Facial mask detection using semantic segmentation. In *2019 4th International Conference on Computing, Communications and Security (ICCCS)* (pp. 1–5). IEEE. (October).
34. Whitehill, J., & Omlin, C. W. (2006) Haar features for FACS AU recognition. In *7th Inter-national Conference on Automatic Face and Gesture Recognition (FGR06)* (pp. 5). IEEE. (April).
35. Abeyrathne, W. S. L., Madushanka, B. G. D. A., & Priyankara, H. D. N. S. (2020). Vision-Based Fallen Identification and Hazardous Access Warning System of Elderly People to Improve Well-Being.
36. Madhusanka, B. G. D. A., & Sureswaran, R. (2020). Recognition of Daily Living Activities Using Convolutional Neural Network Based Support Vector Machine.

Big Health Data: Cardiovascular Disease Prevention Using Big Data and Machine Learning

Salma Lbrini, Abdelhamid Fadil, Zakaria Aamir, Mohamed Khomali, Hassane Jarar Oulidi, and Hassan Rhinane

Abstract By 2030, almost 23.6 million people will die from cardiovascular heart diseases (CVDs). In Morocco, deaths by CVDs represented 38% in 2018. Using machine learning and tracking patient health indicators can reduce this mortality. Indeed, the aim of this study is developing an application that collects and process a stream of geolocation and heart rate data, stores the data and predicts on cardiovascular heart diseases risk. We first construct the machine learning model, define the architecture then we developed and tested the data pipeline. Samsung smartwatch was used to collect heart rate and location, Kafka and Spark were used to collect the streamed data received from the Smartwatch, the Data was then stored in MongoDB. This work produced a development of a complete real time data pipeline from data production to alerts and reports generation using big data and machine learning technologies.

Keywords Big data · Health · Real time · Machine learning · Geolocation · Cardiovascular heart disease

1 Introduction

The World Health Organization defines the cardiovascular heart diseases (CVDs) as the group of disorders of heart and blood vessels, and include.

- hypertension (high blood pressure);
- coronary heart disease (heart attack);
- cerebrovascular disease (stroke);
- peripheral vascular disease;
- heart failure;

S. Lbrini (✉) · A. Fadil · H. Jarar Oulidi
Geoscience Laboratory, Aïn Chock Faculty of Science, Km 8 El Jadida Road, B.P: 5366, Casablanca, Morocco

Z. Aamir · M. Khomali · H. Rhinane
Hassania School of Public Works, Km 7, El Jadida Road, B.P.: 8108, Casablanca, Morocco

© The Author(s), under exclusive license to Springer Nature Switzerland AG 2021
U. Ghosh et al. (eds.), *Machine Intelligence and Data Analytics for Sustainable Future Smart Cities*, Studies in Computational Intelligence 971,
https://doi.org/10.1007/978-3-030-72065-0_17

- rheumatic heart disease;
- congenital heart disease;
- cardiomyopathy.

By 2030, almost 23.6 million people will die from CVDs. In Morocco, based on the report of the national survey on common risk factors of non-communicable diseases of 2017–2018, deaths by CVDs represented 38% in 2018. Smoking, harmful use of alcohol, physical inactivity, unhealthy diets, obesity, hypertension, diabetes and hyperlipidemia are the established risk factors of CVDs [1]. In Morocco, efforts are done to reduce CVDs morbidity, the multisectoral national strategy of prevention and control of non-communicable diseases 2019–2029 stated two main actions [2].

(a)   Improve the response to neuro-cardiovascular emergencies;
(b)   Develop a program for the prevention and control of cardiovascular illnesses.

Big Data and machine learning can help on these two actions. On the one hand, combined with geolocation, monitoring vital signs in real time can ensure interventions in case of emergencies by sending alerts at the right time. On the other hand, Big Data and machine learning proved their efficiency in prevention and prediction CVDs by using Big Data storage capacity and machine learning power of prediction and risk factors determination. This prediction will prevent CVDs mortality by providing the right advice to the patient before it is too late.

In the literature, to prevent CVDs, many studies use machine learning, using mostly as parameters or risk factors the attributes of the UCI Cleveland standard [3]. In 2018, Dinesh and Al proposed a prediction model to predict and prevent heart disease by comparing the accuracies of five algorithms: Support Vector Machine, Gradient Boosting, Random forest, Naive Bayes classifier and logistic regression using UCI machine learning repository [4]. In 2019, Nagaraj and Al found that SVM with radial kernel offers better accuracy than Naïve Bayes classification using UCI Cleveland data set [5]. In 2019, Senthilkumar and Al, using Cleveland dataset, proved that a hybrid approach combining the characteristics of Random Forest (RF) and Linear Method (LM) is quite accurate in the prediction of heart disease with an accuracy level of 88.7% [6]. In 2020, Hossain and Al demonstrated that combining network analytics and machine learning techniques could be used successfully for disease risk prediction [7].

Many studies combine the use of machine learning with the use of big data by monitoring vital signs of the human body. Malti and Al presented in 2019 a system architecture where, with the help of IoT and various communication protocols, the ECG information is sent to the doctor as well as the cloud-based storage system [8]. Samiul and Al proposed in 2019 an algorithm to predict the risk factors of the CVDs with a novel Intelligent Healthcare Platform for continuous data collection and patient monitoring system [9]. In 2018, we proceeded a systematic study that shows a lack in studies interested in Big Health Data technologies used in real time remote sensing, in fact from the 15% articles studying real time remote sensing only 29% are citing the architecture [10].

In this work, we will develop and experiment a system tracking health data in real time, alert on emergencies and use machine learning to predict CVDs. This system can reduce CVDs morbidity by personalized patient monitoring. This connection with the patient will help in giving the right advice at the right time, preventing CVDs and alimenting the system continuously with Big Data that will be used to upgrade machine-learning algorithm. In this chapter, we will first construct the machine learning model, and then proceed studies to define the complete framework architecture from data production to alerts generation. We will conclude with results and overview of this data pipeline application.

2 Methods

This study aims to develop a real-time ingestion, storage, treatment and then alert system for CVDs. To do so, we follow the steps illustrated in Fig. 1.

2.1 Machine-Learning Model Construction to Predict CVDs Risk

The aim of this machine learning is to classify whether the patient has a CVDs risk or not.

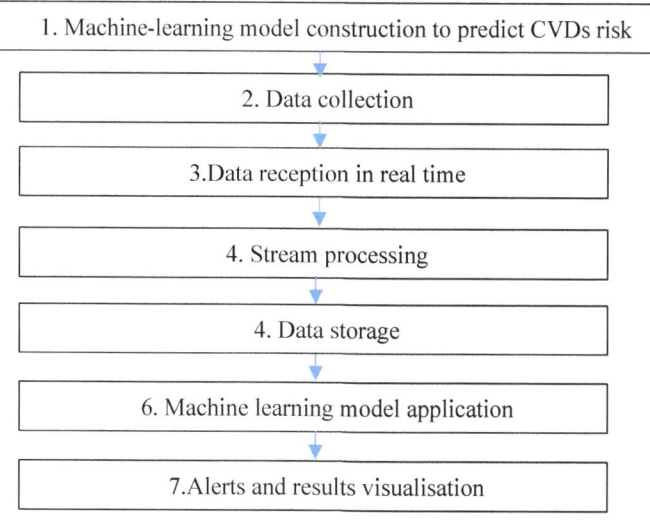

Fig. 1 Steps followed to develop a real-time ingestion, storage, treatment and then alert system for CVDs

We used as risk factors, the14 risk factors listed below.

Age, gender, heart rate, fasting blood sugar, resting blood pressure, serum choles-terol, exercise-induced ST depression, serum cholesterol, chest pain type, resting electrographic results, exercise induced angina, slope of peak exercise ST segment, thalassemia and number of major vessels colored by fluoroscopy.

Using an available online Jupyter notebook [11], we experimented five classifi-cation algorithms on Cleveland dataset heart disease public data.

70% of data was used for test and 30% for validation after cleaning and preparation. The five classification algorithms used are listed below.

Gradientboostingclassifier, randomforestclassifier, svc, extratreesclassifier and logistic regression.

A set of parameters combination for each algorithm was tested. The Table 1 summarizes the parameters' combination that provided the best accuracy.

To choose the suitable model for our study, we evaluated the performance of the five algorithms with the best parameters on the 30% data kept for validation. The evaluation was based on:

The accuracy, the precision and the AUC, which represents the probability of getting accurate results from a classification model [12].

After applying the algorithms on the validation dataset, we obtained the results below (Table 2).

Table 1 Accuracy of selected algorithms applied on Cleveland Data

Algorithm	Best parameters	Accuracy (%)
GradientBoostingClassifier	loss = "deviance" learning rate = 0.01 n estimators = 100 max depth = 3 max features = "log2"	82,1
RandomForestClassifier	n estimators = 100 max features = log2 criterion = "entropy"	81,2
SVC	C = 0.01 gamma = 0.1 kernel = "linear" degree = 3 coef0 = 10.0	83,1
ExtraTreesClassifier	n estimators = 1000 max features = "log2" criterion = "entropy"	80,2
LogisticRegression	C = 0.01 penalty = "l2" fit intercept = False	83,1

Table 2 Precision and accuracy of algorithms applied on validation dataset

Algorithm	Accuracy	Precision
ExtraTreesClassifier	0.922	0.947
GradientBoostingClassifier	0.889	0.919
LogisticRegression	0.889	0.919
SVC	0.878	0.917
RandomForestClassifier	0.867	0.854

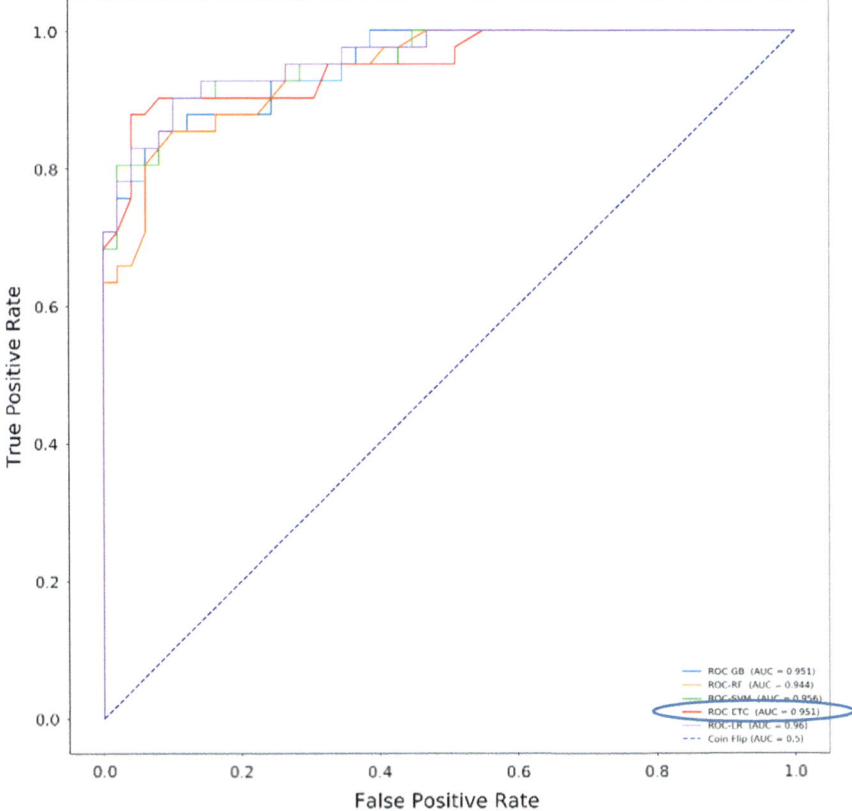

Fig. 2 Receiver Operating Characteristic Curve (ROC)

We then generated the graph that shows that ExtraTreesClassifier's Receiver Operating Characteristic (ROC) curves had Excellent Area under the Curves (AUC): 95.1% (Fig. 2).

Based on these results, the machine-learning model that we will use is the extratreesclassifier with parameters below.

n estimators = 1000; max features = "log2" and criterion = "entropy".

2.2 Data Collection

For this study, we need to collect in addition to geographic position, the 14 risk factors used in Sect. 2.1.

For this work, we need a device that fulfills those requirements.

Table 3 Comparatif of devices sending data options ordered by price

The watch	API	Sending data directly to server
Fitbit Charge HR	Fitbit API	Necessite paired phone
Fitbit Charge 3	Fitbit API	Necessite paired phone
Samsung gear sport	Tizen	Yes
Apple watch series 3 (GPS + Cellular)	Watch os 4	Yes
Apple watch series 4 (GPS + Cellular)	Watch os 5	Yes

- Open source API (Application programming interface) for development,
- sending directly data to servers without necessary connection to the phone,
- Wifi access,
- accurate GPS and accurate health sensors such as heart rate, blood pressure…

A research entitles "Accuracy of Heart Rate Watches: Implications for Weight Management" compared the heart rate accuracy of four wrist-worn devices (Apple Watch, Fitbit Charge HR, Samsung Gear S and Mio Alpha), it was found that none of the devices performed significantly better overall and these devices accurately measure heart rate [13].

Since Apple Watch, and Fitbit Charge HR and Samsung gear sport have GPS sensors, a comparative study based on APIs and prices in 2018 was done to choose the suitable one for this study. The following Table 3 shows the different devices sending data options ordered by price.

For development need we can use apple Series 3 and Apple Series 4 and Samsung Gear Sport but not only those devices can be used. Day after day other wearables that support more features appear.

For our application, we used **Samsung Gear Sport** taking in consideration the price and the features.

In fact, this Smartwatch provides Tizen witch is an open and flexible operating system [14].

This device fulfils also application requirements especially [15].

- HR Sensor
- GPS Battery Time Up to 20 h
- Location Technology GPS, Glonass
- Wi-Fi 802.11b/g/n 2,4 GHz
- NFC available.

Since this device doesn't allow blood pressure and blood sugar level measure, the indicators collection will be as below.

- The age and the gender: inserted once while subscribing patient information.
- The geographic position and the heart rate in rest and maximum heart rate: automatically detected in real time.
- Other risk parameters: updated manually by the patient using a web application form.

However, doctors can suggest other indicators and may need more patient data for this type of studies, for this reason, the architecture must allow doctors to add more indicators flexibly.

2.3   Data Reception in Real Time

To handle emergencies, we choose to receive data in real time using Kafka which is used to build real-time streaming data pipelines [16]. Kafka run as a cluster on one or more servers and stores data streams in categories called topics [17].

2.4   Stream Processing of Received Data

The data generated from Kafka represents a stream of data that should be processed in real time to alert on emergencies. Based on a comparative study between different stream processing frameworks (see Fig. 3) [18], Apache Spark was chosen due to the availability of Streaming query and diverse programming languages especially the Spark Python API (PySpark) that exposes the Spark programming model to Python [19].Spark has also an active user and developer community. It has a list of companies that use it [16].

Many sources including Kafka can be used as a source of Spark Streaming.

This component receives live input data streams to be processed by the Spark engine [21] (see Fig. 4).

2.5   Data Storage

The concept of big data is described according to several levels. The four most commonly recognized dimensions represented by 4 V are Volume, Variety, Velocity, and Veracity [22].

- The volume represents the amount of data stored [23].
- The variety describes the diversity [23].
- Velocity is the manner and speed at which data is stored and analyzed [23].
- The veracity describes the approach that ensures the reliability and quality of data [22].

This study verifies the 4 V dimensions.

- The volume: The stream of data generated by each patient every year can attend more than 10 Go if data is collected every second.
- Variety: Doctors should have the ability to suggest any indicator any time.

	Spark	Storm	Flink	Samza
Data format	DStream	Tuples	DataStream	Message
Data sources	HDFS, DBMS, and Kafka	Spoots	HDFS, DBMS, and Kafka	kafka
Programming model	Transformation and action	Bolts	Actions functions (map,groupby...)	Mapreduce Job
Programming languages	Java, Scala and Python	Java	Java	java
Cluster manager	Hadoop YARN, Apache Mesos	Zookeeper	Hadoop YARN, Apache Mesos	YARN
Latency	Few seconds	Sub-second	Sub-second	Sub-second
Messaging	Exactly once	At least once	Exactly once	Exactly once
Machine learning compatibility	SparkMLLIB	Compatible with SAMOA API	FlinkML	Compatible with SAMOA API
Elasticity	Yes	Yes	No	No
Sliding windows/Windowing	time based	time based and count based	time based	time based and count based
Auto-parallelization	On demand	Pipelined processing	Pipelined processing	On demand
Streaming query	SparkSQL	No	No	Yes (Samza-SQL API)
Data Partitioning	Yes	No	No	Yes
API	Declaratif	Copositionnel	Declartaif	Copositionnel
Data transport	RPC	RPC	RPC	Kafka

Fig. 3 A comparative study between different stream processing [20]

Fig. 4 Illustration of Spark Streaming data communication [21]

- Veracity and Velocity: In addition to predictions, the aim of our study is urgent interventions in real time after receiving alarms and location of patient in danger.

Compared to relational databases, NoSQL (Not Only SQL) databases show higher performance in handling large volumes of structured, semi structured and unstructured data due to its schema less data model [24].

Taking into considerations the parameters below, we used in this study the NoSQL database MongoDB, to store received data.

1. Lower execution times: MongoDB provides lower execution times than MySQL in all four basic operations, which is essential when an application should provide support to thousands of users simultaneously [25].
2. Document Database: in this study, the architecture must allow doctors to add more indicators. MongoDB is a suitable choice because it is a document database built on a scale-out architecture. The document database model is a powerful way to store and retrieve data that allows developers to move fast [26].
3. MongoDB is an open source technology.
4. Geospatial support: MongoDB uses GeoJSON witch is an open-source format containing simple geographical features that is based on JavaScript Object Notation [27].
5. Availability of information: MongoDB has also an active user and developer community, it's the most popular NoSQL database for the last years. The Fig. 5 shows that MongoDB is the most document Database used in the last years [28].

The figure below represents a document of received data stored in MongoDb (Fig. 6).

2.6 Machine Learning Model Application on Data

The aim of this part is to predict CVDs risk presence to alert patients and doctors at early stage. Using pyspark, the python API for spark, we call the machine-learning model constructed in Sect. 2.1. If the risk exists, doctors must receive a report and provide diagnostic and advices to avoid patient health complications.

2.7 Alerts and Data Visualization

Due to its ability to process data stream in real time, in case of emergency, the system allows sending alerts to doctors and a list of person set by the patient.

There are two types of alerts.

- Alerts based on detection of abnormal value of one of the attributes below.

Heart rate, fasting blood sugar, resting blood pressure, serum cholesterol.

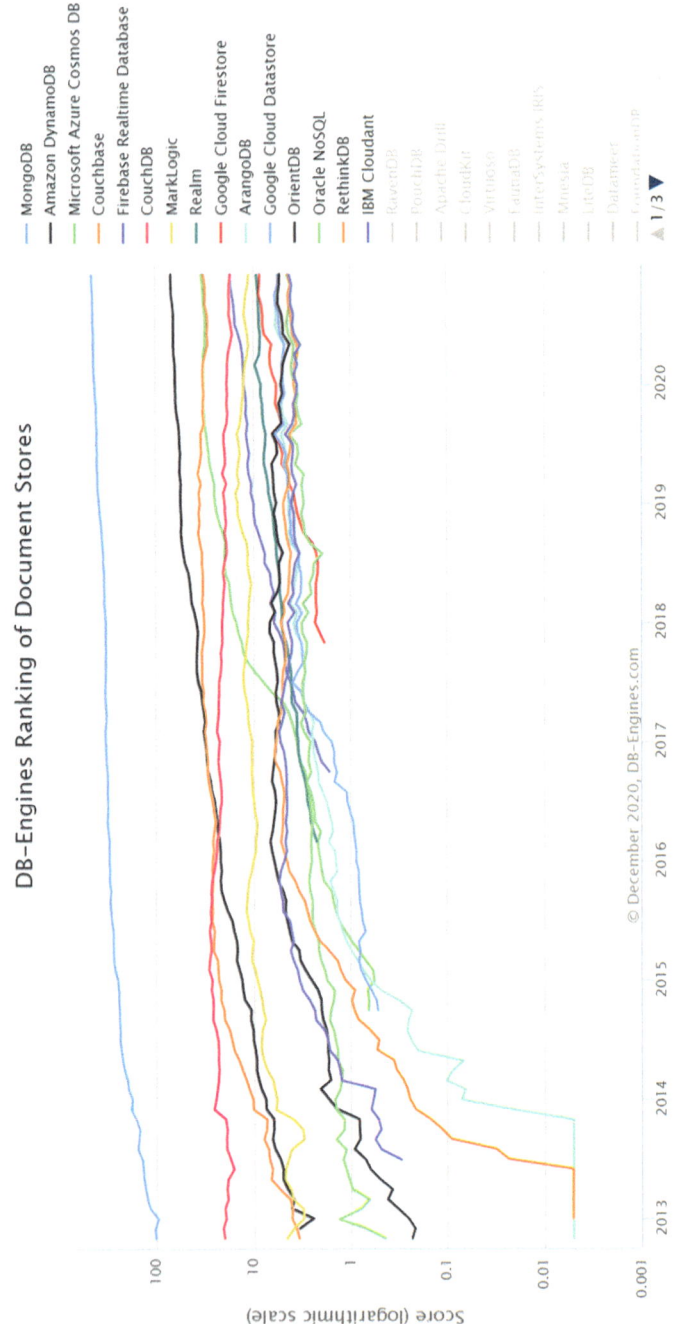

Fig. 5 DB-engines ranking of document stores [28]

```
_id: ObjectId("5f0306673e33e25dfd324976")
name: "Sara "
TIMEPOST: "2020-07-06 11:08:47"
HOUR: "00"
MIN: "10"
SEC: "06"
date: "2020-07-06"
date3: 2020-07-06T23:10:06.000+00:00
RATE: "81"
Longitude: "-7.60282"
Latitude: "33.58751"
Age: "62"
SEX: "0"
FBS: "1"
RBP: "140"
SC: "268"
ST: "3.6"
CPT: "4"
RE: "1"
EIA: "1"
Slope: "3"
Thal: "7"
nbrv: "1"
```

Fig. 6 Json Document structure in MongoDb

Only doctor can specify the normal range based on patient data like age, gender, weight…

- Alerts based on detection of CVDs risk after machine learning application.

  The system allows sending alerts by Short Message Service (sms) or mails.
  A Spring boot web application is also developed to.

- Visualize on map patient location;
- Allow manual integration of patient data;
- Visualize machine learning results…

3   Results

In this work, Samsung smartwatch was used to collect heart rate and location.

Kafka and Spark were used to collect the streamed data received from the Smartwatch. The Data is then stored in MongoDB. Other parameters are integrated directly by the doctor using a spring boot application.

This work needed a development of a complete real time data pipeline from data production to alerts and reports generation.

Based on studies in Sect. 2, the architecture is summarized in Fig. 7.

There's a lack in studies interested in Big Health Data technologies used in real time remote sensing [10], so this work covers a part less treated in this type of studies, furthermore it combines Big Health Data with geolocation and machine learning.

In next paragraph, we will show some application results like the latency and the overview.

3.1   The Latency

Using an intermediate http server to send the data from the smartwatch to Kafka, we calculated the latency, it's estimated to 2 s between the time the data is posted and the time it's subscribed to Kafka topic. Kafka is open source but a Kafka MQTT proxy can be used under license to send directly the data from the smartwatch without an intermediate, which can decrease the latency (Table 4).

3.2   Application Overview

The results are a stream of data generated from every connected device (see Table 5). Processing data in real time generates updated charts and maps as soon as the data arrives: (see Figs. 8 and 9).

We integrated the machine-learning algorithm ExtraTreesClassifier to predict the patient cardiovascular heart disease. If the prediction is positive, we advise patient to take more care of his health as shown in Fig. 10.

4   Conclusion

In Morocco as in the entire world, CVDs are the cause of a huge number of deaths. Big Data and machine learning can reduce this morbidity by decreasing emergencies intervention time and preventing health complications by making people aware of

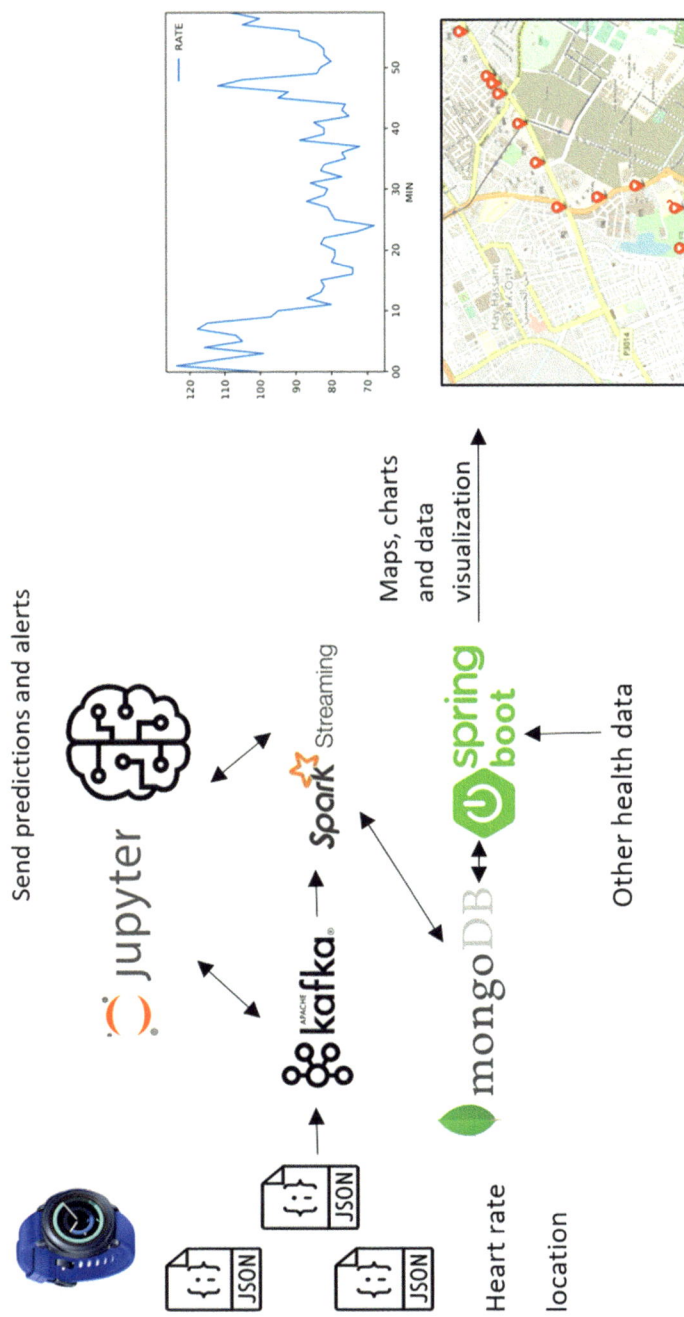

Fig. 7 Application architecture

Table 4 Latency of 10 received dataset from the smartwatch

Difference between timepost/timewatch (in ms)	Difference between timewatch/Kafka (in ms)
130	854
189	1431
150	1029
92	237
101	1396
201	526
101	534
83	1312
313	182
108	1085

Table 5 Data generated from the Smartwatch

ID	Name	Timepost	Rate	Longitude	Latitude
2	SARA	2019-05-06 23:28:01	71	−6.79694	34.06936
2	SARA	2019-05-06 23:29:00	81	−6.79694	34.06936
2	SARA	2019-05-06 23:30:02	81	−7.69543	33.54321
2	SARA	2019-05-06 23:31:00	80	−7.69543	33.54321

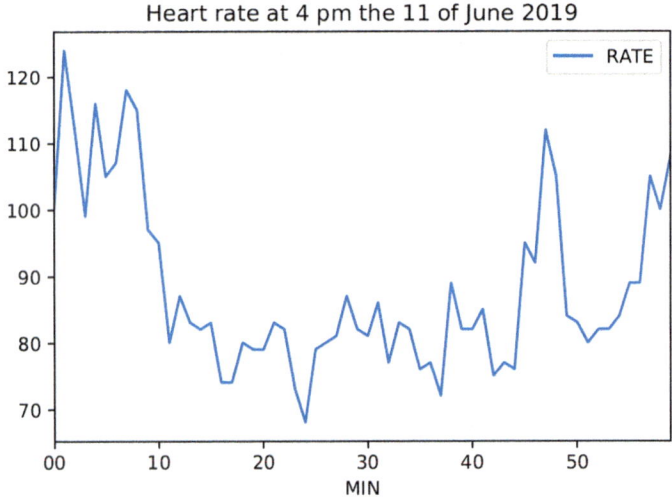

Fig. 8 Data Example of Heart rate chart generated in real time from Smartwatch Data

Hay Hassani
حي الحسني
P3014
Boulevard Sidi Abderrahmane
CAFC
Anfa Clubs
Anfa Sport
Cité de l'Artisanat
Main Street Sud
Anfa Park
Anfa Cité des Arts
Anfa Halles
Université

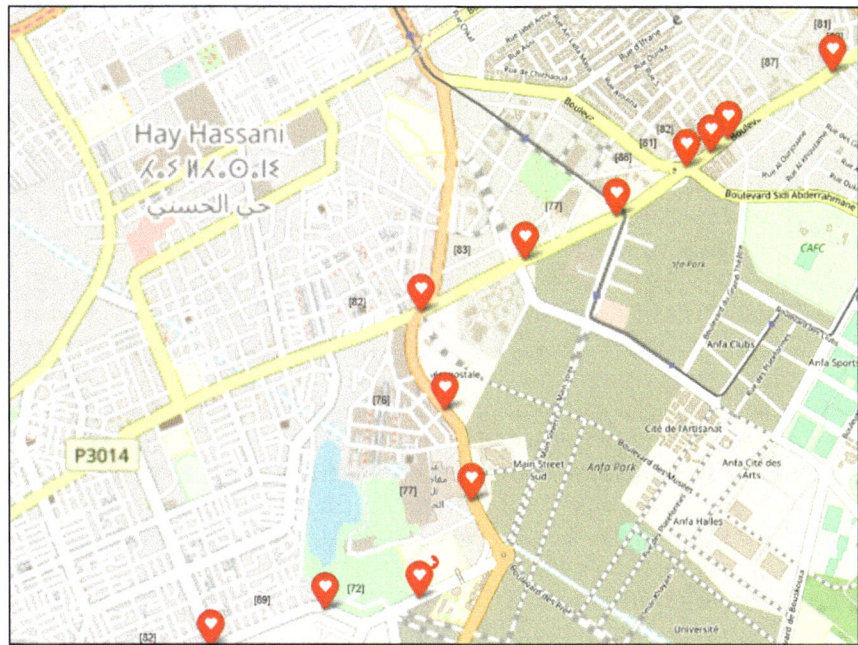

Fig. 9 Example of map with Heart rate values generated in real time from Smartwatch Data

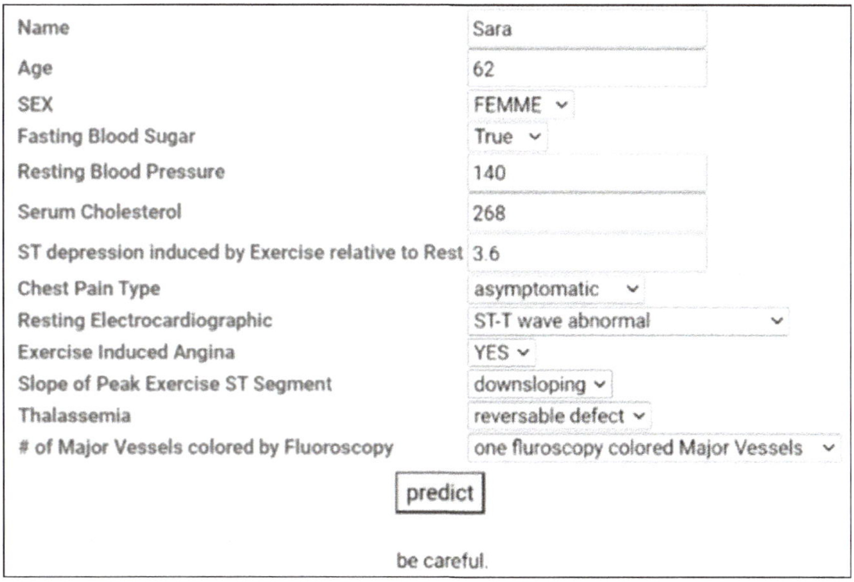

Name
Sara
Age
62
SEX
FEMME
Fasting Blood Sugar
True
Resting Blood Pressure
140
Serum Cholesterol
268
ST depression induced by Exercise relative to Rest
3.6
Chest Pain Type
asymptomatic
Resting Electrocardiographic
ST-T wave abnormal
Exercise Induced Angina
YES
Slope of Peak Exercise ST Segment
downsloping
Thalassemia
reversable defect
# of Major Vessels colored by Fluoroscopy
one fluroscopy colored Major Vessels
predict
be careful.

Fig. 10 Cardiovascular risk prediction

risk factors. To benefit from this technological advance, we needed a development of a complete real time data pipeline from data production to alerts and reports generation.

Indeed, in this work, we developed an application that collects and process a stream of geolocation and heart rate data stores the data and predicts cardiovascular heart disease risk. We first built the machine-learning model, defined the architecture then we developed and tested the data pipeline.

We installed a Tizen application in Samsung gear sport to collect the data, but many other devices can be used by developing such application in their own API.

We found that with using Kafka as a data receiver we had nearly 2 s of latency, this value might decrease by using directly mqtt proxy, which is a product under license that receives data directly from the smartwatch without an intermediate.

To benefit from the analytical and preventive aspects of this application, we suggest experiments involving medical specialists.

In fact, many published works demonstrated a relationship between work related stress and the development of cardiovascular diseases [29], but real time datamining of big geolocated health data can lead to detect dangerous relations between psychology and cardiology like children fear of school, jobs stress…making the patient aware before it's too late.

Two difficulties were noticed. The first one is the battery capacity since the application needs an active consumption in long hours especially to get the geolocation using the Samsung smartwatch. The second is the Wi-Fi availability.

The use of this application by patients will produce a huge quantity of data that can be used in deep learning.

The study didn't cover the security aspect, indeed we propose a dedicated study in this subject.

References

1. Prevalence of cardiovascular disease risk factors: A community-based cross-sectional study in a peri-urban community of Kathmandu, Nepal, https://www.sciencedirect.com/science/article/pii/S0019483217305783, https://doi.org/10.1016/j.ihj.2018.03.003.
2. National Multisectoral Prevention Strategy and Control of Non-Communicable Diseases report, https://www.sante.gov.ma/Docments/2019/02/Plan%20Strate%CC%81gique.pdf.
3. Fatou, N., Ibrahima, F., Mamadou, C., & Alassane, B. (2020). A study on predicting and diagnosing non-communicable diseases: case of cardiovascular diseases, 1–8, https://doi.org/10.1109/iscv49265.2020.9204022.
4. Dinesh, K. G., Arumugaraj, K., Santhosh, K. D., & Mareeswari, V. (2018). Prediction of cardiovascular disease using machine learning algorithms. *International Conference on Current Trends towards Converging Technologies (ICCTCT)*, Coimbatore, pp. 1–7, https://doi.org/10.1109/icctct.2018.8550857.
5. Nagaraj, M. L., Chethan, C., Basavaraj, C, & Pol, S. (2019). Prediction of heart disease using machine learning. *International journal of Recent Technology and Engineering, 8*(2S10), 474–477, https://doi.org/10.35940/ijrte.b1081.0982s1019.

6. Mohan, S., Thirumalai, C., & Srivastava, G. (2019). Effective heart disease prediction using hybrid machine learning techniques. In *IEEE Access*, vol. 7, pp. 81542–81554, https://doi.org/10.1109/access.2019.2923707.
7. Hossain, Md. E., Shahadat, U., & Khan, A. (2020). Network analytics and machine learning for predictive risk modelling of cardiovascular disease in patients with type 2 diabetes. 113918, https://doi.org/10.1016/j.eswa.2020.113918.
8. Malti, B., & Bani, G. (2019). IoT & Big Data in Smart Healthcare (ECG Monitoring), 390–396. https://doi.org/10.1109/comitcon.2019.8862197.
9. MD, I., Haider, U., Samir, U., Mohammed, K. (2019). Intelligent Healthcare Platform: Cardiovascular Disease Risk Factors Prediction Using Attention Module Based LSTM, 167–175, https://doi.org/10.1109/icaibd.2019.8836998.
10. Lbrini, S., Fadil, A., Rhinane, H., & Oulidi, H. J. (2019). Big Health Data a systematic mapping study. The *International* Archives of the *Photogrammetry*, *Remote Sensing* and *Spatial Information* Sciences, XLII-4/W12, pp. 113–119, (2019). Published online 2019 Feb 21, https://doi.org/10.5194/isprs-archives-XLII-4-W12-113-2019. Accessed 9 May 2019.
11. https://towardsdatascience.com/diagnostic-for-heart-disease-with-machine-learning-81b064a3c1dd.
12. Ekta, M., Bondu, V., Arbind, G. (2019). Applying machine learning algorithms to develop a universal cardiovascular disease prediction system, https://doi.org/10.1007/978-3-030-03146-6_69.
13. Wallen, M. P., Gomersall, S. R., Keating, S. E., Wisløff, U., Coombes, J. S. (2016). Accuracy of heart rate watches: implications for weight management. Published online 2016 May 27, https://doi.org/10.1371/journal.pone.0154420. Accessed 12 July 2019.
14. TIZEN Homepage, https://www.tizen.org/about. Accessed 9 May 2019.
15. SAMSUNG Homepage, https://www.samsung.com/latin_en/wearables/gear-sport/SM-R600NZBATPA/. Accessed 9 May 2019.
16. APACHE SAMZA Homepage, http://samza.apache.org/learn/documentation/0.7.0/comparisons/spark-streaming.html. Accessed 9 May 2019.
17. APACHE KAFKA Homepage, https://kafka.apache.org/intro.html. Accessed 9 May 2019.
18. Inoubli, W., Aridhi, S., Mezni, H., Maddouri, M., & Mephu-Nguifo, E (2018). A Comparative Study on Streaming Frameworks for Big Data: Latin America Data Science Workshop. Published online, http://ceur-ws.org/Vol-2170/paper3.pdf. Accessed 9 May 2019.
19. APACHE SPARK Homepage, https://spark.apache.org/docs/0.9.0/python-programming-guide.html. Accessed 9 May 2019.
20. Dendane, Y., Petrillo, F., Mcheick, H., & Ben Ali, S. (2019). A quality model for evaluating and choosing a stream processing framework architecture.
21. APACHE SPARK Homepage, https://spark.apache.org/docs/2.2.0/streaming-programming-guide.html. Accessed 9 May 2019.
22. Ben Salem, A. (2015). Qualité contextuelle des données: Détection et nettoyage guidés par la sémantique des données, PhD thesis in Computer Science, Paris 13 Sorbonne University, Paris, France.
23. Zikopoulos, P., Eaton, C., Deroos, D., Deutsch, T., & Lapis, G. (2012). *Big data: From the business perspective* (pp. 5–7). New York: Understanding Big Data. McGraw-Hill.
24. Bhogal, J., & Choksi, I. (2015). Handling big data using NoSQL. Proceedings of the 2015 29th Interna- tional Conference on Advanced Information Networking and Applications Workshops., Gwangju, pp. 393–398.
25. Győrödi, C., Gyorodi, R., Pecherle, G., Olah, A. (2015). A Comparative Study: MongoDB vs. MySQL. https://doi.org/10.13140/rg.2.1.1226.7685.
26. https://www.mongodb.com/why-use-mongodb.
27. https://kb.objectrocket.com/mongo-db/geospatial-in-mongodb-1183.
28. https://db-engines.com/en/ranking_trend/document+store.
29. Price, Anne E. (2004). Heart disease and work. *Heart (British Cardiac Society), 90*(9), 1077–1084. https://doi.org/10.1136/hrt.2003.029298.

Artificial Intelligence Techniques in Smart Cities Surveillance Using UAVs: A Survey

Narina Thakur, Preeti Nagrath, Rachna Jain, Dharmender Saini, Nitika Sharma, and D. Jude Hemanth

Abstract The security and urbanization challenge is expected to rise to 90% by 2050, and to leverage existing resources, technology is the solitary means to cope with this anticipated raise in entail. The Smart City is focused on the smooth convergence of Information and Communication Technology with the most technological innovations like well-connected home and equipment. Smart city augments the lifestyle of its residents by providing efficacious infrastructure and enhanced security. Surveillance is a recurring and monotonous assignment that descends the performance of human guards when continued for a longer period of time. Unmanned Aerial Vehicles (UAVs) or Drones can be deployed as security cameras to augment human guards. It can be deployed to track intruders, monitor unusual activities such as theft, violence and unprecedented corona-virus pandemic scenarios. UAV based visual surveillance in Smart cities, produces a huge amount of multimedia data. The need to process and analyze the data automatically in real-time is critical. Artificial Intelligence and Deep learning imitates human intelligence and provides excellent analytical capabilities to learn about complex data obtained in real environments. The integrated solution of Deep learning technology with the UAVs an electronic eye-in-the-sky has leveraged the capability of detection, recognition and deterrence in a scalable surveillance system. A comprehensive review on the potential benefits of UAVs and its applications for surveillance in smart cities has been presented. This chapter elaborates seamless integration of UAVs and Deep Learning technologies solutions for smart city surveillance. The paper concludes with a description of main challenges for the application of UAVs in deep learning solutions.

N. Thakur (✉) · P. Nagrath · R. Jain · D. Saini · N. Sharma
Department of Computer Science and Engineering, Bharati Vidyapeeth's College of Engineering, New Delhi, India

R. Jain
e-mail: rachna.jain@bharatividyapeeth.edu

D. J. Hemanth
Department of Electronics and Communication Engineering, Karunya University, Coimbatore, Tamil Nadu, India

Keywords Smart cities · Unmanned aerial vehicles · Deep learning ·
Surveillance · Artificial intelligence

1 Introduction

Today, 54% of the world's population lives in cities, which is projected to hit 66% by
2050. In progression with the rising population expansion, over the subsequent three
decades urbanization will bring another 2.5 billion residents to the urban areas. In the
sense of this rapid expansion that burdens the capital of our cities, environmentally,
socially and economically, sustainability is a must [1, 2]. This rapid growth would
place a burden on capital and pose a major challenge in many aspects of the everyday
life in metropolitan cities including quality service in the medical, educational, environmental, housing, public welfare and defense sectors. The addition of protective
and emergency assistance to the people is therefore an indispensable requisite. This
exigency can be congregated by connected cameras, intelligent transport systems
and public safety monitoring systems.

Currently, the demand for video monitoring costs over 35 billion dollars and is
expected to rise to over 65 billion dollars in five years. Video monitoring can be used
in many different applications, such as traffic control, public safety, parking system,
identification of instability and prevention of crime [3].

The strengths of video surveillance systems have been enhanced by structural
developments in the processing, analysis, distribution and storing of digital content
[4, 5]. Video Surveillance Systems also have a vital role to play in moving towards
Smart cities and the emerging Industrial Internet of Things (IIoT) [6, 7]. Safety and
security are one of the main growth sectors in the rapidly growing Unmanned Aerial
Vehicle (UAV) industry [10]. While the use of drones to provide aerial observations
of operational activities is a comparatively recent addition to industrial development tools in many sectors, the technology has once again emerged, with some of its
most ambitious advancement in surveillance applications evolving. An aerial vehicle
can cover far more area than slower, more complex ground surveillance devices, so
it has been a significant component of military deployment and law enforcement
for years [11–13]. The latest range of safety drones illustrates a number of situations, ranging from fixed-wing versions that are able to easily cover wide areas to
compact quadcopters that search restricted perimeters and build environments with
numerous sensors and tracking systems. Deep Learning and Artificial Intelligence
(AI) has become more widespread as cameras are able to collect data more specifically and make projections based on advanced analytical applications. In the literature, several image processing algorithms have been proposed by the researchers
and academicians for object detection or identifying events in visual data. The use
of profound expertise of surveillance through computer vision has recently evolved
tremendously, and Deep Learning methods for video surveillance systems have been
incorporated. This development has been accomplished, as Deep Learning solutions

have performed better than traditional solutions in many video surveillance applications, in which detection or identification is usually carried out on the basis of certain crafted features developed for particular applications and different scenarios [8, 9]. In using Deep learning techniques, the creation of customized features is not required and unprocessed video frames can be directly fed into a deep learning network to detect objects. Artificial Intelligence in conjunction with drone technology and cloud data management system forms the new technology i.e. Internet of Drones. The essential constituents of Internet of Drones technology is pictorially depicted in Fig. 1.

Figure 2 illustrates the research trends in Unmanned Aerial Vehicle (UAVs) deployed for surveillance using artificial intelligence in smart cities through total number of publications in the past decade. The growth in trend of research in this domain was gradual till the year 2016 and the publications almost doubled in the subsequent year. The major escalation in the development of research articles in this sphere has been seen in the year 2019 with around 1800 publications.

The chapter further follows the following sequence: Sect. 2 offers a detailed research summary on state-of-the art surveillance methods. In this section we discussed conventional methods and video based surveillance methods. The Sect. 3 focuses on the use of Artificial Intelligence in Smart cities. Various Deep Learning architectures used for image and video processing are contained in Sect. 4. The Deep Learning based surveillance in various applications has been summarized in Sect. 5. The discussion on surveillance through UAVs for public safety and security is carried out in Sect. 6. Section 7 highlights the challenges in surveillance through UAVs followed by conclusions drawn in Sect. 8.

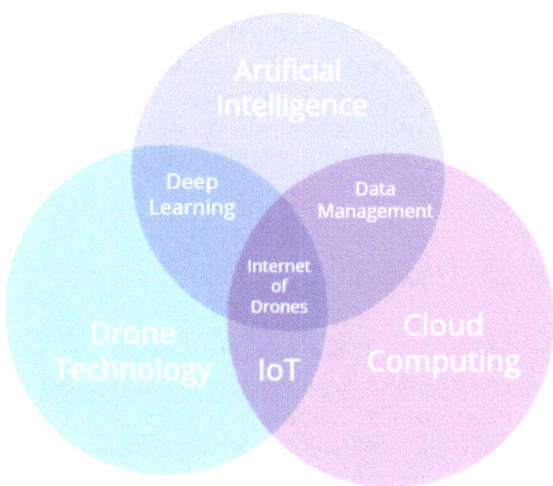

Fig. 1 Essential constituents of Internet of Drones technology

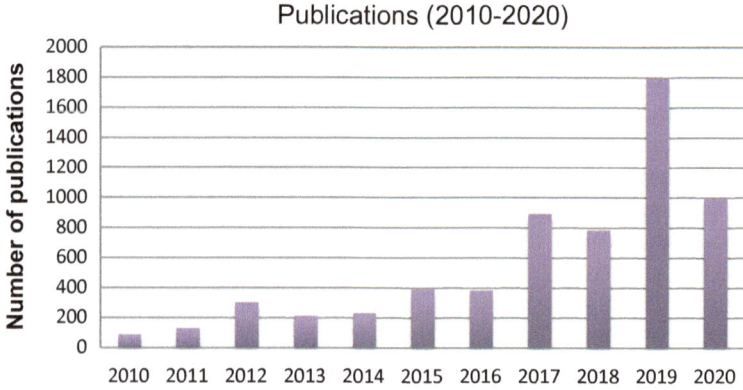

Fig. 2 Total number of publications in Unmanned Aerial Vehicle (UAVs) based surveillance using artificial intelligence in smart cities

2 Research Summary on State-of-the-Art Surveillance Methods

There are several conventional methods that have been deployed for the purpose of surveillance. Human Guards and Law enforcement is still prevalent, however, cameras are included to aid and enhance the surveillance process [14]. Camera systems can gather confidential data requiring cautious and difficult data protection and privacy monitoring. Additionally, the processing of 3D information, e.g. in human height, is difficult and, in dim or foggy smoky conditions, the technological capability of optical sensors is significantly diminished. There are also commonly used other sensor instruments or detectors for tracking public areas. The section is divided into five sub-sections discussing the state-of-the-art surveillance methods from non-visual surveillance to video based surveillance methods which further developed to the incorporation of IoT and UAVs in this realm.

2.1 Conventional Methods

The traditional methods of surveillance consisted of human involvement. The human guards or law enforcement were directly deployed at certain locations to monitor. Later human surveillance was aided with sensor equipped security devices that proved to be an efficient method to scan and detect any malicious activity or weapons. The surveillance methods were further advanced with development of biometric monitoring methods. The conventional surveillance methods have been mapped in the Fig. 3.

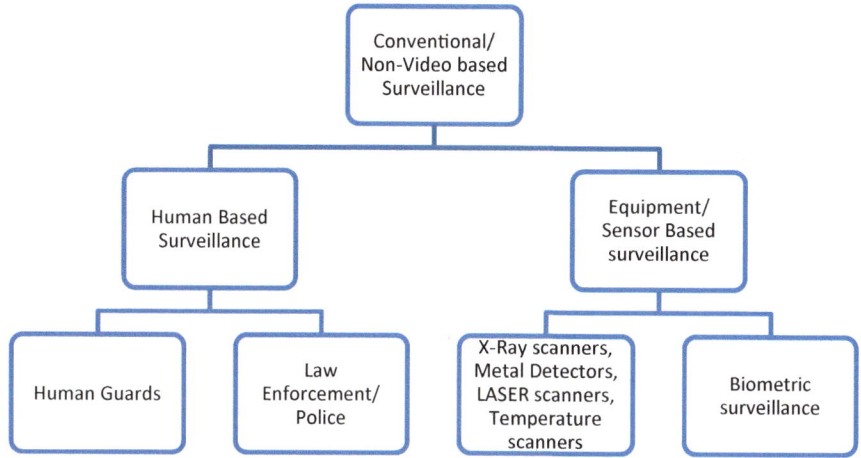

Fig. 3 Conventional surveillance methods based on non visual techniques

- *Human Based Surveillance*

The use of human security guards remains a conventional approach that can adapt to other problems which cannot be remotely controlled such as electricity, gas or chemical contamination, or illegal dumping in hidden spots on large areas [15]. The additional advantages of providing physical guards are high visual disruption; fewer false alarms are created as the incidents can be specifically scrutinized; the capacity for lowering total costs by increasing guard personnel with other services needed, such as administration services; the provision of a more auditory response as the guards can hear break-ins immediately prior to eviction [16]. However, the need for 24 h protection can be extremely expensive, long before such other concerns as recruitment, expenditure to ensure continuity, productivity and employee quality, and shift management.

Law enforcement is just part of the operations of the police department. In various cases, policing has involved numerous tasks, but the primary one being the maintenance of law & order [17, 18].

- *Equipment/Sensor Based Surveillance*

The scan equipment field includes a number of end devices ranging from laser and image barcode readers to baggage or cargo checks and human control [18–22]. These may include the Vehicle X-ray Scanner, X-ray baggage scanners, Biometric surveillance, Hand-held Metal Detector, Full Body Scanner, Under Vehicle Surveillance System, Door Frame Metal Detector, Cargo Scanners, Under Vehicle Search Mirror, and Walk-Through Temperature Scanner etc. [23, 24].

2.2 Closed Camera Television (CCTV) Technology

Since the 1940s and after 1970s, CCTV technology has been a significant player in the safety industry [25, 26]. With almost any surveillance program, there are CCTV security camera versions. Cities have begun to introduce integrated CCTV monitoring networks that include road intersections, government buildings, public places and facilities and that separate whole cities into regions and subzones, so that they have regional monitoring infrastructure as well as relay information to a central location in conjunction [27]. In an automated approach, city's CCTV technology operates effectively and is reliable in wider context, strengthening traffic control, technical monitoring for public protection, increasing response time and capacity, lowering costs for crime while even minimizing investigation time to improve emergency service capability for incidents, natural catastrophe and insurgency [28].

One of CCTV's main drawbacks is secrecy, especially when used at work. Although it will help ensure that staff and clients are secure, they will object to continuous supervision. CCTVs expense is another downside; cameras, monitors, devices for recording and other equipment adding up the costs [29]. Then it's necessary to upgrade the technology, which goes on. CCTV cameras are only able to view and record footage and not much else. Cameras must be able to identify and perceive certain events on their own in order to prevent theft, abuse, vandalism or fire effectively [30]. They ought to be able to communicate with other programs including warning devices.

Here, Internet of Things plays a major role in upgrading the technology. It integrates network-enabled cameras with other devices and applications performing various operations and converts video tracking into smart surveillance and information control.

2.3 IoT Based Surveillance Solutions

CCTVs were initially a great aid in keeping watch on any movement across the protected area. But as the technologies progressed and people became more familiar with extricating these cameras, there was a need for more complex monitoring systems [10]. IoT is built via advanced WiFi communication network, focused on Video Capturing and Continuous Live Streaming of the video to Remote Control System.

In the event of any unauthorized behavior within the premises, it contributes to send alerts to the remote observers. Recovery behavior can be performed over remote devices, against system warnings [31–33]. IoT extends certain features of IP Surveillance systems that enable data to be transmitted in real time to a central server. The method of tracking and managing the remote location facilitates and provide information and reporting through the use of IoT. Combining IoT with Internet technologies and conventional Remote Surveillance Systems builds an efficient security infrastructure [34].

2.4 UAV Based Surveillance Solutions

UAV or Drone-based monitoring is a specialized platform for police officers and emergency responders to capture unrestricted aerial data to provide real-time security risks to extreme circumstances [35]. The fully automatic platform deploys and lands the drone autonomously, and gathers aerial data for processing and analysis [36]. Real-time aerial imagery and images are transmitted directly to field staff, allowing for more proactive decision-making in times of emergencies or regular patrols. In addition to protection and monitoring, industrial inspection drones can be of great benefit for a range of industrial operations.

The drones are designed and categorized based on its deployment application such as surveillance, agriculture, aerial mapping, photography etc. The four major drones on the basis of design aerial platforms are broadly classified in Table 1.

This chapter focuses on how the surveillance methods are being evolved and designed for Smart Cities instituted with Artificial Intelligence. With understanding of state-of-the-art surveillance methods further sections discusses the realm of surveillance scenario in the near future of Smart Cities.

3 AI in Smart Cities and Surveillance

Artificial Intelligence empowers computers or machines to make decisions and to think like humans. Machine learning is further enhanced by the addition of components for training and learning [37]. The access to large datasets and machines with high throughput leverage deep learning methods that automatically extract features or variation factors that differentiate discrete objects [38]. The surveillance video analysis involves a range of modules, like the recognition of an object, recognition of actions and classification of actions identified into anomalous or normal categories. This study focuses primarily on deep learning methods. CNN, auto-encoders, and their variation are commonly used models for tracking activities [39, 40], among the different deep learning architectures.

Smart Cities built and sustained by artificial intelligence, will not only boost people's lifestyle, but also enhance and support society [41]. The evolving idea of Smart Cities actively promotes this possibility by encouraging the convergence of sensors and Big Data through the Internet of Things (IoT). Allam et al. examined the urban capacity of AI and introduced a new system for linking AI technologies and communities, while maintaining the convergence of core aspects of community,

Table 1 Comparative study of drone designs

Drone type	Features	Applications	Drawbacks
Multi rotor	• Easiest to manufacture • Cheapest of all types • Variants: tricopter, quadcopter, heaxacopter, octocopter	Aerial photography Aerial video surveillance	• Limited flight time, limited power and speed • Not suitable for large scale applications • Flight time 20–30 min
Fixed wing	• Wings are the special feature • Wings do not require much energy to fight gravity and stay floating in the air	Ideal for long distance applications like aerial mapping and surveillance	• High cost • Skill training required for flying the drone • Requires runway or catapult launcher to set flight
Single rotor helicopter	• Similar in design to actual helicopter • Comprises one large sized rotor and one small sized rotor (on the tail)	Carrying heavy payloads Applications involving long hovering time	• Involves much complexity and more operational risks • Expensive • Skilled training required
Fixed wing hybrid VTOL	Combined features of fixed and rotor based drones Comprises gyros and accelerometers	Agriculture, infrastructure planning, mapping and land surveying, mining, surveillance and security	• High cost

ecology and governance. The authors are directed at policy makers, developers and data scientists looking to promote the convergence of artificial intelligence and big data in smart cities with the goal of increasing the competitiveness of the urban system and at the same time enhancing economic development and prospects [42, 43]. In a smart city system, one key issue is the capability of traditionally end-IoT devices that are vulnerable to security attacks and susceptible to data breaches of confidentiality and integrity. Blockchain will theoretically overcome these security issues due to the inherent properties of the distributed ledger. Ghosh et al. suggested a decentralised Blockchain architecture that provides stable and scalable smart city infrastructure that can execute a ledger service through a distributed network. [44, 45]. A. Voda and L. Radu addressed their study with the goal of examining the popular opinion towards the effect of AI on the attributes of smart cities and determining whether there are major variations in the viewpoint by age and gender. The mathematical data analysis was conducted using a two-way estimation through Analysis of Variance (ANOVA). Differences between classes have been evaluated using statistical analysis. This paper adds to the awareness of the value of AI strategies for enhancing urban living [46]. Khan et al. have explored a scenario for simulation and useful data sources to construct AI capability within utilities. A brief overview is provided of the BluWave-ai scalable architecture that utilizes deep learning in the data centre and AI deference is exploring edge computing nodes and IoT sensors to maximize the benefits of micro-grids on residential, neighborhood, school, business and community levels [45]. The thesis by Chun Lin et al. analyzed Taiwan's smart society's potential growth using data from separate research organizations [48]. Misra et al. addressed the AI-based concept in the agriculture and food sectors in their article. AI uses external IoT information and other big-data sources, uses knowledge based instructions provided by developers, or uses machine learning to define the basic principles and trends to direct structures to set targets [49] using rule-based AI [50].

The areas discovered are marked as domains of use. Most of the research works offers unique context-based solutions. The smart city surveillance applications include: crowd pulling events, traffic signals, key junctions, office buildings and industrial areas, celebrations and gatherings as a part of religious organizations, etc. Table 2 enlist some of the publications.

Table 2 Artificial intelligence in smart cities

References	Title	Application area	Description
De Paz et al. [51]	Intelligent system for lighting control in smart cities	Centralized public lighting	Presented adaptive architecture for smart lighting systems Uses AI, artificial neural network (ANN), Multi agent System (MAS), EM Algorithm etc. for modular architecture
Mathur and Modani [52]	Smart city—a gateway for Artificial Intelligence in India	Role of AI in smart cities	The role of AI in public transportation, power supply, waste management, healthcare
Srivastava et al. [53]	Safety and security in smart cities using Artificial Intelligence—a review	Surveillance	Presents the already in place smart solutions for smart cities a hybrid system of AI and human is concluded as the smarter system
Misra et al. [49]	IoT, big data and artificial intelligence in agriculture and food industry	Agriculture and food industry	Advances in intelligent farm machinery; Low altitude spectral imaging
Nikolopoulos et al. [54]	Embedded intelligence in smart cities through multi-core smart building architectures: research achievements and challenges	Smart building architectures	Important difficulties that cities face in seeking to become smart with an emphasis on the unique field of smart buildings

Without a full infrastructure system and processes to protect its residents, enterprises, activities and utilities, a smart city is incomplete. The authors give a description of the protection of smart cities and an insight into the concept of secure smart cities. They speak about the authenticity, access protection, encryption and firewalls of current authentication security mechanisms and their adequacy to protect a smart city. In particular, data storage, the internet, water sources, power, the city brain and other vital public facilities, as well as potential animosity threats against a smart city have been addressed and debated. Finally, best practices for securing the smart cities were discussed [55]. When quad copters fly into the sky and film everything at a wide angle, it's easy to see a large field from one drone. It's easier to quickly identify the intruder (human & other animal) from the video taken by the drone, with the aid of machine learning. Just one drone will do the job of 10–30 border guards very easily if we fly a drone around a wide field, which is quite attainable [56].

4 Deep Learning Methods and Architectures

This section discusses the state-of-the-art deep leaning methods and architectures with a brief discussion on the applied applications domains.

4.1 Deep Learning Methods

Artificial Intelligence is vast area with Machine Learning algorithms as its subset and Deep Learning as super subset. Deep learning usually requires learning interpretations at various levels of hierarchy, such that abstract structures can be built from simplistic ones. This section encompasses the most significant deep learning methods categorized in Fig. 4.

Supervised Learning Methods

Algorithms with a dataset that contains a series of features are addressed in supervised learning. Each sample is often supplied with labels or target values. This translation of the features to the target values labels is where the information is encoded. After learning, the algorithm can find mapping for its right labels or target values from the features of unseen samples.

Makantasis et al. applied deep supervised learning through convolutional neural networks for hyperspectral data classification in [57]. In another work [58] Partially-supervised learning was used in video surveillance for facial recognition from facial trajectories. Semi-supervised Learning was used in [59] by Sillito et al. for Anomalous Trajectory Detection. Techniques like LSTM, BiLSTM and three dimensional CNN has been used for violence detection in real time, activity recognition, prediction of human trajectory in crowded spaces etc. [60–64].

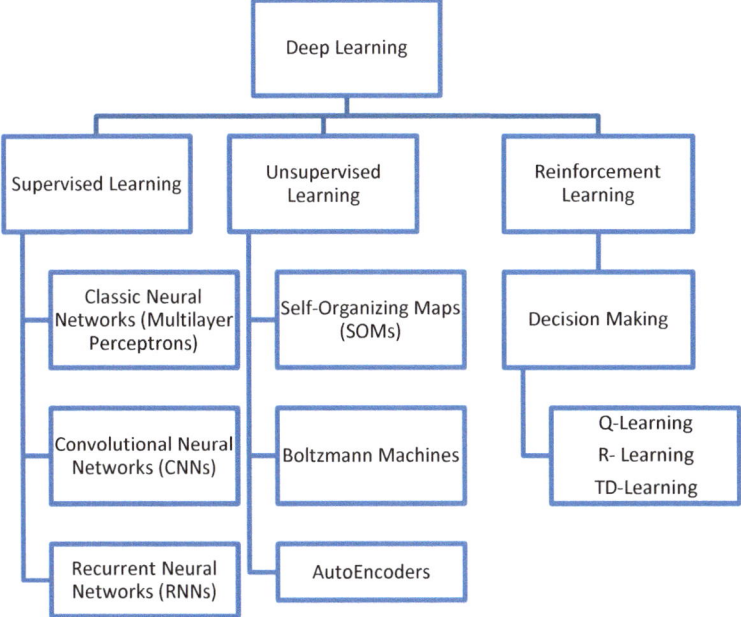

Fig. 4 Categorization of deep learning algorithms

Unsupervised Learning

The objective of unsupervised learning is to collect concrete observations and to illustrate key data characteristics. In this case there is no need for labels or target values to benefit from the results.

Unsupervised learning [65] has been majorly employed in temporal anomaly detection and learning the patterns through video surveillance [66–68]. Anomalies have also been detected in industrial robots using unsupervised learning method [69]. This method is also used for aerial planning through UAVs [70].

Reinforcement Learning

Reinforcement learning algorithm aims at the method that learns by trial and error process. It is different from both supervised and unsupervised machine learning. While supervised learning can predict labels for complex inputs, and unsupervised learning can group together related items, reinforcement learning predicts the action that will yield the best result. An AI agent communicates with a specific or virtual world in reinforcement learning algorithms.

Distributed wildfire surveillance with autonomous UAVs [71, 72], vehicle classification [73], traffic monitoring and control [74], aerial planning [75], real-time anomaly detection [76] and video face recognition [77] are some of the applications where deep reinforcement learning has been applied and found effective yielding high accuracies.

4.2 Deep Learning Architectures

This section addresses different architectures of Deep Learning technologies as shown in Fig. 5. CNNs are amongst the best learning algorithms for interpreting image content and have demonstrated exemplary performance in the segmentation, classification, detection and retrieval of image based tasks [78, 79]. CNNs influence has attracted interest outside academia. Google, Microsoft, AT&T, NEC, and Facebook have established active research groups to discover new CNN frameworks [80]. Most of the frontrunners of image processing and computer vision (CV) world championships are currently using Deep CNN-based models.

CNNs were applied for visual tasks as of the late 1980s. LeCun et al. have proposed the first Multilayer CNN called ConvNet in1989 and the name source is rooted in Neocognitron [81] of Fukushima. LeCun introduced a supervised learning of ConvNet using back-propagation algorithm as compared to the unsupervised Reinforcement learning model used by Neocognitron [82, 83]. Thus the work of LeCuN established a foundation for the modern 2D CNNs. This ConvNet showed better results for problems related to the handwritten recognition of digits and recognition of zip code. In 1998, LeCuN presented an upgraded version of ConvNet, known as LeNet-5, and it continued utilizing CNN in the classification of characters in applications related to document recognition [84]. Because of CNNs profound performance in optical character recognition and fingerprint recognition, it was commercially used in ATM and Banks began in 1993 and 1996 [85]. LeNet-5 achieved many accomplishments throughout this time frame.

From the late 1990 to 2000s, various improvements throughout the research methods and architecture of CNN learning were aimed at making CNN scalable to major, heterogeneous, complex, and multiclass problems. Innovations in CNNs include diverse facets such as processing unit improvements, optimization strategies for constraints and hyperparameters, design patterns and layer connectivity, etc. After AlexNets outstanding performance on the ImageNet dataset in 2012 CNN-based applications became prevalent [86]. Since then, new breakthroughs have been

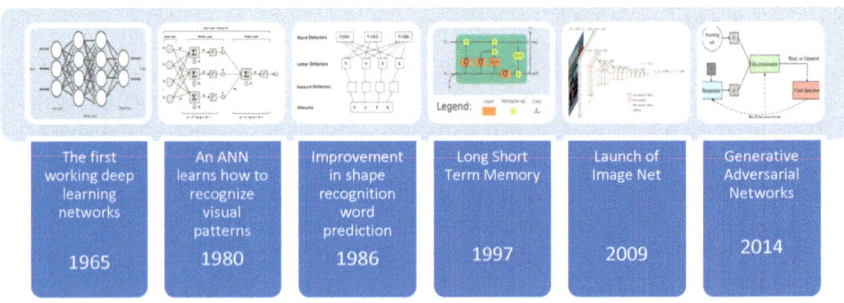

Fig. 5 Evolution of deep learning techniques over the years

suggested at CNN and are basically due to processing unit redesign and the construction of new blocks. Zeiler and Fergus provided CNNs formulation of layer-wise visualization to enhance the reliability of feature extraction stages, which shifted the paradigm towards the extracting features in deep architecture to low spatial resolution as demonstrated in VGG [87]. The initial block offers the concept of branching into a layer for the first time, allowing features to be abstracted at different spatial scales. The ResNets definition of skip links for deep CNNs was popular in 2015. Most successor networks such as Inception-ResNet, Wide ResNet, ResNeXt etc. have consequently used this term. The influence of multi-tiered transformations on CNNs' learning capacities through implementation of cardinalty or enhancing width is explored by various architectural concepts such as Big ResNet, ResNeXt, Pyramidal Net, Xception, PolyNet and several others.

You only look once (YOLO) is an object detection technology that is able to process in real time. YOLO is released in five versions i.e. YOLO, YOLOv2 and YOLO9000 and YOLOv3 and YOLOv4 [88]. With 43.5% average precision at 65 FPS on a Tesla V100, YOLO v4 significantly outperforms state-of-the-art performance at a real time speed on the MS COCO dataset.

5   Deep Learning in Surveillance

Deep learning has largely evolved the surveillance system through computer vision from detection of target to extracting features to tracking and predicting the position of the target. This section discusses the literature of application of deep learning algorithms particularly in monitoring tasks.

Figure 6 shown below illustrates the real-time surveillance mechanism based on deep learning methods. The key constituent of surveillance task is obtaining data through the sensing element which can be a static camera (static sensor) or a UAV equipped with camera (dynamic sensor). Using various deep learning methods (as discussed in Sect. 4) the data is processed and critical information is mined through feature extraction. An appropriate deep learning model is applied to learn from the extracted features followed by model optimization and validation on real-time data.

The section consists of a deeply ingrained study that begins with object identification, behavior recognition, crowd awareness and eventually identification of aggression in a crowd environment. The authors explored the underlying deep learning technologies used in different multi-group video analysis procedures in [89]. In their article, Yazdi et al. present a survey on the new approaches to detect objects in motion from the moving camera's video sequences [90]. While several studies and excellent works looked at object detection methods and background subtraction for a fixed camera, a thorough description of the latest various approaches for the moving camera was never published [90]. The demonstration of the autonomous nano-drone navigation engine capable of closed loop end-to-end visual navigation by DNN was introduced by Bah et al. [91] Pradeep et al. focused on combining deep learning methodology with a real-time flight camera feature descriptor to correctly define the

Fig. 6 Process-flow of surveillance method using deep learning techniques for continuous visual monitoring

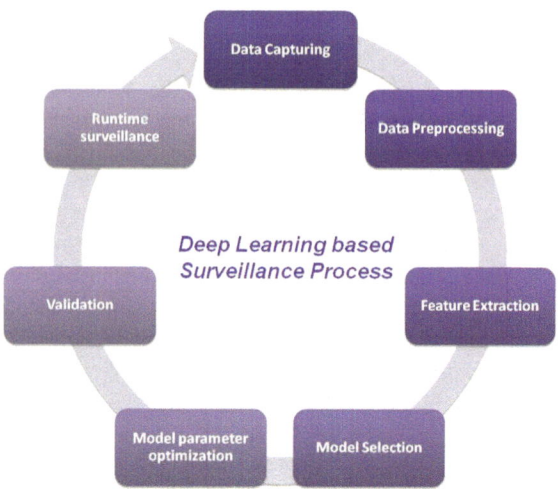

specific target [92]. The suggested solution is described as being successful in both military and civil defense in identifying suspected persons' actions or even in finding the flying objects emitted by an opposing nation.

Multi-object monitoring for moving cameras is yet another technology which has recently gained significant attention with the use of several mathematical models or more sophisticated filtering prediction models. Due to the availability of huge annotated data set for its training [93] deep neural networks have also been integrated into visual tracking. The authors intend to define the most relevant and essential architecture problems in their paper for Cyber Physical Systems (CPS) implementations in multi UAV systems. Kim et al. integrated differentiated implementations for monitoring smart cities and vast oceans and suggested a tight plane-based framework utilizing current public transportation, including public buses, city trains and their routes, to provide time-sensitive monitoring [94]. Nawaratne et al. [9] found that Face recognition in drone crowd surveillance is yet to be proven to enhance the effectiveness of UAVs equipped with IoT devices to provide high-level IoT services. It illustrated how UAVs can be used for Face Recognition based crowd surveillance. The Open Source Computer Vision Local Binary Pattern Histogram approach is used to perform the face recognition [95].

The deep learning surveillance can be deployed several domains, majorly four domains are pedestrian detection [41, 43, 96], border surveillance [46, 47], perimeter monitoring [91, 92] and traffic monitoring [42, 97].

6 UAV Surveillance for Public Safety and Security

In the coming years, with the increasing proliferation of unmanned aerial vehicles (UAVs) [98], large quantities of knowledge will be shared between edge devices and UAVs. This is supposed to make traffic management using UAVs and edge computing devices an integral part of smart transport systems of the next decade. Monitoring, however, involves continual data exchange, collective decision-making, and stable network structure as depicted in Fig. 7.

This section includes a description of possible new paths and concepts that may affect future research in this domain. Approaches to networks of embedded systems, such as drone swarms, where each system has the same processing capability (in terms of energy and speed), need to be expanded. In order to optimize either the lifespan of such machines or life of the entire swarm, the computational burden can be distributed among the entire swarm. The framework will establish the metrics to optimally reduce and reallocate these networks, validate them by measuring power and articulate various compromises which must be addressed.

Taking reference from the taxonomy proposed in [99, 100] Aerostack architecture, the process of surveillance backed by deep learning algorithms can be understood at different abstract stages. The architecture describes the usage of UAVs and processing of data acquired by systematically applying deep learning methods for the purpose of

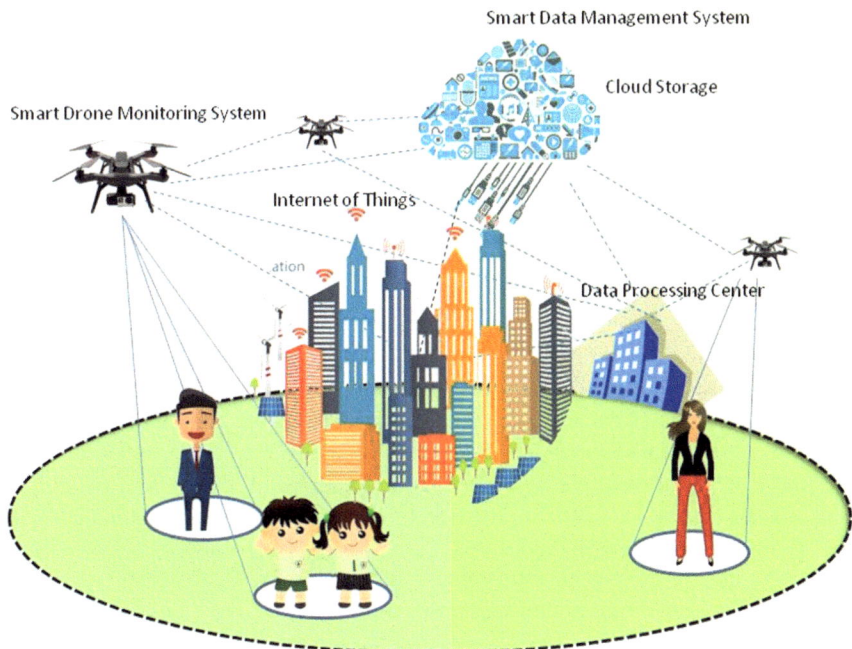

Fig. 7 Framework of Internet of Drones in Smart city surveillance

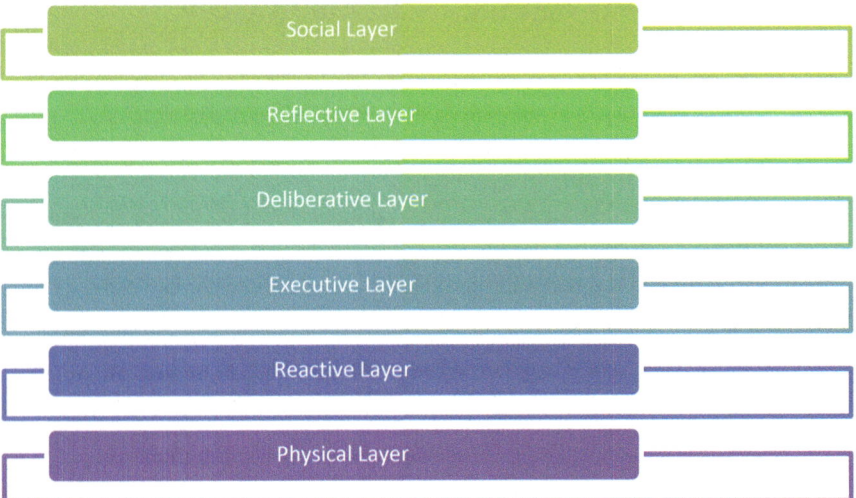

Fig. 8 A six abstraction layered Aerostack architecture

classification deployed for various applications. It is defined in different abstraction levels or layers. This UAV based architecture accentuated the process of observation, management, routing, and controlling of the aerial rotorcraft systems at these abstract layers. The elements of an unmanned robotic aerial vehicle can be divided into the following structures and interfaces, according to Aerostack as shown in Fig. 8.

6.1   Deep Learning Mechanism Through Aerostack Architecture

The *Physical layer* comprises of the hardware interface that includes the UAV sensors and actuators. The data is sensed and retrieved at this stage. Feature extraction from the raw data collected is processed using deep learning algorithms in the *Reactive Layer*. This layer also constitutes the motor control system which is responsible to control motion and visual servos connected to the actuators present in the physical layer. The extracted features are fed to the *Execution Layer* for self-localization and mapping of the features in the processed images and videos with the help of multi sensor fusion state estimator the features are combined and estimated again making use of deep learning algorithms. The situational awareness system present in this stage creates the environmental understanding from the visual data by generating state variables from the compiled sensor information. The Deep Learning algorithms also find its significance in the *Deliberative Layer* that binds the entire planning system. Based on the complex actions and behaviour sequences detailed, the action plan is revised and optimized using high-level learning algorithms assisted by strong feedback system thus providing global solutions to the intricate activities in path planning of UAVs. Monitoring of the entire surveillance mission takes place in the *Reflective Layer* that processes the signals from rest of the stages and the signals generated are communicated through robot interface present in *Social Layer*. Deep Learning algorithms can exemplify the supervision system in top two layers of the architecture in order to check the progress of the autonomous robotic system towards its goals, detect faults and generate recovery actions.

UAVs would facilitate a range of applications spanning from video streaming to medical allocation of resources. Drones are limitlessly effective due to its compact size and ability to fly with no pilot onboard, for example in agriculture, surveillance and various other utilities. Their use, however, poses many challenges to national entities when concerning public security, such as nuclear installations, historic landmarks and the residences of public officers because they are capable of holding explosives and other harmful substances. The ever-increasing use of monitoring drones includes rigorous identification, recording, intruder positioning and jamming. The identification precision is a fundamental device prerequisite. Generally, effective identification is time-consuming.

The prompt identification of a suspicious drone is already an open challenge for security providers. Recently, a multitude of schemes have tackled the issue. The strategy however has a drawback, such as adverse weather conditions and massive data requirements. Several other challenges encountered by drone surveillance are discussed in the next section.

7 Challenges in Surveillance

CCTV surveillance is amongst the most popular methods of monitoring presently available. Also, while CCTV cameras are deployed predominantly for video surveillance, and to dissuade theft and certain other illegal acts, they also contribute to enhancing the efficiency and job performance of the staff. UAVs are perpendicular to the pedestrian travel path and provide a top view position. Their position, speed and distance may also be obtained in addition to the actual height and width. Useful statistics can be extracted for several reasons of course dependent upon the data obtained: guided count-volume for pedestrians, traffic detection, flow distribution, speed determination and perhaps more insight. Researchers have made considerable efforts to overcome these difficulties, often producing incredible results; however, important challenges remain.

Challenges Due to Capturing Device

Occlusion is one of the major challenges in surveillance is occlusion [93, 95, 96]. Occlusion is the effect of blocking an object from the 3D view. Lighting conditions is an added challenge in surveillance which may vary all day long [97]. The viewpoint variation is a serious distortion due to degree changes in image viewpoints. It is difficult to identify persons/objects of interest from different angles and has a direct influence on their precision [98]. Another significant challenge in Surveillance applications is multiple spatial scales and aspect ratios, which may appear in a wide range of sizes and aspect ratios.

UAV Handling

Criminals can find new methods for infringing drone safety. The protection of these portable safety devices would be another challenge. The biggest driving force in using surveillance drones is to look out for unauthorized drones that can cause major disasters if there is no prompt precautionary action. The ever-increasing use of monitoring drones includes rigorous identification, recording, intruder positioning and jamming.

Privacy Challenges

As most IoD apps using the drones are in real time, users (external parties) are generally interested in accessing real-time services for the drones belonging to a certain fly zone from the deployed drones. It involves collecting sensitive information about human beings, thus privacy issues are raised. It is challenging to define new methods where everything from smart city sensors to homes is linked to a network [101, 102].

Data Management Challenges

Large time lags in real-time detection due to high computational complexity are generated to process and analyze the humongous amount of data that has been recorded. The massive volume of data course between various drones, control centers and end device components running in implicit arrangement of IoD. Moreover, it is difficult to achieve efficient delivery ratio as secure data forwarding is not realized because of the possibility of data leakage. There is a need to devise a secure and steadfast data distribution system for the IoD environment. An efficient data processing system needs to be set-up to ensure on-board processing interfaced with inertial sensors with high precision and high computational speed.

Technological Challenges

Another challenge faced is by the city workers who need to learn how to use this new technology and be comfortable with. A special workforce needs to be trained to ensure effortless working of Internet of Drones in civil applications and surveillance. Keeping the surveillance drones safe from amateur drones and regular repairing of these flying objects pose an inconceivable challenge.

Environmental Challenges

UAVs can be used to study the structure of the atmosphere, air quality and habitats owing to its potential to enter harmful conditions, such as thunderstorms, hurricanes and volcanic fumes. The use of UAVs indoors is a daunting task because it requires higher demands than outdoors. To prevent collisions in an indoor environment, GPS is very difficult to use, typically indoors is a GPS negated environment. Optical sensor-based camera collision-prevention approaches struggle from intense image processing computing operations. These cameras and optical sensors are light

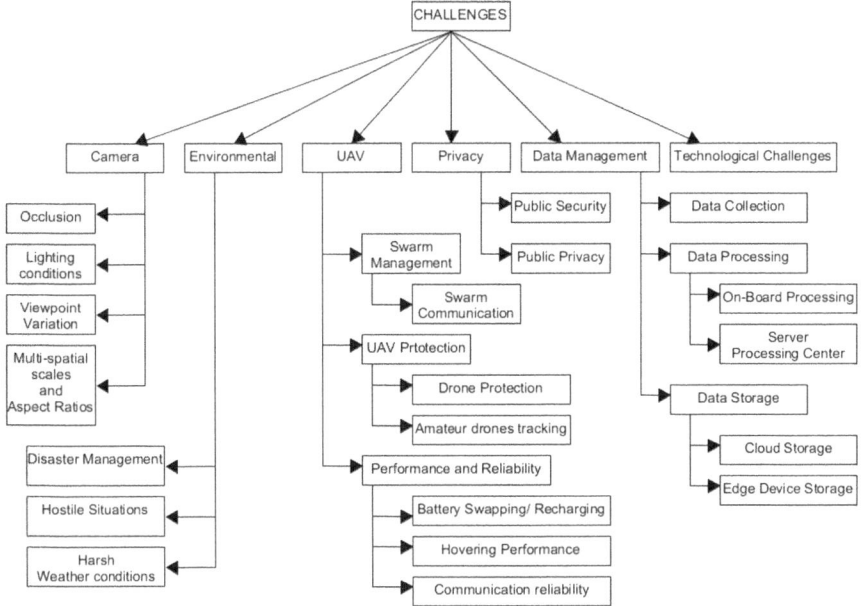

Fig. 9 Mapping of challenges in implementation of Internet of Drones

sensitive, which means that the steam and smoke will cause the collision avoidance device to malfunction (i.e. adequate illumination to operate correctly). Harsh weather condition can also be significant such as heavy rain falls, hurricanes, tornados etc.

All the above discussed challenges are mapped in Fig. 9.

8 Conclusion and Future Work

A detailed analysis has been presented on state-of-the art of surveillance methods including Sensor Based surveillance, CCTVs, IoT based surveillance solution, drone-based surveillance solution, state-of-the art deep learning solutions and potential benefits of UAVs in smart cities surveillance. In recent years, crime rates have risen in urban communities, such as violent robberies, arson and terrorism. Therefore it is an effective method to predict crimes by identifying and distinguishing offenders within crowds of people. Many security guards and an immense amount of human activity are required in conventional patrol structures to provide the requisite protection for people. UAVs may be used in this way to support the security guards by tracking individuals at points of concern remotely. UAVs can offer protection from any threat and aid not only in monitoring but in locating, identifying and recognizing offenders that incorporates methods of face recognition. Using UAVs with compatible IoT systems, such as video cameras, can provide an effective crowd monitoring system;

identify any unusual activity and illegal action; and recognize the faces of criminals. Using this tool offers a panoramic view for crowd control and facial recognition. So public protection and defense can be improved and at the same time reducing the number of security guards stationed on the field. The face recognition process consists of well-defined steps: the extraction of facial features, the development of identified faces in the database and the identification of faces comparing videotaped faces with profiled faces can be accomplished by combining UAV technology with deep learning techniques to create artificial intelligent surveillance systems. The UAV based surveillance system equipped with deep learning methods prevent and reduce chances of theft, provide real-time video surveillance, gather useful criminal evidence, yields better employee productivity, assists in cutting down on security-related costs, enables to monitor high-risk areas and provides fool-proof coverage.

Acknowledgments This work titled "Artificial Intelligence and Deep Learning Technologies in Smart Cities Surveillance using UAVs" is supported by the grant from Department of Science and Technology, Government of India, against CFP launched under Interdisciplinary Cyber Physical Systems (ICPS) Programme, DST/ICPS/CPS-Individual/2018/181(G).

References

1. Silva, B. N., Khan, M., & Han, K. (2018). Towards sustainable smart cities: A review of trends, architectures, components, and open challenges in smart cities. *Sustainable Cities and Society, 38,* 697–713.
2. Shahidehpour, M., Li, Z., & Ganji, M. (2018). Smart cities for a sustainable urbanization: Illuminating the need for establishing smart urban infrastructures. *IEEE Electrification Magazine, 6*(2), 16–33.
3. Yin, C., Xiong, Z., Chen, H., Wang, J., Cooper, D., & David, B. (2015). A literature survey on smart cities. *Science China Information Sciences, 58*(10), 1–18.
4. Srinivasan, S., Latchman, H., Shea, J., Wong, T., & McNair, J. (2004). Airborne traffic surveillance systems: Video surveillance of highway traffic. In *Proceedings of the ACM 2nd International Workshop on Video Surveillance & Sensor Networks* (pp. 131–135).
5. Gurwicz, Y., Yehezkel, R., & Lachover, B. (2011). Multiclass object classification for real-time video surveillance systems. *Pattern Recognition Letters, 32*(6), 805–815.
6. Alam, M. F., Katsikas, S., Beltramello, O., & Hadjiefthymiades, S. (2017). Augmented and virtual reality based monitoring and safety system: A prototype IoT platform. *Journal of Network and Computer Applications, 89,* 109–119.
7. Malik, P. K., Sharma, R., Singh, R., Gehlot, A., Satapathy, S. C., Alnumay, W.S., Pelusi, D., Ghosh, U., & Nayak, J. (2020). Industrial Internet of Things and its applications in industry 4.0: State of the art. *Computer Communications.*
8. Chen, J., Li, K., Deng, Q., Li, K., & Yu Philip, S. (2019). Distributed deep learning model for intelligent video surveillance systems with edge computing. *IEEE Transactions on Industrial Informatics.*
9. Nawaratne, R., Alahakoon, D., De Silva, D., & Yu, X. (2019). Spatiotemporal anomaly detection using deep learning for real-time video surveillance. *IEEE Transactions on Industrial Informatics, 16*(1), 393–402.
10. Motlagh, N. H., Bagaa, M., & Taleb, T. (2017). UAV-based IoT platform: A crowd surveillance use case. *IEEE Communications Magazine, 55*(2), 128–134.

11. Semsch, E., Jakob, M., Pavlicek, D., & Pechoucek, M. (2009). Autonomous UAV surveillance in complex urban environments. In *2009 IEEE/WIC/ACM International Joint Conference on Web Intelligence and Intelligent Agent Technology* (Vol. 2, pp. 82–85). IEEE.

12. Kopfstedt, T., Mukai, M., Fujita, M., & Ament, C. (2008). Control of formations of UAVs for surveillance and reconnaissance missions. *IFAC Proceedings Volumes, 41*(2), 5161–5166.

13. Vattapparamban, E., Güvenç, İ., Yurekli, A. İ., Akkaya, K., & Uluağaç, S. (2016). Drones for smart cities: Issues in cybersecurity, privacy, and public safety. In *2016 International Wireless Communications and Mobile Computing Conference (IWCMC)* (pp. 216–221). IEEE.

14. Li, J., Huang, W., Shao, L., & Allinson, N. (2014). Building recognition in urban environments: A survey of state-of-the-art and future challenges. *Information Sciences, 277,* 406–420.

15. Richman, L. (2005). Human guard enhancing multiple site integrated security system. U.S. Patent 6,894,617, issued May 17, 2005.

16. Cuthbert, A. R. (1995). The right to the city: Surveillance, private interest and the public domain in Hong Kong. *Cities, 12*(5), 293–310.

17. Richman, L., Vacaru, A., Zatusevschi, O., Belyshev, A., Oostendorp, M., Denisov, D., & Alexeev, K. (2003). Method and protocol for real time security system. U.S. Patent Application 10/176,565, filed November 6, 2003.

18. Peek-Asa, C., Schaffer, K. B., Kraus, J. F. & Howard, J. (1998). Surveillance of non-fatal workplace assault injuries, using police and employers' reports. *Journal of Occupational and Environmental Medicine, 40*(8), 707–713.

19. Guo, S., He, T., Mokbel, M. F., Stankovic, J. A., & Abdelzaher, T. F. (2010). On accurate and efficient statistical counting in sensor-based surveillance systems. *Pervasive and Mobile Computing, 6*(1), 74–92.

20. Aguilar-Ponce, R., Kumar, A., Luis Tecpanecatl-Xihuitl, J., & Bayoumi, M. (2007). A network of sensor-based framework for automated visual surveillance. *Journal of Network and Computer Applications,* 30(3), 1244–1271.

21. Bellazreg, R., Boudriga, N., & An, S. (2013). Border surveillance using sensor based thicklines. In *The International Conference on Information Networking 2013 (ICOIN)* (pp. 221–226). IEEE.

22. Hu, T., Zhang, H., Zhu, X., Clunis, J., & Yang, G. (2018). Depth sensor based human detection for indoor surveillance. *Future Generation Computer Systems, 88,* 540–551.

23. Shao, X., Zhao, H., Shibasaki, R., Shi, Y., & Sakamoto, K. (2011). 3D crowd surveillance and analysis using laser range scanners. In *2011 IEEE/RSJ International Conference on Intelligent Robots and Systems* (pp. 2036–2043). IEEE.

24. Aliakbarpour, H., Nunez, P., Prado, J., Khoshhal, K., & Dias, J. (2009). An efficient algorithm for extrinsic calibration between a 3d laser range finder and a stereo camera for surveillance. In *2009 International Conference on Advanced Robotics* (pp. 1–6). IEEE.

25. Norris, C., & Armstrong, G. (1999). *The maximum surveillance society: The rise of CCTV* (Vol. 2). Oxford: Berg.

26. Levine, R. M. (2000). *SIDE and closed circuit television (CCTV): Exploring surveillance in public space.*

27. Nieto, M., Johnston-Dodds, K., & Simmons, C. W. (2002) *Public and private applications of video surveillance and biometric technologies* (Vol. 2, No. 6). Sacramento, CA: California State Library, California Research Bureau.

28. Kruegle, H. (2011). *CCTV surveillance: Video practices and technology.* Elsevier.

29. Norris, C., & Moran, J. (2016). *Surveillance, closed circuit television and social control.* Routledge.

30. McCahill, M. (2013). *The surveillance web.* Routledge.

31. Muhammad, K., Hamza, R., Ahmad, J., Lloret, J., Wang, H., & Baik, S. W. (2018). Secure surveillance framework for IoT systems using probabilistic image encryption. *IEEE Transactions on Industrial Informatics,* 14(8), 3679–3689.

32. Memos, V. A., Psannis, K. E., Ishibashi, Y., Kim, B.-G., & Gupta, B. B. (2018). An efficient algorithm for media-based surveillance system (EAMSuS) in IoT smart city framework. *Future Generation Computer Systems, 83,* 619–628.

33. Hu, L., & Ni, Q. (2017). IoT-driven automated object detection algorithm for urban surveillance systems in smart cities. *IEEE Internet of Things Journal, 5*(2), 747–754.

34. Batra, I., Verma, S., Malik, A., Ghosh, U., Rodrigues, J. J. P. C., Nguyen, G. N., Hosen, A. S. M., & Mariappan, V. (2020). Hybrid logical security framework for privacy preservation in the green Internet of Things. *Sustainability, 12*(14), 5542.

35. Quadri, S. A. I., & Sathish, P. (2017). IoT based home automation and surveillance system. In *2017 International Conference on Intelligent Computing and Control Systems (ICICCS)* (pp. 861–866). IEEE.

36. Sarkar, S., Totaro, M. W., Elgazzar. K. (2019). Intelligent drone-based surveillance: Application to parking lot monitoring and detection. In *Unmanned Systems Technology XXI* (Vol. 11021, p. 1102104). International Society for Optics and Photonics.

37. Sharma, V., You, I., Pau, G., Collotta, M., Lim, J. D., & Kim, J. N. (2018). LoRaWAN-based energy-efficient surveillance by drones for intelligent transportation systems. *Energies, 11*(3), 573.

38. Vannicola, V. C., & Mineo, J. A. (1988). Applications of knowledge based systems to surveillance. In *Proceedings of the 1988 IEEE National Radar Conference* (pp. 157–164). IEEE.

39. Rego, A., Canovas, A., Jiménez, J. M., & Lloret, J. (2018). An intelligent system for video surveillance in IoT environments. *IEEE Access, 6,* 31580–31598.

40. Feldstein, S. (2019). *The global expansion of AI surveillance* (Vol. 17). Carnegie Endowment for International Peace.

41. Bilal, M., Khan, A., Khan, M. U. K., & Kyung, C.-M. (2016). A low-complexity pedestrian detection framework for smart video surveillance systems. *IEEE Transactions on Circuits and Systems for Video Technology, 27*(10), 2260–2273.

42. Wei, H., & Kehtarnavaz, N. (2019). Semi-supervised faster RCNN-based person detection and load classification for far field video surveillance. *Machine Learning and Knowledge Extraction, 1*(3), 756–767.

43. Helbing, D., Buzna, L., Johansson, A., & Werner, T. (2005). Self-organized pedestrian crowd dynamics: Experiments, simulations, and design solutions. *Transportation science, 39*(1), 1–24.

44. Ghosh, U., Chatterjee, P., Shetty, S., & Datta, R. (2020). An SDN-IoT-based framework for future smart cities: Addressing perspective. In *Internet of Things and secure smart environments: successes and pitfalls* (p. 441). CRC Press.

45. Malik, A. A., Tosh, D. K., & Ghosh, U. (2019). Non-intrusive deployment of blockchain in establishing cyber-infrastructure for smart city. In *2019 16th Annual IEEE International Conference on Sensing, Communication, and Networking (SECON)* (pp. 1–6). IEEE.

46. Laouira, M. L., Abdelli, A., Othman, J. B., & Kim, H. (2019). An efficient WSN based solution for border surveillance. *IEEE Transactions on Sustainable Computing.*

47. Shen, C., Xie, R., & Zhang, L. (2015). Temporal recursive pedestrian trajectory identification for multi-camera surveillance system. In *2015 IEEE International Conference on Computer and Information Technology; Ubiquitous Computing and Communications; Dependable, Autonomic and Secure Computing; Pervasive Intelligence and Computing* (pp. 1158–1163). IEEE.

48. Lovas, T., & Barsi, Á. (2015). Pedestrian detection by profile laser scanning. In *2015 International Conference on Models and Technologies for Intelligent Transportation Systems (MT-ITS)* (pp. 408–412). IEEE.

49. Misra, N. N., Dixit, Y., Al-Mallahi, A., Bhullar, M. S., Upadhyay, R., & Martynenko, A. (2020). IoT, big data and artificial intelligence in agriculture and food industry. *IEEE Internet of Things Journal.*

50. Garg, S., Singh, A., Batra, S., Kumar, N., & Yang, L. T. (2018). UAV-empowered edge computing environment for cyber-threat detection in smart vehicles. *IEEE Network, 32*(3), 42–51.

51. De Paz, J. F., Bajo, J., Rodríguez, S., Villarrubia, G., & Corchado, J. M. (2016). Intelligent system for lighting control in smart cities. *Information Sciences, 372,* 241–255.

52. Mathur, S., & Modani, U. S. (2016). Smart city-a gateway for artificial intelligence in India. In *2016 IEEE Students' Conference on Electrical, Electronics and Computer Science (SCEECS)* (pp. 1–3). IEEE.

53. Srivastava, S., Bisht, A., & Narayan, N. (2017). Safety and security in smart cities using artificial intelligence—A review. In *2017 7th International Conference on Cloud Computing, Data Science & Engineering-Confluence* (pp. 130–133). IEEE.

54. Nikolopoulos, B., Dimitrakopoulos, G., Bravos, G., Dimopoulos, A., Nikolaidou, M., & Anagnostopoulos, D. (2016). Embedded intelligence in smart cities through multi-core smart building architectures: Research achievements and challenges. In *2016 IEEE Tenth International Conference on Research Challenges in Information Science (RCIS)* (pp. 1–2). IEEE.

55. Toh, C. K. (2020). Security for smart cities. *IET Smart Cities, 2*(2), 95–104.

56. Mojib, E. B. S., Bahalul Haque, A. K. M., Nafis Raihan, M., Rahman, M., & Alam, F. B. (2019). A novel approach for border security; Surveillance drone with live intrusion monitoring. In *2019 IEEE International Conference on Robotics, Automation, Artificial-intelligence and Internet-of-Things (RAAICON)* (pp. 65–68). IEEE.

57. Makantasis, K., Karantzalos, K., Doulamis, A., & Doulamis, N. (2015). Deep supervised learning for hyperspectral data classification through convolutional neural networks. In *2015 IEEE International Geoscience and Remote Sensing Symposium (IGARSS)* (pp. 4959–4962). IEEE (2015).

58. De-la-Torre, M., Granger, E., Radtke, P. V. W., Sabourin, R., & Gorodnichy, D. O. (2015). Partially-supervised learning from facial trajectories for face recognition in video surveillance. *Information Fusion, 24*, 31–53.

59. Sillito, R. R., & Fisher, R. B. (2008). Semi-supervised learning for anomalous trajectory detection. In *BMVC* (vol. 1, pp. 035–1).

60. Fenil, E., Manogaran, G., Vivekananda, G. N., Thanjaivadivel, T., Jeeva, S., & Ahilan, A. (2019). Real time violence detection framework for football stadium comprising of big data analysis and deep learning through bidirectional LSTM. *Computer Networks, 151*, 191–200.

61. Ullah, A., Muhammad, K., Ser, J. D., Baik, S. W., & de Albuquerque, V. H. C. (2018). Activity recognition using temporal optical flow convolutional features and multilayer LSTM. *IEEE Transactions on Industrial Electronics, 66*(12), 9692–9702.

62. Ko, K.-E., & Sim, K.-B. (2018). Deep convolutional framework for abnormal behavior detection in a smart surveillance system. *Engineering Applications of Artificial Intelligence, 67*, 226–234.

63. Alahi, A., Goel, K., Ramanathan, V., Robicquet, A., Fei-Fei, L., & Savarese, S. (2016). Social LSTM: Human trajectory prediction in crowded spaces. In *Proceedings of the IEEE Conference on Computer Vision and Pattern Recognition* (pp. 961–971).

64. Lu, N., Yidan, W., Feng, L., & Song, J. (2018). Deep learning for fall detection: Three-dimensional CNN combined with LSTM on video kinematic data. *IEEE Journal of Biomedical and Health Informatics, 23*(1), 314–323.

65. Makris, D., & Ellis, T. (2005). Learning semantic scene models from observing activity in visual surveillance. *IEEE Transactions on Systems, Man, and Cybernetics, Part B (Cybernetics), 35*(3), 397–408.

66. Cobb, W. K., & Seow, M.-J. (2012). Unsupervised learning of temporal anomalies for a video surveillance system. U.S. Patent 8,167,430, issued May 1, 2012.

67. Venkatesan, R., Dinesh Anton Raja, P., & Balaji Ganesh, A. (2016). Unsupervised learning based video surveillance system established with networked cameras. In *Advances in signal processing and intelligent recognition systems* (pp. 603–614). Cham: Springer.

68. Seow, M.-J., & Cobb, W. K. (2015). Unsupervised learning of feature anomalies for a video surveillance system. U.S. Patent 9,111,148, issued August 18, 2015.

69. Munawar, A., Vinayavekhin, P., & De Magistris, G. (2017). Spatio-temporal anomaly detection for industrial robots through prediction in unsupervised feature space. In *2017 IEEE Winter Conference on Applications of Computer Vision (WACV)* (pp. 1017–1025). IEEE.

70. Faigl, J., Váňa, P., Pěnička, R., & Saska, M. (2019). Unsupervised learning-based flexible framework for surveillance planning with aerial vehicles. *Journal of Field Robotics, 36*(1), 270–301.

71. Julian, K. D., & Kochenderfer, M. J. (2019). Distributed wildfire surveillance with autonomous aircraft using deep reinforcement learning. *Journal of Guidance, Control and Dynamics, 42*(8), 1768–1778.

72. Julian, K. D., & Kochenderfer, M. J. (2018). Autonomous distributed wildfire surveillance using deep reinforcement learning. In *2018 AIAA Guidance, Navigation, and Control Conference* (p. 1589).

73. Zhao, D., Chen, Y., & Lv, L. (2016). Deep reinforcement learning with visual attention for vehicle classification. *IEEE Transactions on Cognitive and Developmental Systems, 9*(4), 356–367.

74. Wei, H., Zheng, G., Yao, H., & Li, Z. (2018). Intellilight: A reinforcement learning approach for intelligent traffic light control. In *Proceedings of the 24th ACM SIGKDD International Conference on Knowledge Discovery & Data Mining* (pp. 2496–2505).

75. Aberkane, S., & Elarbi, M. (2019). Deep reinforcement learning for real-world anomaly detection in surveillance videos. In *2019 6th International Conference on Image and Signal Processing and Their Applications (ISPA)* (pp. 1–5). IEEE.

76. Yan, C., Xiang, X., & Wang, C. (2019). Towards real-time path planning through deep reinforcement learning for a UAV in dynamic environments. *Journal of Intelligent & Robotic Systems*, 1–13.

77. Rao, Y., Lu, J., & Zhou, J. (2017). Attention-aware deep reinforcement learning for video face recognition. In *Proceedings of the IEEE International Conference on Computer Vision* (pp. 3931–3940).

78. Ciresan, D., Giusti, A., Gambardella, L. M., & Schmidhuber, J. (2012). Deep neural networks segment neuronal membranes in electron microscopy images. In *Advances in neural information processing systems* (pp. 2843–2851).

79. Liu, X., He, P., Chen, W., & Gao, J. (2019). Multi-task deep neural networks for natural language understanding. arXiv preprint arXiv:1901.11504.

80. Deng, H. (2013). Guided random forest in the RRF package. arXiv preprint arXiv:1306.0237.

81. Fukushima, K., & Miyake, S. (1982). Neocognitron: A self-organizing neural network model for a mechanism of visual pattern recognition. In *Competition and cooperation in neural nets* (pp. 267–285). Berlin, Heidelberg: Springer.

82. Linnainmaa, S. (1970). *The representation of the cumulative rounding error of an algorithm as a Taylor expansion of the local rounding errors*. Master's Thesis (in Finnish), University of Helsinki (pp. 6–7).

83. LeCun, Y., Boser, B., Denker, J. S., Henderson, D., Howard, R. E., Hubbard, W., et al. (1989). Backpropagation applied to handwritten zip code recognition. *Neural Computation, 1*(4), 541–551.

84. LeCun, Y., Bottou, L., Bengio, Y., & Haffner, P. (1998). Gradient-based learning applied to document recognition. *Proceedings of the IEEE, 86*(11), 2278–2324.

85. Jackel, L. D., Lecun, Y., Stenard, C., Strom, B., Sharman, D., & Zuckert, D.(1998). Optical character recogntion for automatic teller machines. In *Industrial applications of neural networks* (pp. 375–378).

86. Krizhevsky, A., Sutskever, I., & Hinton, G. E. (2012). Imagenet classification with deep convolutional neural networks. In *Advances in neural information processing systems* (pp. 1097–1105).

87. Zeiler, M. D., & Fergus, R. (2014). Visualizing and understanding convolutional networks. In *European conference on computer vision* (pp. 818–833). Cham: Springer.

88. Bochkovskiy, A., Wang, C.-Y., Mark Liao, H.-Y. (2020). YOLOv4: Optimal speed and accuracy of object detection. arXiv preprint arXiv:2004.10934.

89. Sreenu, G., & Saleem Durai, M. A. (2019). Intelligent video surveillance: a review through deep learning techniques for crowd analysis. *Journal of Big Data, 6*(1), 48.

90. Yazdi, M., & Bouwmans, T. (2018). New trends on moving object detection in video images captured by a moving camera: A survey. *Computer Science Review, 28,* 157–177.
91. Bah, M. D., Hafiane, A., & Canals, R. (2018). Deep learning with unsupervised data labeling for weed detection in line crops in UAV images. *Remote Sensing, 10*(11), 1690.
92. Akhloufi, M. A., Castro, N. A., & Couturier, A. (2018). UAVs for wildland fires. In *Autonomous systems: Sensors, vehicles, security, and the Internet of Everything* (Vol. 10643, p. 106430M). International Society for Optics and Photonics.
93. Zhang, S., Wen, L., Bian, X., Lei, Z., & Li, S. Z. (2018). Occlusion-aware R-CNN: Detecting pedestrians in a crowd. In *Lecture notes in computer science (including subseries Lecture notes in artificial intelligence and Lecture notes in bioinformatics)* (pp. 657–674).
94. Lee, D., Kim, G., Kim, D., Myung, H., & Choi, H.-T. (2012). Vision-based object detection and tracking for autonomous navigation of underwater robots. *Ocean Engineering, 48,* 59–68.
95. Yang, S., Luo, P., Loy, C. C., & Tang, X. (2015). From facial parts responses to face detection: A deep learning approach. *Proceedings of the IEEE International Conference on Computer Vision, 2015*(3), 3676–3684.
96. Yao, L., & Wang, B. (2019). Pedestrian detection framework based on magnetic regional regression. *IET Image Process.*
97. Tian, B., Li, Y., Li, B., & Wen, D. (2014). Rear-view vehicle detection and tracking by combining multiple parts for complex Urban surveillance. *IEEE Transactions on Intelligent Transportation Systems, 15*(2), 597–606.
98. Hayat, S., Yanmaz, E., & Muzaffar, R. (2016). Survey on unmanned aerial vehicle networks for civil applications: A communications viewpoint. *IEEE Communications Surveys Tutorials, 18*(4), 2624–2661.
99. Sanchez-Lopez, J. L., Molina, M., Bavle, H., Sampedro, C., Suárez Fernández, R. A., & Campoy, P. (2017). A multi-layered component-based approach for the development of aerial robotic systems: the aerostack framework. *Journal of Intelligent & Robotic Systems, 88*(2–4), 683–709.
100. Carrio, A., Sampedro, C., Rodriguez-Ramos, A., & Campoy, P. (2017). A review of deep learning methods and applications for unmanned aerial vehicles. *Journal of Sensors.*
101. Lin, C., He, D., Kumar, N., Raymond Choo, K.-K., Vinel, A., & Huang, X. (2018). Security and privacy for the internet of drones: Challenges and solutions. *IEEE Communications Magazine, 56*(1), 64–69.
102. Wazid, M., Das, A. K., & Lee, J.-H. (2018). Authentication protocols for the internet of drones: Taxonomy, analysis and future directions. *Journal of Ambient Intelligence and Humanized Computing,* 1–10.

Machine Learning Based Cybersecurity Defense at the Age of Industry 4.0

Sourabh Kumar Vishavnath, Adnan Anwar, and Mohiuddin Ahmed

Abstract Industry 4.0 is mainly recognized as the digital transformation of the industrial sector which is driven through machine learning and artificial intelligence. It includes the historical collection of information, the capture of live data via sensors, data aggregation and connectivity between routing, gateways and other protocols, PLC integration, the dashboard for analysis and monitoring. The convergence of Machine learning (ML) and Artificial Intelligence (AI) has overcome data integration and decision-making challenges with the adoption of Industry 4.0. This article justifies the context and relevance of data sharing in the industrial sectors and the cyber threats in industry 4.0 and also provides the preventive techniques used via AI and ML. In addition, this book chapter illustrates real use cases and potential prospects for both technologies.

1 Introduction

The security requirement for the digital system has enhanced the availability and scope of the business stakeholders. The technical improvement is highlighted from machine learning and artificial intelligence which has boosted the regular business functions and also assisted to secure a huge volume of data used in the organization. Machine learning refers to the ability of computers to learn without explicit programming which has developed behaviour models before using those models for framing future decisions based on input data [1]. The technology has enabled the

S. K. Vishavnath · A. Anwar (✉)
Center for Cyber Security Research and Innovation (CSRI), School of IT, Deakin University, Geelong, Australia
e-mail: adnan.anwar@deakin.edu.au

S. K. Vishavnath
e-mail: svishavnath@deakin.edu.au

M. Ahmed
Edith Cowan University, Joondalup, Australia
e-mail: mohiuddin.ahmed@ecu.edu.au

© The Author(s), under exclusive license to Springer Nature Switzerland AG 2021
U. Ghosh et al. (eds.), *Machine Intelligence and Data Analytics for Sustainable Future Smart Cities*, Studies in Computational Intelligence 971,
https://doi.org/10.1007/978-3-030-72065-0_19

management to interpret the threats of the digital field and respond to the attacks. On the other hand, artificial intelligence-based infrastructure has simplified the detection of threats of cybersecurity along with prevention and prediction of the failure. Core reasons for the issue and automated mitigation are characterized as the benefits of the company after adopting AI whereas the reduction on the dependency of human and savings cost is executed through the technology [2]. The development of industry 4.0 is justified through the diversified use of technical aspects. In this chapter, the background of industry 4.0 and the importance of data sharing are justified along with the cyber risks in industry 4.0 and the prevention strategies applied through AI and ML are included. Moreover, real-world use cases and future opportunities of both the technologies are demonstrated in the report.

2 Background of Industry 4.0

Industry 4.0 is considered as the intelligent networking of processes and machines for the particular sector with the assistance of information and communication technology. It is reflected as the digital transformation of production and manufacturing-based companies along with value creation approaches [7]. Industry 4.0 is applied interchangeably with the fourth industrial rebellion which has represented new stages in the company and controlling the trade value chain. The contemporary control system along with embedding software systems and IoT would be reflected as the main sign of industry 4.0 [6]. The notation also indicates more effort on automation than the third industrial revolution and proper bridging of the physical and digital system enabled by the industrial internet of things. The shift from the industrial control system to the smart devices defining by production steps and close-loop data models are highlighted as characteristics of industry 4.0. It has also developed upon data mapping and data models across value streams and end-to-end product lifecycle. The integration of technologies is reflected as the main key for industry 4.0.

Artificial intelligence and machine learning are considered as main technologies that drive industry 4.0 whereas big data, cloud computing, IoT, and augmented reality also contribute to technological advancement [21]. The smart factory is developed of hyper-connected production approaches where communication is supported through relying on AI platforms for collection and analysis of various patterns of information including code text and images. Industry 4.0 is mainly flagging the path for the digitization of the manufacturing industry where machine learning and AI-based systems have changed the information interaction manner [15]. The impact of artificial intelligence in industry 4.0 is reflected through yield and predictive quality, predictive maintenance, the collaboration of human and robot, design of generation and supply chain and market adaptation. Yield and predictive quality have mainly applied industrial AI to reveal the hidden reasons for the production losses of the manufacturing sector. This is mainly progressed through consistent and multi-variation of the analysis using the algorithm of machine learning [11]. Supervised learning is mainly integrated for identifying the patterns and trends in information where the involvement of both machine learning and AI is justified.

The involvement of machine learning and AI has changed the manufacturing companies to be incorporated with the framework of industry 4.0 which has created a new business opportunity that has resulted in improvement in business efficiency. The reduction in cost through predictive maintenance leads to limited maintenance activity which indicates reduced labour cost, minimized inventory and wastage of materials [24]. The prediction of Remaining Useful Life has improved business performance at the time of machine health maintenance. It also minimizes the scenarios that have caused unplanned downtime. Autonomous vehicles and equipment have streamlined their operations before proper transportation and controlling of better quality with human-machine collaboration [5]. Customer-focused manufacturing would be executed by adopting machine learning and AI in the context of industry 4.0.

Industry 4.0 has offered several benefits that have raised productivity. It includes productivity, flexibility, quality and speed which has made the construction work effective and up to the mark. The main identified manufacturing conditions are reflected as safety, working conditions, training and collaboration, protection of the environment and innovative capability [3]. Industry 4.0 is mainly recognized as the digital transformation of the manufacturing sector which is driven through machine learning and artificial intelligence. It also includes the historical collection of information, the capture of live data via sensors, data aggregation and connectivity between routing, gateways and other protocols, PLC integration, the dashboard for analysis and monitoring [9]. The integration of machine learning and AI has solved the unique complexity issue after digital modelling for ensuring full traceability. It has also revealed the impact of process inefficiency with having multivariate analysis and multisource data tags.

Improved productivity through automation and optimization are identified as the benefit of industry 4.0 which has reflected through cost-saving, rising profitability, waste reduction, automation for error prevention and boosted completion of the business process [22]. The impact of artificial intelligence in industry 4.0 is reflected through yield and predictive quality, predictive maintenance, the collaboration of human and robot, design of generation and supply chain and market adaptation. Maintaining the proper flow of the supply chain and balance between demand and need of the customers would be simplified through machine learning. Moreover, the understanding of regional demands is effectively addressed through artificial intelligence which would assist the management to produce particular quantity and quality products.

3 Importance of Data Sharing

Data is considered a critical business tool for organizations that prefer to operate business operations with the help of different data applications. Modern-day use of data becomes more vital in every area of business. The organizational business has entered a new data economy, where organizations need to involve in data sharing for utilizing valuable data insights in terms of better decision making [4]. Data become one of the most valuable assets in every business aspect. All possible opportunities have lied under the data, where proper utilization and application of the data can only offer such opportunities. Data sharing also plays a vital role in data applications. Not only the business, different other areas such as societal aspects and educational research work, investigation, etc...., may also be facilitated with the advancement in data sharing facilities [19]. Data sharing options allow individuals to access more amount of information and get sufficient knowledge regarding any expected topic or works.

Data application undergoes certain evaluation, where immense growth in it promotes the development and better utilization of different predictive data technologies such as artificial intelligence (AI) and machine learning (ML). Data gets such demand and acceptance on a global scale due to its ability to make predictive decisions from collected data [19]. The information system is quite compatible to collect real-time data about any business operation, condition of the individual, markets, and others. Such collected information is utilized in AI and ML techniques for producing different predictive analysis and decision making. Without data sharing facility, such data utilization would be restricted enough within the organization or individual [12]. Data sharing facility, make the data available for a larger application. The collection of large amounts of data increases the accuracy of decision-making with the information. The larger the amount of data represents more knowledge. It may offer better and effective data utilization in different business and individual areas.

Data sharing brings lots of facilities to be means of data transparency, collaboration, research acceleration, reproducibility, etc. Machine learning and artificial intelligence upgrade the requirement of data sharing to another level, where data not only holds high value but also become essential in different applications. Before focusing on data sharing one may focus on data collection, analysis processes. Modern-day information technologies are capable enough to collect information in a large amount [16]. Such a large amount of information has enough role to bring more accurate data in front to be utilized for AI and ML applications. Machine learning would also assist in developing the security framework against the particular threat as the programing comprises an algorithm for preventing the threats. The machine learning application is operated based on the data input and its behaviour to get outputs [5]. Without proper data sharing facilities, insufficient data may offer limited possibilities for data uses. Machine learning technology is preferred for extracting organized data from unorganized data set through a supervised approach . Data sharing makes it possible to access such a large amount of data, where ML may be deployed for proper data utilization. On the other hand, AI technology also gets enough acceptance in every aspect of the business. It seems that AI technology offers better data output rather than a human being [19]. Such data predictive technology requires a sufficient amount of data input to be experienced and further provides autonomous data output for specific data input. One industry or individual can't access a sufficient amount of information until it access to shared data from others. The business organization gets enough facilitation from it in terms of better data efficiency, collaboration, flexibility, and lots more. Accessing a large amount of data become simpler and cost-effective [20]. ML and AI become one of the essential technology that is preferred within the organizational process. At the same time, data sharing facilitates both the technology by supplying sufficient data and the outputs of those technologies can also be made available for everyone through the data sharing facilities.

In the present time, closed data is recognised as the major player which has contributed to the economy of European data which is projected to be € 739 billion by 2020. The valuation of data has enhanced rapidly and the application of AI and ML has insisted on the authority to treat the resources as assets [3]. One of the important data sharing approaches includes publishing data into the domain of specific research platform concerning the usage and types of data that has supported data centres. It includes cancer imaging archive, PubChem and ChEMBL. It has offered good access given the data submission nature along with the capability for linking data (Fig. 1).

Using the trusted partner or alliance model for storing data is accessible based on agreed rules for data depositors [22]. The partners of the consortium model are required to sign the agreements and reflect their capability for preserving the security and privacy of data. It has also taken excess time for establishing legal agreements to allow the appropriate sharing of data along with restricting the shared data type (Fig. 2).

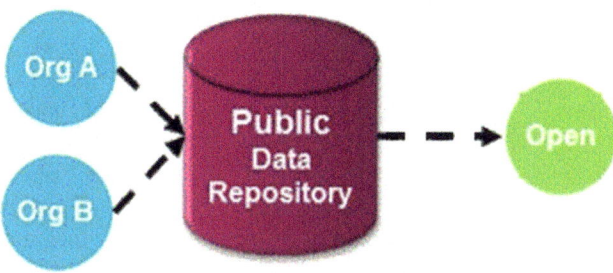

Fig. 1 Publishing data into broad domain agonistic public research source

Fig. 2 Trusted partner
model source

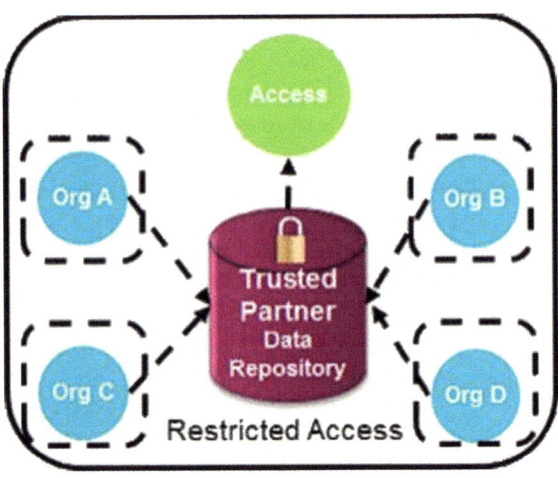

Fig. 2 Trusted partner
model source

4 Cyber Risk at the Age of Industry 4.0

The revolution in the industrial field has brought once again the advancement in the way. The fourth-generation revolution has caused the world to see the new advancement in the technology but along with that, it brought a new operational risk [23]. The advancement in the interconnected system of the network and industry 4.0 has brought the cyber risk with a better pace and means to have a more extensive effect on the networks more than the previous. The manufacturer and the suppliers or the providers may lack in supporting the technology with the counter-attack to the cyber risks. In this age of industry 4.0, cybersecurity should be strategies to have better security, vigilant, and resilient. They should be fully integrated into the organizational and information technology strategies. The strong connectivity of the smart machinery in the advancement of the system as the industry 4.0 raises the stake of having cyber risks [21]. This advancement has initiated a better version of the attack and even has a lack of ideas to counter-attack the threats to protect the system.

Industry 4.0 is a combination of all the previous systems as they are going to give a tailored amalgamation of the digital world, technologies, and the physical actions of the smart factories with advanced manufacturing. While the idea of the industry 4.0 is to bring the best of the digital capabilities along with the manufacturing and the supply processes to drive a revolutionary change but it also brings the risk of new cybercrimes [19]. These industries were not prepared for this risk. For example, energy grid has been the target for a wide number of cyber incidents [13, 14]. It is fundamental to have a system working with the integrated working of the systems and digital connections. New risks that are associated with the industry 4.0 need to be addressed so that the threat can be taken care of. Cybersecurity should become one of the integral parts of the system now as these threats of the cyber risks are pouncing in the system now and then [1]. There have been several aspects of the cyber risks that are associated with industry 4.0.

The data sharing through the digital connections of the industry 4.0 increases the access to the cyber risks in the working. The organization should know properly how to share the data in the system and how to overcome the threats of cyber risks. The organization also pay attention to vendor processing [15]. This will, in turn, enhance the transactions of the payment in the system and can also cause a lot of danger to it. The organization switch the system into a better and big group of suppliers to avoid the danger of cyber risks [2]. Some of the known cyber threats that are affecting the system are notably as malware, phishing, Spear Phishing, Man in the middle, Trojans, Ransomware. These have started coming in the front in a new way and are getting initiated with the involvement of the industry 4.0. It also includes the historical collection of information, the capture of live data via sensors, data aggregation and connectivity between routing, gateways and other protocols [21]. Modern-day information technologies are capable enough to collect information in a large amount. Such a large amount of information has enough role to bring more accurate data in front to be utilized for AI and ML applications.

The software that is performing the malicious threats to the system is unable to be handled by this new advancement [4]. This new advancement is giving way to such threats. The malware is the software-related risks that target the device and the network to damage the data in them. The contemporary control system along with embedding software systems and IoT would be reflected as the main sign of industry 4.0. Yield and predictive quality have mainly applied industrial AI to reveal the hidden reasons for the production losses of the manufacturing sector. The information system is quite compatible to collect real-time data about any business operation, condition of the individual, markets, and others. Phishing is the one that works in the emails of the users in the network [9]. The more sophisticated one in this field is the spear phishing. All these risks got increased in rate with the advent of industry 4.0. This advent of the system in the digital world has encouraged the world to take a better way of involving in the business but lacks in protecting the system with the counter solutions of the attacks of the cyber risks.

In present times, the companies and the user have faced particular threats which have caused the loss of confidential data. The key issues faced by the companies include geographically aloof IT systems, detection of manual threats, and reactive nature of cybersecurity and frequent cover of IP address along with modifying. Geographical distance has made the communication difficult for tracking the accidents [24]. The cybersecurity experts are required to overcome the differences in infrastructure for monitoring the incidents which are occurred at the regional level. The detection of manual threats would be beneficial for the company whereas it is expensive and time-taking with unexpected. The authority of the business fixes the issues after rising the threat only which is identified as the major challenge over the digital medium. The hackers mainly apply a diversified range of programs like a proxy server, virtual private network, Tor browser and others. Using these sites would make the identity of hackers as undetected and anonymous [12]. The cyber-security is also impacted by a diversified range of challenges which include vast attack surface, thousands or hundreds of devices per organisation, big shortfalls in the skilled security professional number and data masses that moved beyond the human-scale problem.

5 Combating Cyber-Attacks Using Advanced Analytics

There are several ways to combat cybersecurity against the cyber risks that are imposed in the network. Cybersecurity is one of the most important features of this world. The increasing rate of cyber risks in the system is putting an alarm in the mind of the organizations. Cybersecurity is going through massive changes in its form. The cyber securities are the set of the technological processes which are designed to protect the system from the cyber risks [7]. In the recent world, there has been the implementation of machine learning which helps in the processes of cybersecurity. Machine learning is a tool that is significantly changing cybersecurity and data science. This is a pattern in which the technologies are changing with each passing day (Sharma et al. 2020). The integration of machine learning and artificial intelligence would also support the accuracy in design which is identified as the important criteria.

There is a huge emphasis on cybersecurity and is processed through the machine learning method. The machine learning method talks about the relationship and the usages of the context of cybersecurity. The ultimate goal of cybersecurity is data-driven and is imposed to protect the data in the network from the cyber risks that are imposed on the network from several malicious elements [22]. The machine learning models show how cyber risks are prevented. The machine learning method emphasis the comprehensive study of the model which protects the data in the system. Machine learning would also assist in developing the security framework against the particular threat as the programing comprises an algorithm for preventing the threats [3]. It mainly applies the algorithm which is induced from the old dataset along with statistical analysis for formulating decision based on customer buying. The security of customer data is also supported through technology integration.

The pattern of cybersecurity is maintained in this flexible manner. The learning techniques are capable of malicious anomalies and the data-driven pattern which is associated with the security incident to make the intelligent decision to process the cybersecurity [16]. Depending on the data-driven pattern the cybersecurity is used to emphasize the concept of cyber securities based on it. The machine learning techniques include their relations with the cybersecurity tasks and summarise the concept of the number of machine learning methods based on the cybersecurity model. The multi-layered smart technology of the cybersecurity services helps to provide better cybersecurity to the network (Sharma et al. 2020). Protection of the data in the network in the form of cybersecurity is helping the network to protect the information in a better and useful way. Machine learning methods help in restoring the data and to stop the inculcation of the cyber risks in the system.

As the patterns of cyber-attack have changed rapidly, traditional security approaches have struggled for detecting malware and threats. The hackers have put their consistent effort on developing smarter ways to breach data, control access and compromising the secured network [17]. Artificial intelligence assists the under-resourced system of security which has broadened the current solution scope of cybersecurity after paving the approaches of developing a new one. AI has enabled the authority to detect the threats, conduct vulnerability management, data centre and network security. Traditional safety process has employed compromise indicators or signatures for identifying the threats whose success rate is 90%. On the other hand, the adaptation of AI has raised the threat detection rate to 95% [18]. The companies use AI for enhancing the threat hunting process through the integration of behavioural interpretation. As an example, the AI model has been leveraged to develop application policies by processing huge end-point data within the network of the organisation.

Machine learning is a pattern in which the technologies are changing with each passing day. In 2019, the newly reported vulnerability was 20,362 which is reportedly higher than 17.8%. The organisation has failed to handle and identify a diversified range of vulnerability which are combatted through AI techniques which are identified as user and event behavioural analytics, the endpoint and behaviour of server baseline. It also ensures protecting data by integrating AI. The technology has mainly optimised and monitored the processes in the data centre which include cooling filters, backup power, internal temperature and bandwidth application. The calculative power and consistent monitoring capability would develop efficiency and security infrastructure. The technology mainly minimises the hardware maintenance cost after altering the threats induced in the business. The warnings would assist the authority to fix the potential issue smartly. Google had announced 40% minimisation of cooling cost at the facility along with 15% savings in power consumption.

Machine learning is recognised as the mere application of the algorithm which is applied for interpreting the cyber threats in a better manner. The benefits of integrating machine learning include detecting the malicious activities with the prevention of cyber-attacks, analysing mobile end-points for the threats which are applied by Google. In addition, improvement in the human analysis is executed by ML from the malicious threats to the protection of end-point [4]. The advantages of integrating AI is demonstrated through IT asset inventory, exposure of threats, the effectiveness of controls, prediction of breach risks and explainability which has ensured an ultimate level of security. The hackers follow the trend whose identification has become simplified by AI framework. The cybersecurity system provides up-to-date data regarding industrial and global threats for framing the effective decisions related to the context. Gaining accurate and complete inventory of users, devices and applications with the access of information systems would be identified as the benefits of using AI as the solution of cybersecurity. It is also very useful to identify the influence of security processes and tools for maintaining strong posture [1]. The identified strengths and gaps of infosec are executed through the AI framework which also justifies the correct prediction of risks.

AI-based systems have offered improved context for the prioritization and security alert response which has simplified the mitigation process of the threats. Thus, the technical development programmed by AI and ML justifies the security approaches fulfilment. The integration of machine learning and AI has solved the unique complexity issue after digital modelling for ensuring full traceability [7]. Maintaining the proper flow of the supply chain and balance between demand and need of the customers would be simplified through machine learning. The information system is quite compatible to collect real-time data about any business operation, condition of the individual, markets, and others.

6   Real-World Use Cases

In this section of the critical research, real cases of using machine learning and artificial intelligence against cyber-attacks have been presented. For the first real case scenario, Hink et al. [8] using machine learning techniques to reduce power system disturbances caused by cyber-attacks and malware. The study applied numerous learning calculations to Modbus RTU information to show their feasibility as interruption location devices on a straightforward gas pipeline framework. Condition of-the-practice arrangement calculations were applied to show a capacity to separate order and information infusion assaults for basic and little scope SCADA frameworks. For building up the security framework the analysts utilized an unpredictable blend of administrative control frameworks interfacing with different brilliant electronic gadgets supplemented by network checking gadgets, for example, SNORT and Syslog frameworks. The organization is made out of 4 breakers constrained by astute electronic transfers. These IEDs hand-off data back through a substation switch through a switch back to the administrative control and information securing frameworks. Assault situations were fabricated and reproduced with the supposition that an entertainer had just accessed the substation organization and represents an insider danger by giving orders from the substation switch. Digital assault situations like a short out issue in an electrical cable and can happen in different areas along the line, the area is demonstrated by the rate range and Line support where at least one breakers are opened utilizing the far off transfer trip order for upkeep. The situation additionally included hand-off setting change that is arranged with separation insurance conspire and the aggressor changes the setting to handicap the transfer capacity with the end goal that hand-off won't trip for a substantial shortcoming or a legitimate order. The information was drawn from 15 informational indexes which included a huge number of individual examples of estimations all through the force framework for every occasion type. The datasets were arbitrarily tested at 1% to lessen the estimate and assess the viability of little example sizes. For this examination, there was a normal of 294 'No occasion' occurrences, 3711 assault examples and 1221 common occasions cases utilized over the arrangement plans. The date and time data were eliminated since situations were run successively and time and date would consummately arrange the information. In light of the discoveries, it was set up that the underlying benchmarks for applying AI ways to deal with power framework unsettling influence grouping on a shrewd force lattice system. Utilizing the JRipper + Adaboost strategy over a three-class like Attack, Natural Disturbance, and No Event order conspire, the specialists had the option to dependably characterize power framework unsettling influences with low bogus positive rates. In light of the aftereffects of applying learning strategies to this force framework information, we reason that AI is a suitable way to deal with giving solid choice help to control framework administrators on whether the framework is enduring an onslaught.

A similar real case study was conducted to develop countermeasures against load redistribution assaults as identifiers that dissect the deliberate burdens in the force business. The rate circulation of heaps of the PJM framework is near that of the heaps in the 30-transport framework, which legitimizes the planning from each zone to the comparably estimated load in our test framework. The planning has been performed by first deciding a planning proportion between the PJM and the 30 transport framework and afterwards increasing each hourly heap of the 20 zones by this factor. Along these lines, the examination made new burden designs comparing to all year long for the test framework. The PJM information was utilized to make a stacked dataset for the IEEE 30 transport framework for a very long time, relating to the period from January 2012 to December 2016 which was thought to be the ordinary information. Concerning show assault plan, scientists planned a bi-level streamlining issue to discover the assault vector c that amplifies the stream on a particular objective line by redistributing loads and enhances this detailing making it more effective and versatile to enormous scope frameworks. After figuring the assault vector, the arrangement of bogus estimations is passed to the state assessor which computes the relating loads. The investigation proposed three finders which examined the deliberate heaps of a force framework during any hour of the day and decide whether they are ordinary burdens or if they have been noxiously altered through a digital assault. Every finder depends on an alternate AI strategy, however, the overall methodology in deciding the adequacy of a lot of estimations is comparable. The best identifier as far as recognition capacity and computational productivity are the one dependent on the closest neighbour calculation, which had the option to reliably identify the assaults with little paces of bogus alert.

On the other hand, Komar et al. [10] aimed to increase the security of intrusion detection system, the summed up structure of an elite versatile framework for digital assaults discovery was created. The model contained a fluffy location analyzer, a lot of neural finders and a subsystem of dynamic and actualizing. The subsequent one is executed in programming and spoke to by a devoted PC, which is utilized for current assaults examination and formation of suitable insurance implies. In this part, as indicated by the recently characterized highlights, the subsystems for examination and neural organization preparing, control and measurable handling of cyberattacks information can be recognized. This part incorporates additionally a committed cushion where dubious code is recorded for its investigation. What's more, inhabitant programming of the subsequent part monitors the support codes to treat them just as information. Any of the guidelines, which are remembered for the code, contained in a committed cushion, can't be performed. The framework for discovery and balance of cyber-attacks runs in the accompanying manner. Fluffy location analyzer leads a nonstop examination of the current message source. Collected information about the historical backdrop of cyberattacks, distinguished beforehand, permit to lessen the ideal opportunity for investigation of information, contained in a specific message. For this reason, the consequences of measurable preparing of the recurrence of digital assaults and their sorts are utilized. Fluffy analyzer correspondingly orchestrates a lot of neural indicators to lessen the ideal opportunity for examination of cradle information. Each neural organization indicator produces two codes for the dynamic subsystem. One of them decides the probability of a digital assault. The subsequent one decides the likelihood of the nonappearance of such an assault. Such neural organization locators association builds the proficiency of the dynamic subsystem. The methodology of improving of the security level of the interruption location framework is offered by the usage of the neural organization identifiers on the programmable rationale clusters and presentation of the decision-making framework because of Mamdani fuzzy surmising rules.

7 Future Opportunities

Artificial intelligence has mainly played a crucial role in the automation of the operation which has driven the business process without facing any particular concern. The technology mainly applies facial recognition to track and spot the suspects for catching the criminals which are recognised as the future development of technology (Tanczer et al. 2018). On the other hand, AI is also able to predict the outcomes which are based on data interpretation. It has evaluated the patterns of customer data for inclining the deliverable into that direction. The tail off of the demands is also analysed by the technology which is useful in assisting the company for the purchase of current stock in proper volume [21]. It is also applied for improving the recruitment process which has indicated the incorporation of smart techniques for completing the work process. Personalised marketing, operational automation, inventory management and customer service would be boosted through integrating AI into business operations. In the healthcare context, the technology is applied for introducing virtual personal healthcare assistant whereas the healthcare bots also assist the patients regarding the effective service deliverables. Data sharing makes it possible to access such a large amount of data, where ML may be deployed for proper data utilization [5]. On the other hand, AI technology also gets enough acceptance in every aspect of the business.

8 Conclusion

From the above study, it has been concluded that the integration of machine learning and artificial intelligence has benefitted the user and companies to secure the information and progress in the environment of industry 4.0. The development of industry 4.0 is justified through the diversified use of technical aspects. Industry 4.0 is mainly recognized as the digital transformation of the manufacturing sector which is driven through machine learning and artificial intelligence. The information system is quite compatible to collect real-time data about any business operation, condition of the individual, markets, and others. ML and AI become one of the essential technology that is preferred within the organizational process.

The smart factory is developed of hyper-connected production approaches where communication is supported through relying on AI platforms for collection and analysis of various patterns of information including code text and images. In this age of industry 4.0, cybersecurity should be strategies to have better security, vigilant, and resilient. They should be fully integrated into the organizational and information technology strategies. Some of the known cyber threats that are affecting the system are notably as malware, phishing, Spear Phishing, Man in the middle, Trojans, Ransomware. The cybersecurity experts are required to overcome the differences in infrastructure for monitoring the incidents which are occurred in regional level. The machine learning techniques include their relations with the cybersecurity tasks and summarize the concept of the number of machine learning methods based on the cybersecurity model.

References

1. Beier, G., et al. (2020). Industry 4.0: How it is defined from a sociotechnical perspective and how much sustainability it includes–A literature review. *Journal of Cleaner Production,* 120856.
2. Carrasquilla, J., & Melko, R. (2017). Machine learning phases of matter. *Nature Physics, 5*(13), 431–434.
3. Chen, B., et al. (2017). Smart factory of industry 4.0: Key technologies, application case, and challenges. *IEEE Access, 6*(1), 6505–6519.
4. Churpek, M., et al. (2016). Multicenter comparison of machine learning methods and conventional regression for predicting clinical deterioration on the wards. *Critical Care Medicine, 14*(2), 17.
5. Copeland, B. J., & Proudfoot, D. (2007). Artificial intelligence: History, foundations, and philosophical issues. *Philosophy of Psychology and Cognitive Science North-Holland, 1*(1), 429–482.
6. Erwin, A., et al. (2020). Machine learning and data analytics for the IoT. *Neural Computing and Applications.*
7. Guzmán, V., et al. (2020). Characteristics and skills of leadership in the context of industry 4.0. *Procedia Manufacturing, 43*(1), 543–550.
8. Hink, R. C. B., Beaver, J. M., Buckner, M. A., Morris, T., Adhikari, U., & Pan, S. (2014, August). Machine learning for power system disturbance and cyber-attack discrimination. In *2014 7th International Symposium on Resilient Control Systems (ISRCS)* (pp. 1–8). IEEE.

9. Kerr, R., & Szelke, E. (Eds.). (2016). *Artificial intelligence in reactive scheduling.* London: Springer.
10. Komar, M., Kochan, V., Dubchak, L., Sachenko, A., Golovko, V., Bezobrazov, S., & Romanets, I. (2017, September). High performance adaptive system for cyber attacks detection. In *2017 9th IEEE International Conference on Intelligent Data Acquisition and Advanced Computing Systems: Technology and Applications (IDAACS)* (Vol. 2, pp. 853–858). IEEE.
11. Komura, D., & Ishikawa, S. (2018). Machine learning methods for histopathological image analysis. *Computational and structural biotechnology journal, 1*(16), 34–42.
12. Kozma, R., Alippi, C., Choe, Y., & Morabito, F. (Eds.). (2018). *Artificial intelligence in the age of neural networks and brain computing.* London: Academic Press.
13. Le, T. D., Anwar, A., Beuran, R., & Loke, S. W. (2019). Smart grid co-simulation tools: Review and cybersecurity case study. In *icSmartGrid: IEEE Proceedings of the 7th International Conference on Smart Grid.*
14. Le, T. D., Anwar, A., Loke, S. W., Beuran, R.,& Tan, Y. (2020). GridAttackSim: A cyber attack simulation framework for smart grids. *Electronics, 9,* 1218.
15. Li, D., Landström, A., Fast-Berglund, Å., & Almström, P. (2019). Human-centred dissemination of data, information and knowledge in industry 4.0. *Procedia CIRP, 84*(1), 380–386.
16. McKinnel, D., Dargahi, T., Dehghantanha, A., & Choo, K. (2019). A systematic literature review and meta-analysis on artificial intelligence in penetration testing and vulnerability assessment. *Computers & Electrical Engineering, 75*(1), 175–188.
17. Minnaar, A. (2014). 'Crackers', cyberattacks and cybersecurity vulnerabilities: The difficulties in combatting the 'new' cybercriminals. *Acta Criminologica: African Journal of Criminology & Victimology,* 127–144.
18. Moses, S., & Rowe, D. C. (2016). Physical security and cybersecurity: Reducing risk by enhancing physical security posture through multi-factor authentication and other techniques. *International Journal for Information Security Research (IJISR), 6*(2)
19. Murray, J., Hughes, G., & Kreutz-Delgado, K. (2005). Machine learning methods for predicting failures in hard drives: A multiple-instance application. *Journal of Machine Learning Research, 6*(1), 783–816.
20. O'Connor, C., Calkin, D., & Thompson, M. (2017). An empirical machine learning method for predicting potential fire control locations for pre-fire planning and operational fire management. *International Journal of Wildland Fire, 26*(7), 587–597.
21. Patel, P., Ali, M., & Sheth, A. (2018). From raw data to smart manufacturing: AI and semantic web of things for industry 4.0. *IEEE Intelligent Systems, 33*(4), 79–86.
22. Russo, D., et al. (2018). Comparing multiple machine learning algorithms and metrics for estrogen receptor binding prediction. *Molecular Pharmaceutics, 15*(10), 4361–4370.
23. Safdar, N., Banja, J., & Meltzer, C. (2019). Ethical considerations in artificial intelligence. *European Journal of Radiology,* 108768.
24. Sanders, A., Elangeswaran, C., & Wulfsberg, J. (2016). Industry 4.0 implies lean manufacturing: Research activities in industry 4.0 function as enablers for lean manufacturing. *Journal of Industrial Engineering and Management (JIEM), 3*(9), 811–833.

Security Systems for Smart Cities Based on Acoustic Sensors and Machine Learning Applications

Giuseppe Ciaburro

Abstract Safety in urban areas assumes a central position in the requalification of modern cities or in the planning of future cities. Security problems have a different nature: problems deriving from the mobility of citizens, micro-crimes, terrorism are just a few examples. A smart city equipped with an infrastructure of sensors capable of alerting security managers to a possible risk becomes fundamental for the safety of citizens. Often in cities monitoring systems based on video cameras are used, adding to these acoustic sensors connected to an automatic identification system based on machine learning can significantly improve the security of smart cities.

Keywords Smart city · Deep learning · Machine learning · Sounds classification · Data mining · Artificial intelligence · Convolutional neural networks · Audio events detection · Spectral signature · Mel-frequency cepstral coefficients

1 Introduction

Modern society, in continuous transformation and evolution, places ever more tight rhythms and puts people in a position to have to maximize resources to automate procedures as much as possible. Cities are undergoing profound changes regarding their demographic, social, environmental, and economic structure [1]. In this context, Smart Cities are proposed as a solution model of current urban problems, placing citizens at the center of their organization as the main users of resources and developing ideas and services in a collaborative way, aimed at satisfying and improving the real needs of the community [2].

A Smart City is a city that combines technology with government actions and the demands of society, driven by a growing demand for smart and sustainable actions that reduce environmental impact and offer citizens a high-quality life. New technologies

G. Ciaburro (✉)
Department of Architecture and Industrial Design, University of Campania Luigi Vanvitelli, Caserta, Italy
e-mail: giuseppe.ciaburro@unicampania.it

© The Author(s), under exclusive license to Springer Nature Switzerland AG 2021
U. Ghosh et al. (eds.), *Machine Intelligence and Data Analytics for Sustainable Future Smart Cities*, Studies in Computational Intelligence 971,
https://doi.org/10.1007/978-3-030-72065-0_20

represent the enabling factor which, together with traditional techniques, can carry out numerous interventions to obtain a more efficient and sustainable city [3].

The smartness to which the Smart City idea refers is a distributed, shared, horizontal, social intelligence that favors the participation of citizens and the organization of the city with a view to optimizing resources and results [4].

To undertake the transformation of a city towards the Smart City model, the following aspects are crucial:

- introduction of new technologies in daily life and in the various fields of work
- communication between citizens and administration for a continuous constructive dialogue towards the problems to be faced and the ways to do it better
- improvement of mobility towards sustainability and the development and promotion of new renewable energy sources

Smartness does not correspond to the degree of technology used in the urban context, but to the internal management of the city, aimed at the construction of a technological and intangible infrastructure, capable of making people and objects communicate, integrating information, generating inclusion and intelligence, improving life quality [5].

Social and environmental sustainability are core components of smart cities. Today, urban realities rely more and more on natural resources, so the goal is to avoid the depletion of these riches, ensuring their protection in the long term. The protection of the environment, the reduction of waste and more generally sustainability are issues that in recent decades have seen growing attention from civil society and policy makers, with a growing attention to the opportunities associated with the new economic-social-productive models. The origin of this process is mainly attributable to the increased attention paid to two closely interrelated phenomena: the scarcity of natural resources and population growth [6].

Strategic sectors through which to promote the development of the sustainability of smart cities are:

- efficiency and energy saving
- efficient uses of resources
- the prevention and recycling of waste
- sustainable mobility
- eco-innovation

The idea of Smart City is based on the awareness and participation of citizens in public life. This comparison represents for the administrators the possibility of continuously detecting concrete and real needs and any problems connected to them, making the response efficient and effective. By promoting a continuous dialogue and interaction between citizens and policy makers, the city promotes a peaceful coexistence between various cultures and currents of thought. Citizens are involved in decision making and the city is designed thanks to its citizenship, which becomes co-author of public policies [7].

In the Smart City, the role of Information and Communication Technology (ICT) is crucial, therefore the presence of infrastructures capable of connecting different devices and applications to each other becomes indispensable, offering support to connectivity that must be widespread, dynamic and scalable. The technologies are functional to make the objectives of Smart Cities operational, starting from those linked to citizen participation and their training, through those dedicated to environmental sustainability and the development of smart and sustainable mobility, up to those connected to the economy of the territory [8].

All these activities use ICT in an instrumental way, by exploiting some technologies that more than others lend themselves to the achievement of the objectives of Smart Cities. Among these, the Web plays a decisive role, which makes it easier to access services based on the Internet of Things (IoT), which allows objects to become intelligent collectors and distributors of information on mobility, energy consumption, services and assistance to the citizen, the cultural and tourist offer and much more. The massive use of sensors to measure the characteristics of the environment becomes essential for the proper management of the services that a smart city must guarantee to citizens [9].

In this chapter, we will show how the widespread installation of sensors and the use of technologies based on artificial intelligence can significantly improve the security systems of smart cities. Modern sensors can detect physical and chemical parameters with a high degree of reliability, with an economic investment that has been significantly reduced over time. Today it is possible to install sensor systems for the detection of physical parameters interconnected with each other and with a server for data collection. These data represent a wealth of information which, given the amount of data (Big Data), can hardly be analyzed with traditional statistical procedures. In these cases, technologies based on machine learning algorithms come to our aid. Using such algorithms, we can extract knowledge from the large amount of data that sensor installations return to us. In this work we will see how to identify the presence of drones in confined environments.

The Chapter is organized as follows: Sect. 2 discusses the security needs required by Smart Cities and how they are currently being addressed. In Sect. 3 Machine Learning-based technologies are introduced: Crucial concepts such as Supervised, Unsupervised, and Reinforcement learning are also covered. Section 4 proposes a case study relating to the use of Deep Learning to detect Unmanned Aerial Vehicles: First the basic concepts of the Convolutional Neural Network (CNN) are introduced, and then move on to analyze the characteristics of Unmanned Aerial Vehicles. Subsequently the proposed methodology is analyzed in detail with an accurate discussion of the results obtained. Finally, Sect. 5 summarizes the essential points of the work.

2   Smart Cities Security Management

The rise of the Internet of Things (IoT), where more and more devices are interconnected to exchange data and interact with minimal human intervention, combined with intensive software analysis and extensive connectivity, allows us to monitor and control complex systems present in a city at an unprecedented level, through the inclusion of connected sensors and actuators [10]. The huge growth of connected devices within a Smart City leads to an increase in the access possibilities for a potential attacker. In the absence of clear indications, standards and regulations, connected devices often lack basic security measures and allow attackers to easily access systems, potential sources to be hijacked for Denial of Service (DoS) attacks and targets vulnerable critics in the very infrastructure of the city [11].

There is a global shortage of specialized information security skills, which is accentuated in smart cities where governments struggle to attract enough suitable people to provide them with effective security.

The smart city is subject to a wide range of potential threats, from criminal activities such as ransomware, to politically motivated hacktivists to acts of terrorism. There have been reports of security breaches, ransom demands, personal data leaks, denial of service attacks in government and municipal departments and systems, as well as attacks on public services and critical infrastructure that pose a real threat to people's safety. As smart cities evolve and grow, the above issues will become more pressing [12].

In recent times, cyber-attacks have been reported on the Smart City infrastructures of the main cities of the world. Often the attack takes place through ransomware and with a series of potential impacts ranging from compromising data, to Denial of Service, up to service interruptions that are potentially dangerous for citizens' lives. In fact, the most critical services have been able to restore an alternative backup system. However, the number and severity of these attacks are increasing, making a life-threatening attack increasingly likely. The most widespread and serious attack vectors are through "Advanced Persistent Threats" in which hackers aim to access networks in a covert way and manage to remain undetected for long periods until they reach their goals [13].

Smart cities security is not just a technical problem for ICT teams to solve. It involves all the people and all the systems we use throughout our lives. The simplest way to reduce security threats is not to adopt a new technology, because a huge opportunity would be missed. While there is no single solution to make a city "safe", there are many things that can be done to enable cities and communities to reap the benefits of connected digital technology while minimizing risks. This should be part of any organization's resilience strategy. The balance between being defensive and being progressive must be carefully evaluated [14].

We tend to think of cybersecurity in terms of controlling access to our networks. The main vulnerabilities are at both ends of the system. Whether you manage a city, a community, a company or an organization, the basic rules are the same. Cyber security starts with data. Most attacks continue to enter through phishing, social engineering, and the like. At the other end of the system are the many networked IoT devices, often with very little security and little system-level knowledge or control providing a large and vulnerable attack surface [15].

It becomes essential to ensure that the Information Security policy adopted is complete, and that it covers all risks. Security is not easy and it's not a single step in the process. It is essential for the design, commissioning, use, maintenance and decommissioning of the project. If planned and executed effectively during the life of a project, it can be the basis for a successful 'smart city', where the benefits of smart technology can be realized by the whole community [16].

The problem of security has always characterized the urban realities. The city was born as a security system for people and for commercial exchanges. In less recent history, insecurity was outside the cities: beyond the city walls there were insecure spaces and territories, at the mercy of crime and removed from the control of the public authorities. In the modern city, insecurity has moved into the urban fabric [17]. The space of cities tends to become alien and, therefore, unsafe for the inhabitants: it is in the city that criminal episodes mainly occur and one is exposed most to risks and dangers, now increasingly fueled by global terrorism. In cities, therefore, the causes of insecurity are multiplying. These can be found both in criminal conduct and in behaviors of incivility, both in situations of social hardship and in the difficult coexistence between different social groups, both in phenomena of urban decay and in landscape and hydrogeological instability [18].

In cities, the extent of the security problem is changing no longer just prevention and repression of crimes, but above all promotion and guarantee of better living conditions. This is an objective that is sought to be achieved through the integration of security policies with other public policies: urban planning, territorial protection, social and cultural integration, conflict resolution, education for legality. To these problems are added those deriving from the considerable flow of people who move between the suburbs and the city center for work, study, and more generally personal reasons. In fact, urban mobility, especially during peak hours, represents a great challenge to the safety of citizens that administrators must necessarily face [19].

As part of a capillary control of the territory, the widespread use of sensors can help stakeholders in the management of emrgenze in order to miglioarre considerably the security of the smart cities (Fig. 1).

A tool widely used in the security management of urban centers makes use of systems based on video surveillance. Although today several thousand cameras are already installed in our cities, in fact they support the emergency service in an inefficient way, since in most cases they are not interconnected to a single service center, limiting themselves to making simple recordings, between the other difficult to consult. Hence the need to trace the safety of a city not so much to the numerical explosion of the installed cameras, but to their integration, management, and analysis of images, in favor of a more effective use of new technologies [20].

To improve the safety of a city, fighting crimes and promptly identifying possible emergency situations, it is necessary to intervene in a capillary way starting from the streets, using integrated systems based on multiple technologies and connected to each other and to emergency systems, as a tool real-time prevention and identification of real problems. The mere use of the visual channel to identify an emergency is no longer enough, but it is necessary to combine this tool with the analysis of the sounds that a smart city offers. For the identification of an emergency it becomes crucial to exploit the automatic recognition tool of the characteristic sounds of that

Fig. 1 Examples of sensors that can be installed in a smart city: sound recognition, video surveillance, smoke detection, and weather station

problem to add it to the contributions deriving from video surveillance. A joint management of these tools can bring added value in the management of the security of smart cities [21].

The urban sound environment has undergone many transformations over time thanks to new sound characters that have been added to the old ones, from the faint sounds coming from the increasingly widespread computer tools to the heavy noise resulting from modern transport infrastructures. The buzz due to the substantial anthropogenic presence thus approaches the noise of traffic, penetrating the urban environment with different wavelengths and in opposite ways. Urban space, full of sounds, thus becomes influenced and its perception by its users increasingly difficult (Fig. 2). Due to the predominance of nearby sound sources, it is therefore increasingly difficult to perceive sounds from long distances and the acoustically recognizable urban space appears reduced. To study the urban sound environment, it is therefore necessary to discover its significant characteristics, particular sounds, important for their individuality and their dominant presence [22].

Traditionally, the recognition of sounds has been dealt with using classical instruments such as in situ measurements for the sound component and frontal interviews for the perceptual component. To these tools today others are added that allow us to analyze the phenomenon through an innovative approach. We refer to the machine learning algorithms widely used in other research fields, which in my view are well suited to the characteristics of sound study. In particular, algorithms based on the use of artificial neural networks in recent years have been used to train systems that make use of artificial intelligence, as well as to build prediction and classification models in various fields of medicine and engineering [23].

Fig. 2 Sounds event detection using acoustics sensors

3   Machine Learning-Based Technologies

Machine learning is the branch of Artificial Intelligence (AI) that develops the ability of machines to learn something without being explicitly given the instructions to do so. We refer to algorithms that, based on probability and statistical studies, autonomously develop their knowledge thanks to the data patterns received, without the need to have specific initial inputs from the human developer [24].

This means setting aside the classic explicit programming, a system in which the human component programs a model based on conditional commands, in favor of a method in which the machine is able to establish by itself the patterns to follow to obtain the desired result. Therefore, the real factor that distinguishes artificial intelligence is autonomy [25].

We can divide any machine learning process into three main phases:

- the learning of data, in the various forms in which they can manifest themselves
- the evaluation of the data itself, in which the computer system hypothesizes statistical models that describe the observed reality
- the optimization of the estimated models and the consequent formulation of a response/action strategy based on the feedback collected with experience

First, let's consider the concept of data, elements with which the machine learning system comes into contact. They are not all the same, and they certainly don't all have the same value. In the first phase of the learning process, the system receives a set of data necessary for training (training dataset), at which time the computer estimates the relationships between the input and output data [26].

These identified links describe the parameters (or weights) of the model estimated by the system (Fig. 3). Sometimes the input data are supplied with the respective output data, such as the system receives initial data which is given a precise future realization determined by the actions taken. What the computer sees is already tagged or classified data. The presence of labels in the initial data is what distinguishes some machine learning methodologies: supervised learning. More precisely, the label in machine learning is nothing more than a target, what in statistics is defined as a dependent variable. Remaining in the statistical field, what is usually called an explanatory variable, in machine learning is called feature. Once the system has received the training data, what it does is an estimate of the recurring structures and logical rules that it observes in the examples provided, which is nothing more than the determination of models and parameters that describe reality. Wanting to resort to statistics again, these data patterns attempt to describe the so-called Data Generating Process (DGP2) of the data set [27].

The model estimated by the computer does not necessarily have to be static, or determined a posteriori, but rather it is constantly evolving: after an initial training, the machine begins to apply the strategy of actions that it considers optimal in order to achieve the set goal; this sequence of operations can prove to be as effective as not, and this is described by a value function that returns feedback to the machine on which it can base its future moves. As the training progresses, and therefore the feedback received, the computer improves its moves in response to the situations it faces [28]. At the end of the analysis of the training dataset, the system, based on the accumulated experience, will have a first estimate on how to behave when dealing with new data but similar to past ones; however, what interests human subjects is to evaluate the efficiency of future information that may be different from that already

Fig. 3 Training procedure in a Machine Learning-based algorithm

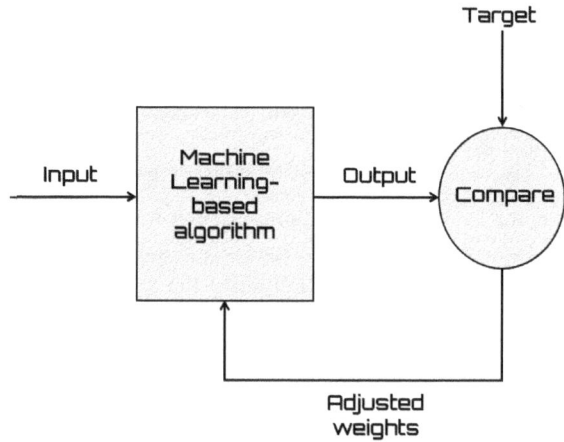

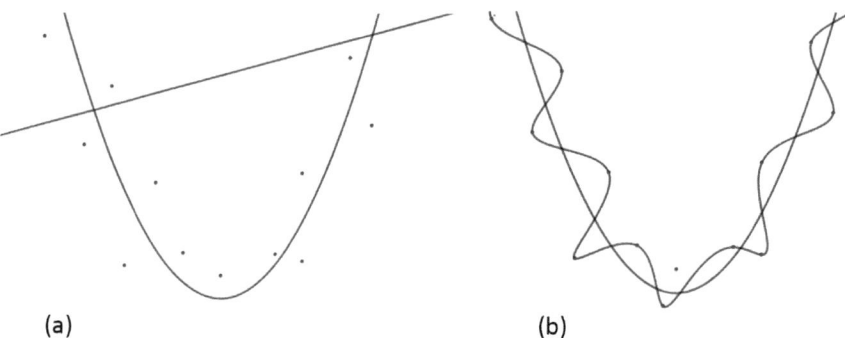

Fig. 4 **a** Underfitting occurs when the algorithm cannot capture the underlying trend of the data; **b** Overfitting occurs when a model begins to memorize training data, rather than learning to generalize from a trend

absorbed by the system, and to receive future forecasts. A further category of data called test data is inserted, which serves precisely to test the results that the "machine" returns after the learning phase on the training data [29] (Fig. 4).

The set of rules, formulas, actions and logical operations seen allow to define the concept of algorithm: through experience, the machines create general rules that convey into learning models, which in turn allow to build learning algorithms necessary to solve a particular problem. The algorithm indicates to the system the actions/operations it can perform in each context. If the system can do this, it will use this information for subsequent tasks [30].

Although studies on machine learning algorithms are in continuous development, the learning theory does not provide any guarantee on the effective performance of the algorithms. The finiteness of the data and the lack of knowledge on the future evolution of the model on the other hand represent a limit in themselves. For the modeling, to provide the best possible performance, the complexity of the inductive hypothesis must be equal to that of the model underlying the training dataset. Problems of underfitting and overfitting arise precisely from this disparity: in particular, the problem of overfitting arises when a complex statistical model over-fits the observations because it has an excessive number of parameters compared to the number of sample data (Fig. 4b). Conversely, it is true for underfitting [31]. Overfitting occurs when a model trained on a training set is then unable to generalize the pattern learned on a larger set of data: it is very precise as regards training data, but is not useful on large series of unknown information. Underfitting (Fig. 4a), on the other hand, occurs in cases where there is not enough data to estimate an adequate number of parameters and therefore the model does not adhere to the real PGD. To avoid these problems, it is important to perform training also using unexpected data or using gimmicks. In any case, to date, there is no single solution to problems of this type: each application, each task to be solved, requires different models and algorithms. It is also for this reason that various machine learning proposals have been developed in the last twenty years, the result of experimentation and study [32].

The learning process can be implemented in very different ways, depending on the tasks required of the machine or the data that can be collected. The functions of machine learning are typically classified into three main macro-categories, which are distinguished by the nature of the inputs on which the learning is based or the value function with which the system judges the actions taken. These subsets of machine learning, also called paradigms, are supervised learning, unsupervised learning, and reinforcement learning [33].

Parallel to these, deep learning now plays a fundamental role, a method that arises from the study of neural networks and which, to date, seems to be the driving force behind the development of machine learning science, given its large applications. In addition to these four main forms, both hybrid forms, the point of contact between the paradigms of machine learning, and alternative forms that originate in different scientific approaches have arisen over the years [34].

3.1 Supervised Learning

In Supervised Learning, the model is provided with both input datasets and information on the related outputs, with the aim that the computer identifies patterns and general rules that associate the initial information x with the results named y. Thus, it should be possible to reuse that logical link also used for other similar tasks, that is for new inputs do not present in the training set. The goal of supervised learning is to provide an inductive hypothesis, a function that is able to learn from the input-output pairs provided in the training phase, and then obtain the desired results in a new test dataset [35].

Each input data xi represents a feature, and can be, for example, a numeric, textual attribute or a more sophisticated object, such as an image or an audio file. The output yi instead represents the result, the response variable, and can take any form. The latter is what distinguishes the different application approaches of supervised learning (Fig. 5). From a logical point of view, supervised learning starts from the assumption

Fig. 5 Supervised Learning dichotomous classification

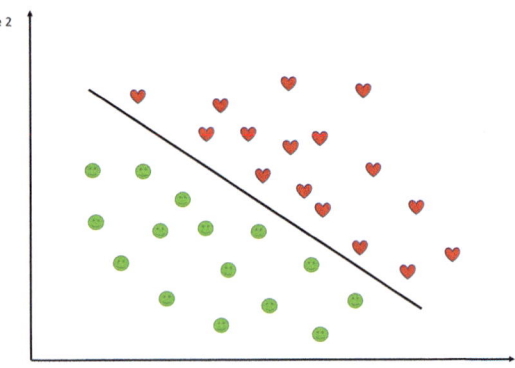

Fig. 6 Unsupervised
Learning grouping objects

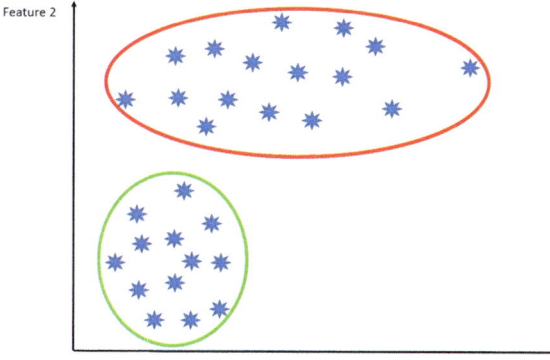

that, if the system has an adequate number of information, it will have sufficient experience to allow it to determine a function that can approximate the objective function [36].

3.2 Unsupervised Learning

Unlike the supervised version, in unsupervised learning the system receives only the input data sets without any indication of the outputs to be obtained. The purpose of this learning method is to trace hidden patterns, that is, to identify a logical structure in the input data without having previously been classified. These algorithms represent the best solution in cases where the problem insists on numerical data, since they can use the techniques of statistics and distribution studies. Conversely, they are ineffective methods for the analysis of non-numerical data [37].

The main problems faced by unsupervised learning systems are clustering, that is the search for groups and classes within the data (Fig. 6), association rules, estimating the distribution of data, displaying data in graphs or tables and reducing of the volume of data in summary representations [38].

3.3 Reinforcement Learning

In Reinforcement Learning, software interacts with a dynamic environment and with a system of rewards and punishments that guides it towards improving its performance.

Once the input data is obtained, the computer must reach a goal that is established based on a value function that only tells it if the goal has been achieved or not [39].

In this way, thanks to the feedback, the behavior of the system progresses through a trial and error process. It is basically the same process that man goes through when

Fig. 7 Reinforcement learning interaction with the environment

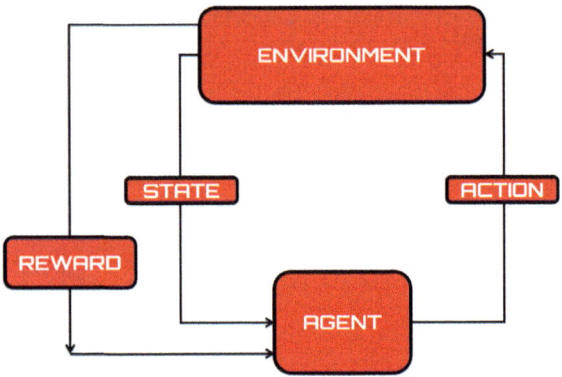

he has to learn to ride a bike for example: at the beginning you often fall, until, a little at a time, you learn what the wrong moves are, how to balance the weights, which the position to maintain (Fig. 7). Over time, evaluating what has worked and what leads to a fall, allows you to learn to ride a bike. The same is true for computers that use reinforcement learning algorithms: they try out different actions, evaluate which of them gave the best result, reinforce those that worked, and so on [40].

4 Using Deep Learning to Detect Unmanned Aerial Vehicle

Artificial neural networks (ANN) are a computational model that is inspired by the circuits of biological neurons present in the animal brain. They are in fact made up of units, called artificial neurons, connected to each other through synapses that allow the transmission of signals from one neuron to another [41]. What is formed is a network in which the neurons represent the nodes of the network and the edges the connections between the latter: the receiving neuron (called postsynaptic) processes the signals and transmits them to the downstream neurons connected to it, and so for the whole network. Usually these signals have a numerical state that oscillates between 0 and 1; the output that comes out of each neuron is obtained through some non-linear function that "sums" the different input data it receives from upstream neurons [42]. Neurons and synapses usually have different weights that can vary over time, or, as learning progresses, they can increase or decrease the strength of the signal they send downstream. Furthermore, there may also be thresholds, such that, if the signal arriving at the neuron does not reach a certain value, it is not transmitted to subsequent neurons [43].

Typically, neurons are organized in layers. Different layers can perform different types of transformations on the input data. The signals therefore travel from the first input layer to the last output layer, sometimes after crossing the intermediate layers several times (Fig. 8). In these cases, ANNs are defined as deep neural networks (DNNs) precisely to underline the presence of different learning "plans", exactly what happens in deep learning [44].

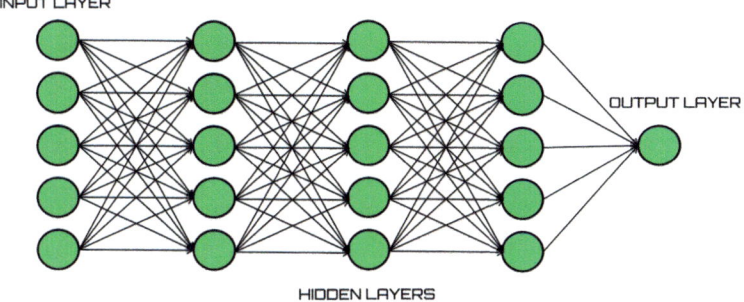

Fig. 8 Deep Learning-based architecture with layers and nodes

Further elements, then, which affect the functioning of neural networks, are:

- the learning rates
- the cost function which evaluates the results and provides feedback
- back-pay, a method that regulates and determines the weights of the connections to compensate for errors encountered during learning, and which works through the stochastic descent of the gradient relating to each previous state

A Neural Network is therefore made up of three layers: input layer, series of hidden layers, output layer. The nodes receive input data, process it, and can send the information to other subsequent nodes. Through more or less numerous input-processing-output cycles, the networks become able to generalize and provide correct outputs associated with inputs that are not part of the training set [45].

4.1 Convolutional Neural Network (CNN)

Deep Learning is a sub-category of Machine Learning that can be defined as a path to achieve advanced machine learning. To achieve this goal, learning models stratified on several levels are created that allow the transformation of raw input data into classified output data. Tele technique takes the name of representation learning. The data, however, are not provided by humans, but are learned through the use of statistical calculation algorithms that try to understand the functioning of the human brain and the way in which it interprets the various basic inputs. Learning, therefore, is modeled according to a pyramid, in which the highest concepts are learned starting from the lowest levels. Deep Learning allows computational models composed of multiple levels of processing to learn representations of data with multiple levels of abstraction [46].

Deep Learning is used for both supervised and unsupervised learning and consists of two main phases:

- Training: which consists in the transformation of the basic levels, having raw data, into higher more abstract levels using the backpropagation algorithm, this works by varying the parameters used to calculate the representation in each level by modifying that of the previous level. The set of methods that allows you to feed a machine with raw data and automatically discover the representations necessary for detection or classification is called learning of representations
- Test: consists of training the model, created in the previous phase, using data of a different nature than the input ones

Convolutional Neural Network (CNN) is one of the most common Deep learning algorithms. It is used for processing data characterized by a particular grid topology: a CNN is able to emphasize local relationships starting from adjacency structures present in the data, through automatic and adaptive learning of low to high level patterns [47].

CNNs therefore represent the main model used in the field of computer vision and in general in applications that require object recognition and artificial vision.

It is an architecture inspired by the biological structure of the visual cortex, in which there is a hierarchy of two basic types of cells: simple and complex cells. Simple cells react to primitive patterns present in sub-regions of the visual field, called receptive fields, while complex cells synthesize information from the former to identify more complex structures. Similarly, the neurons present in a convolutional layer are connected to sub-regions of the previous level and are not affected by signals located outside that area. The receptive fields can also overlap: the neurons of a CNN therefore produce spatially correlated results [48].

It is therefore possible to identify the main difference with a fully connected neural network: while in a fully connected neural network the number of parameters to be learned increases as the input size increases, a convolutional neural network reduces the number of parameters thanks to the reduced number of connections, shared weights and sub-sampling.

A convolutional neural network can have tens or hundreds of layers, each of which consists of different filters used to detect different features and build the corresponding feature maps. In the case of images, in fact, filters (called kernels) are applied to each input and, through a convolution operation, feature maps are generated. The latter will be used as input for the next layer (Fig. 9). The filters of the first layers look for very simple features (edges), to assume increasingly complex shapes, able to uniquely define the object [49].

The architecture of a convolutional neural network includes several blocks, such as, for example, convolutional layers, pooling layers and fully connected layers.

The architecture of the convolutional neural network can be divided into two parts:

- a part of feature detection which then deals with the extraction of features through operations such as convolution, pooling and ReLU
- a classification part that generates the predicted output using fully connected layers and operations, or SoftMax

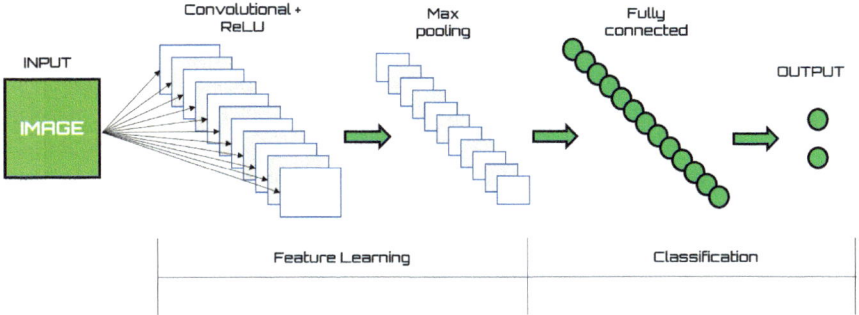

Fig. 9 Convolutional Neural Network-based architecture for binary classification

The optimization of the network parameters takes place in the following ways: the function signal is propagated forward in the network up to the final layer, where the gradient of the loss function is computed and retro-propagated to allow the weights to be updated through an descent of the gradient.

The feature extraction layers perform operations that alter the data with the intent of learning the specific characteristics of the data. Three of the most common layers are: convolution, activation or ReLU, and pooling.

4.1.1 Convolutional Layer

Convolution is a particular type of linear operation in which a filter, or kernel, is applied to a multidimensional array of numbers, called the input tensor, i.e. a multidimensional parameter mask limited in height and width but which extends throughout the depth. of the input volume [50].

The core scans the input tensor in height and width: in each position, the scalar product between the kernel and the covered input portion is calculated, to obtain the output value in the current position. The output volume obtained once the process is complete is called the activation map or feature map.

4.1.2 ReLU Layer

The convolution layer is typically followed by a non-linear activation function: only the activated features are passed to the next layer. The most used function is the rectified linear unit (ReLU), which performs a thresholding operation on each element by mapping the negative values to zero and maintaining the positive values. It has been demonstrated how the use of the ReLU function allows for faster and more effective training than traditional units such as the hyperbolic tangent [51].

4.1.3   Pooling Layer

The pooling layer performs a non-linear subsampling operation that reduces the transversal dimension of the activation maps, leaving their depth unchanged, with the aim not only of simplifying (reducing its parameters) the output of the previous convolutional layer, but also to introduce translational invariance to small displacements and distortions and, consequently, providing greater robustness with respect to the localization of the features [52].

Taking up the analogy with the visual cortex, the pooling unit was inspired by the behavior of complex cells: by capturing a growing visual field, they are able to learn spatial hierarchies of feature-patterns resulting less sensitive to slight shifts in the position of salient features.

The most popular form of the pooling operation is max pooling, which performs non-linear subsampling by dividing the input into rectangular regions and returning the maximum value within each window.

4.2   Unmanned Aerial Vehicle

In recent years, technological development has led to significant growth in the field of robotics and artificial intelligence. The use of robots to improve the lifestyle by applying these sciences in various fields such as civil, industrial, or medical is increasingly common. Especially thanks to the reduction of costs and safer technologies, robotics is also taking its first steps inside homes, even in terms of simple entertainment [53].

Different types of robots can be used, essentially distinguishable in ground, water and flying robots. Ground robots can move on wheels, tracks, or legs. Water robots are boats of various types. Autonomous flying robots fall into the category of drones. We are witnessing an enormous spread of multirotor, specifically quadcopters, which with their ease of use, attract a lot both hobbyists and professionals in various sectors, from video shooting, to monitoring, to agricultural treatment. The quadcopter is a multirotor equipped with four motors, placed at the ends of the arms that form a cross. Adjacent motors rotate in the opposite direction (Fig. 10). It is an unmanned robotic system with vertical take-off and landing. They use a control system based on the reading of the data coming from the IMU (Inertial Measurement Unit) thanks to which a stabilized flight is possible [54].

UAVs can fly at low altitudes and having no human pilot on board, they can be used in inaccessible areas or in missions dangerous to human life. They also have low operating and management costs and can be equipped with various types of sensors. UAVs use a radio control for remote piloting or a control station. The control station allows the programming of a flight plan with the list of geographical coordinates of the places to fly over [55].

UAVs represent added value for the advancement of smart cities. The use of UAVs can make many routine operations easier: home delivery of goods and infrastructure

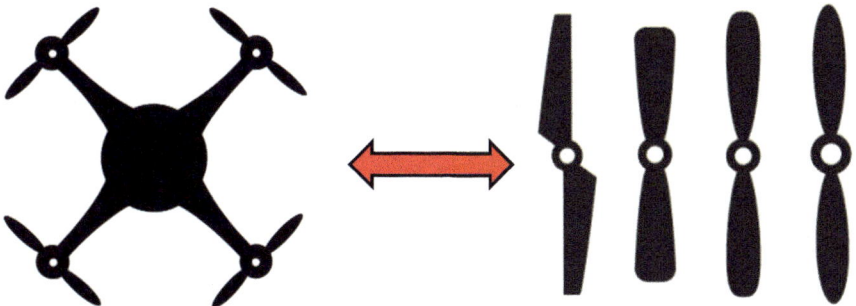

Fig. 10 Structure of a quadcopter with different types of blades

monitoring, just to name a few. The use of UAVs imposes in-depth considerations relating to the safety of citizens: already several states around the world use UAVs to monitor crowds during mass events. The recent period characterized by the pandemic situation due to the spread of covid-19, has imposed periods of limitation of citizens' freedom all over the world. Lockdown periods were imposed in several states in which citizens were forced to stay at home. In these contexts, UAVs have been widely used to control the restrictions imposed on citizens.

There are many possible harmful activities that can be carried out with the help of UAVs. A UAV can be operated on specific radio channels avoiding countermeasures based on electromagnetic disturbances, it can carry illegal cargoes such as drugs and it can be used to transport goods in prohibited areas such as prisons. Finally, UAVs can be used by terrorist organizations to perform attacks: the attacks on the Venezuelan prime minister and the one on the oil extraction plant in Saudi Arabia are the latest reported cases. The identification of the presence of UAVs in smart cities represents a necessary commitment to guarantee the safety of citizens. The small size of the UAVs and the use of rotating parts make UAVs difficult to identify with traditional systems. An alternative to such methods uses the beep they emit when they are in motion.

4.3 Methodology

In this section, a new methodology is proposed for the detection of UAVs in environments with anthropogenic background noise. The UAV identification procedure can be generalized to other sounds to improve safety in a smart city. Specific acoustic sensors are used to record sounds which are then used to train a model based on artificial neural networks. In this way an emergency alarm signal with identification of the location can be sent. Figure 11 describes in detail all the steps of the procedure [56].

The recordings were made inside a shopping center in the period of maximum crowding. In such conditions any problem can cause panic in the crowd that can generate escapes to exits with potential emergency conditions. In this scenario, the

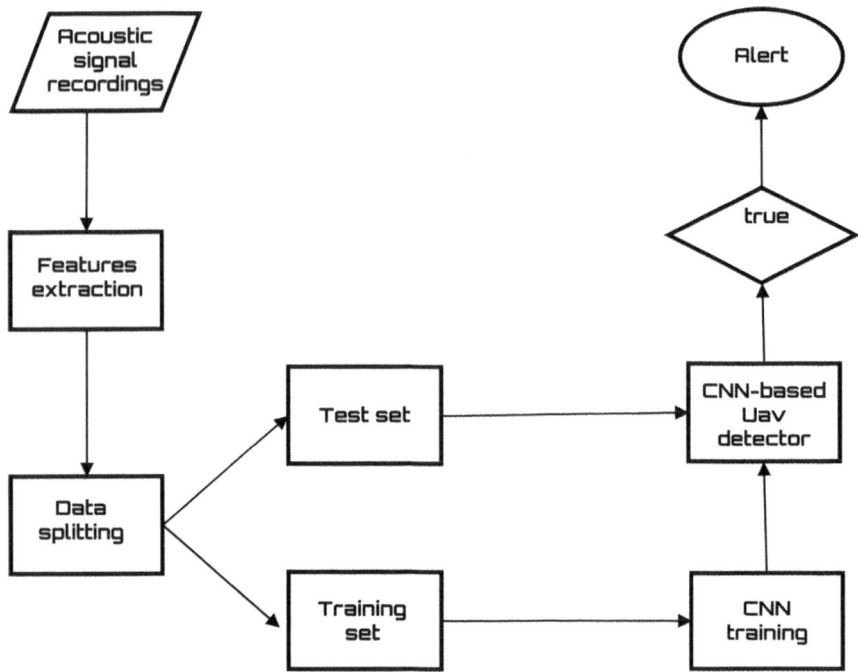

Fig. 11 Flow chart for UAV detection procedure

automatic identification of the presence of a hazard becomes crucial. In the presence of large numbers of people, video surveillance alone is not enough, but it is necessary to support innovative technologies [57].

Two scenarios were simulated: anthropogenic noise, and anthropogenic noise with the presence of an active UAV (Fig. 12). To generalize the measurement conditions, several measurement sessions were carried out by placing the UAV at different distances from the acoustic sensor and with different crowding conditions. A high-quality Zoom H6 Handy Recorder portable recorder with X-Y microphone was used for sound collection.

An acrylonitrile butadiene styrene (ABS) quadcopter with a 6-axis gyroscope was used to simulate the potential hazard. This is radio-controlled UAV with a frequency of 2.4 GHz and a control distance of 80 m. The dimensions of the UAV are: 310 × 310 × 120 mm. The UAV has four propellers, each equipped with two blades. The UAV's propellers move alternately according to the following pattern: two clockwise while the other two moves counterclockwise. The main noise source of the UAV is represented by the blades: shape and speed of rotation characterize the sound emission. The propeller consists of the hub which is the supporting structure and transmits the rotation to the blade, and the blade that generates the thrust for the flight of the drone [58].

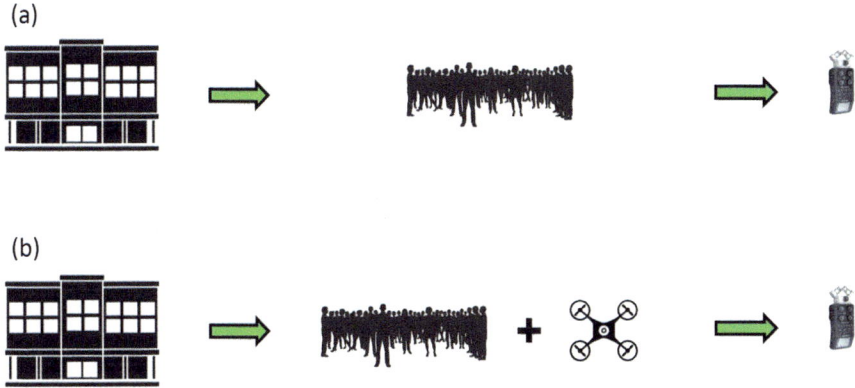

Fig. 12 Simulated recording scenarios: **a** anthropogenic noise; **b** anthropogenic noise with the presence of an active UAV

Table 1 Recorder and microphone specifications	Type	Specifications
	Microphone	Sensitivity: –41 dB, 1 kHz at 1 Pa
		Input gain: –∞ to 46.5 dB
		Maximum sound pressure input: 136 dB SPL
	Recording device	Sampling rate: 48 kHz
		Bit depth: 24 bit

The recordings were made with a high-quality portable recorder Zoom H6 Handy Recorder with a X-Y microphone (Table 1).

To train the CNN-based model, you need to extract the features to be used as model input. In this work the spectrogram of the recorded sounds was used. The spectrogram describes a signal through the relationship between the following properties of a sound: Frequency, Time, and Intensity. The relationship is guaranteed through a color map. In a spectrogram, the abscissa measures time, the ordinate the frequency and intensity of sound is described by a color map: dark colors represent low-intensity sound, light colors high-intensity.

The images containing the spectrogram characterized by a grid topology are provided as input to the model based on convolutional neural networks. CNN can extract the local characteristics of the adjacency structures present in the data. This is done using the adaptive learning of models that start from the bottom to reach the highest level. For these reasons and CNN are associated with object recognition in computer vision.

Convolutional neural networks perform particularly well on data characterized by strong spatial dependencies in local regions. A two-dimensional image can be represented as a grid: a matrix of values between 0 and 255. The pixels are interlaced

Table 2 Hardware requirements

Central Processing Unit (CPU)	Intel Core i5 6th Generation processor or higher, and AMD equivalent processor
RAM	8 GB minimum, 16 GB or higher is recommended
Graphics Processing Unit (GPU)	NVIDIA GeForce GTX 960 or higher
Operating System	Ubuntu or Microsoft Windows 10

Table 3 Software requirements

Programming platform	Python
Library	TensorFlow Scikit-Learn Keras

with each other and together define a pattern. CNNs associate these characteristics with values called weights, which will be similar for local regions with similar patterns [59].

In Table 2 and 3 are reported the hardware and software used to challenge the problem of sound event detection using CNN-based model.

With the specifications indicated in Tables 2 and 3 it will be possible to carry out the recognition in real time.

4.4 Results and Discussion

After obtaining the registration of the two scenarios, we tried to extract the features. Tracks of the recordings lasting 10 s were used to contain an adequate content of information. About 200 samples equally distributed between the two identification classes (UAV, no-UAV) were extracted. The spectrograms for each sample were then processed, labeling them with the two classes. To increase the number of samples to be subjected to the simulation model, the data augmentation technique was used. Random rotate, flip, and cut operations were performed on the spectrogram images. As a result, 2000 samples were obtained equally distributed between the two classes (UAV, no-UAV). 70% of the samples were used for training and the remaining 30% were used for testing.

Figure 13 compares a spectrogram of the UAV scenario with a NoUAV scenario.

The comparison of the two spectrograms (Fig. 13) highlights a high frequency component (5 kHz) which is present only in graph a) relating to the UAV scenario. This component is the spectral signature of the UAV which therefore characterizes its operation and allows us to identify it.

The convolutional network adopted has three convolutional hidden layers each consisting of the following layers:

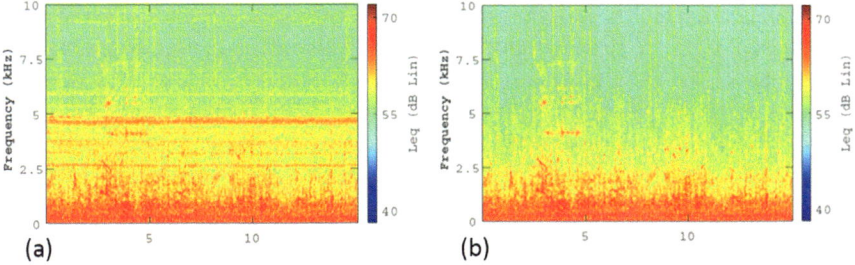

Fig. 13 Spectrograms of two simulated scenarios: **a** Anthropic noise and UAV switched on; **b** Anthropic noise and UAV switched off

- 2D Convolutional layer
- Average pooling operation for spatial data
- Rectified Linear Unit activation

A flattening layer is added to this structure which allows us to flatten the feature map into a single column:

- Flatten operation
- Dropout

A fully connected layer follows to link each input function with the corresponding output function:

- Densely-connected NN layer
- Rectified Linear Unit activation
- Dropout

Finally, a dense layer is added to obtain the probability that an image belongs to one of the two classes (UAV, no-UAV):

- Densely-connected NN layer
- Softmax activation

To evaluate the performance of the simulation model based on convolutional neural networks, the accuracy metric was adopted [60]. It returns a value close to 1 if our model works well. Our simulation model recorded an accuracy of 0.89, demonstrating the validity of the procedure for identifying a UAV in a complex acoustic scenario.

In Table 4 are shown the confusion matrix values.

Table 4 Confusion matrix for the test set (600 samples)

		Detected	
		UAV	NoUAV
Actual	UAV	270	30
	NoUAV	36	264

In a confusion matrix, the comparisons between the values predicted by the model and the real ones are reported in tabular form. This metric is used to evaluate the performance of a classification model. The rows show the current class while the columns show the values predicted by the system. The values present on the main diagonal of the confusion matrix are crucial as they represent correctly classified observations. Off-diagonal cells correspond to misclassified observations.

The values present on the main diagonal are equally distributed between the two classes (UAV, NoUAV). This result is consistent with the starting dataset which was equally represented by the two classes. The classification errors are slightly different, in fact the number of wrong classifications of the presence of UAVs are greater than the wrong classifications of NoUAV.

5   Conclusion

Security is crucial for the evolution of smart cities. Security is not limited to the data collected by sensors, but must necessarily include new methodologies which, using such data, can return better levels of security. In this study a new methodology for the identification of UAVs in a complex urban scenario was developed and verified. Sounds detected by acoustic sensors are transmitted to an expert system based on convolutional neural networks. The spectrograms from the detected sounds are sent to CNN which recognizes the patterns by returning a dichotomous response relating to the two classes (UAV, NoUAV).

The simulation carried out with the data recorded in a shopping center with different levels of crowding returned good results by measuring an accuracy of 0.89. The methodology lends itself to the identification of different types of sounds and can be integrated into a complex surveillance system based on data detected by video and audio sensors.

References

1. Campbell, T. (2013). *Beyond smart cities: How cities network, learn and innovate.* Routledge.
2. Deakin, M. (Ed.). (2013). *Smart cities: Governing, modelling and analyzing the transition.* Routledge.
3. Rodríguez-Bolívar, M. P. (2015). *Transforming city governments for successful smart cities.* Springer.
4. Axelsson, K., & Granath, M. (2018). Stakeholders' stake and relation to smartness in smart city development: Insights from a Swedish city planning project. *Government Information Quarterly, 35*(4), 693–702.
5. Yigitcanlar, T., & Kamruzzaman, M. (2018). Does smart city policy lead to sustainability of cities? *Land Use Policy, 73,* 49–58.
6. Basiago, A. D. (1998). Economic, social, and environmental sustainability in development theory and urban planning practice. *Environmentalist, 19*(2), 145–161.

7. Coletta, C., Evans, L., Heaphy, L., & Kitchin, R. (Eds.). (2018). *Creating smart cities.* Routledge.
8. McClellan, S., Jimenez, J. A., & Koutitas, G. (Eds.). (2017). *Smart Cities: Applications, technologies, standards, and driving factors.* Springer.
9. Schuurman, D., Baccarne, B., De Marez, L., & Mechant, P. (2012). Smart ideas for smart cities: Investigating crowdsourcing for generating and selecting ideas for ICT innovation in a city context. *Journal of Theoretical and Applied Electronic Commerce Research, 7*(3), 49–62.
10. Zanella, A., Bui, N., Castellani, A., Vangelista, L., & Zorzi, M. (2014). Internet of things for smart cities. *IEEE Internet of Things Journal, 1*(1), 22–32.
11. Mohanty, S. P., Choppali, U., & Kougianos, E. (2016). Everything you wanted to know about smart cities: The internet of things is the backbone. *IEEE Consumer Electronics Magazine, 5*(3), 60–70.
12. Ijaz, S., Shah, M. A., Khan, A., & Ahmed, M. (2016). Smart cities: A survey on security concerns. *International Journal of Advanced Computer Science and Applications, 7*(2), 612–625.
13. Elmaghraby, A. S., & Losavio, M. M. (2014). Cyber security challenges in Smart Cities: Safety, security and privacy. *Journal of advanced research, 5*(4), 491–497.
14. AlDairi, A. (2017). Cyber security attacks on smart cities and associated mobile technologies. *Procedia Computer Science, 109,* 1086–1091.
15. Khatoun, R., & Zeadally, S. (2017). Cybersecurity and privacy solutions in smart cities. *IEEE Communications Magazine, 55*(3), 51–59.
16. Nayyar, A., Jain, R., Mahapatra, B., & Singh, A. (2019). Cyber security challenges for smart cities. In *Driving the Development, Management, and Sustainability of Cognitive Cities* (pp. 27–54). IGI Global.
17. Amini, M. H., Arasteh, H., & Siano, P. (2019). Sustainable smart cities through the lens of complex interdependent infrastructures: panorama and state-of-the-art. In *Sustainable Interdependent Networks II* (pp. 45–68). Springer, Cham.
18. Vitunskaite, M., He, Y., Brandstetter, T., & Janicke, H. (2019). Smart cities and cyber security: Are we there yet? A comparative study on the role of standards, third party risk management and security ownership. *Computers & Security, 83,* 313–331.
19. Cugurullo, F., & Acheampong, R. A. (2020). *Smart cities.* Handbook of Urban Mobilities, 344.
20. Naeem, M., Ejaz, W., Iqbal, M., Iqbal, F., Anpalagan, A., & Rodrigues, J. J. (2020). Efficient scheduling of video camera sensor networks for IoT systems in smart cities. *Transactions on Emerging Telecommunications Technologies, 31*(5),
21. Sodhro, A. H., Pirbhulal, S., Luo, Z., & de Albuquerque, V. H. C. (2019). Towards an optimal resource management for IoT based Green and sustainable smart cities. *Journal of Cleaner Production, 220,* 1167–1179.
22. Alves, J., Guerreiro, P., Marques, G., & Paulo, J. (2020). Urban sound event detection and classification. *i-ETC: ISEL Academic Journal of Electronics Telecommunications and Computers, 6*(1), 2.
23. Carleo, G., Cirac, I., Cranmer, K., Daudet, L., Schuld, M., Tishby, N.,… & Zdeborová, L. (2019). Machine learning and the physical sciences. *Reviews of Modern Physics, 91*(4), 045002.
24. Hutter, F., Kotthoff, L., & Vanschoren, J. (2019). *Automated machine learning: methods, systems, challenges* (p. 219). Springer Nature.
25. Nosratabadi, S., Mosavi, A., Keivani, R., Ardabili, S., & Aram, F. (2019, September). State of the art survey of deep learning and machine learning models for smart cities and urban sustainability. In *International Conference on Global Research and Education* (pp. 228–238). Springer, Cham.
26. Ardabili, S., Mosavi, A., & Várkonyi-Kóczy, A. R. (2019, September). Advances in machine learning modeling reviewing hybrid and ensemble methods. In *International Conference on Global Research and Education* (pp. 215–227). Springer, Cham.
27. Zeng, T. (2001). Mean reversion and the comovement of equilibrium spot and futures prices: implications from alternative data-generating processes. *Journal of Futures Markets: Futures, Options, and Other Derivative Products, 21*(8), 769–796.

28. Sarker, I. H., Kayes, A. S. M., Badsha, S., Alqahtani, H., Watters, P., & Ng, A. (2020). Cyber-security data science: an overview from machine learning perspective. *Journal of Big Data, 7*(1), 1–29.

29. Chicco, D., & Jurman, G. (2020). Machine learning can predict survival of patients with heart failure from serum creatinine and ejection fraction alone. *BMC Medical Informatics and Decision Making, 20*(1), 16.

30. Zhong, X., & Enke, D. (2019). Predicting the daily return direction of the stock market using hybrid machine learning algorithms. *Financial Innovation, 5*(1), 4.

31. Van der Aalst, W. M., Rubin, V., Verbeek, H. M. W., van Dongen, B. F., Kindler, E., & Günther, C. W. (2010). Process mining: A two-step approach to balance between underfitting and overfitting. *Software & Systems Modeling, 9*(1), 87.

32. Zhang, H., Zhang, L., & Jiang, Y. (2019, October). Overfitting and Underfitting analysis for deep learning based end-to-end communication systems. In 2019 11th *International Conference on Wireless Communications and Signal Processing (WCSP)* (pp. 1–6). IEEE.

33. Virvou, M., Alepis, E., Tsihrintzis, G. A., & Jain, L. C. (2020). Machine learning paradigms. In *Machine Learning Paradigms* (pp. 1–5). Springer, Cham.

34. Hassanien, A. E. (Ed.). (2019). *Machine learning paradigms: Theory and application*. Cham: Springer.

35. Zhai, X., Oliver, A., Kolesnikov, A., & Beyer, L. (2019). S4l: Self-supervised semi-supervised learning. In *Proceedings of the IEEE International Conference on Computer Vision* (pp. 1476–1485).

36. Zhou, Z. H. (2018). A brief introduction to weakly supervised learning. *National Science Review, 5*(1), 44–53.

37. Celebi, M. E., & Aydin, K. (Eds.). (2016). *Unsupervised learning algorithms*. Berlin: Springer International Publishing.

38. Srinivas, B., & Rao, G. S. (2018, January). Unsupervised learning algorithms for MRI brain tumor segmentation. In *2018 Conference on Signal Processing And Communication Engineering Systems (SPACES)* (pp. 181–184). IEEE.

39. Botvinick, M., Ritter, S., Wang, J. X., Kurth-Nelson, Z., Blundell, C., & Hassabis, D. (2019). Reinforcement learning, fast and slow. *Trends in cognitive sciences, 23*(5), 408–422.

40. François-Lavet, V., Henderson, P., Islam, R., Bellemare, M. G., & Pineau, J. (2018). An introduction to deep reinforcement learning. arXiv preprint arXiv:1811.12560.

41. Livingstone, D. J. (Ed.). (2008). *Artificial neural networks: methods and applications* (pp. 185–202). Totowa, NJ, USA: Humana Press.

42. Rojas, R. (2013). *Neural networks: A systematic introduction*. Springer Science & Business Media.

43. Da Silva, I. N., Spatti, D. H., Flauzino, R. A., Liboni, L. H. B., & dos Reis Alves, S. F. (2017). Artificial neural network architectures and training processes. In *Artificial Neural Networks* (pp. 21–28). Springer, Cham.

44. Braspenning, P. J., Thuijsman, F., & Weijters, A. J. M. M. (1995). *Artificial neural networks: An introduction to ANN theory and practice* (Vol. 931). Springer Science & Business Media.

45. Kacprzyk, J., & Pedrycz, W. (Eds.). (2015). *Springer handbook of computational intelligence*. Springer.

46. Dong, C., Loy, C. C., & Tang, X. (2016, October). Accelerating the super-resolution convolutional neural network. In *European Conference on Computer Vision* (pp. 391–407). Springer, Cham.

47. CireşAn, D., Meier, U., Masci, J., & Schmidhuber, J. (2012). Multi-column deep neural network for traffic sign classification. *Neural networks, 32*, 333–338.

48. Sarıgül, M., Ozyildirim, B. M., & Avci, M. (2019). Differential convolutional neural network. *Neural Networks, 116*, 279–287.

49. Dheir, I. M., Mettleq, A. S. A., Elsharif, A. A., & Abu-Naser, S. S. (2020). *Classifying nuts types using convolutional neural network*.

50. Shi, W., Caballero, J., Theis, L., Huszar, F., Aitken, A., Ledig, C., & Wang, Z. (2016). Is the deconvolution layer the same as a convolutional layer?. arXiv preprint arXiv:1609.07009.

51. Xu, B., Wang, N., Chen, T., & Li, M. (2015). Empirical evaluation of rectified activations in convolutional network. arXiv preprint arXiv:1505.00853.
52. Socher, R., Huang, E. H., Pennin, J., Manning, C. D., & Ng, A. Y. (2011). Dynamic pooling and unfolding recursive autoencoders for paraphrase detection. In *Advances in Neural Information Processing Systems* (pp. 801–809).
53. Mozaffari, M., Saad, W., Bennis, M., & Debbah, M. (2016). Unmanned aerial vehicle with underlaid device-to-device communications: Performance and tradeoffs. *IEEE Transactions on Wireless Communications, 15*(6), 3949–3963.
54. Yu, H., Li, G., Zhang, W., Huang, Q., Du, D., Tian, Q., et al. (2020). The unmanned aerial vehicle benchmark: Object detection, tracking and baseline. *International Journal of Computer Vision, 128*(5), 1141–1159.
55. Yao, H., Qin, R., & Chen, X. (2019). Unmanned aerial vehicle for remote sensing applications— A review. *Remote Sensing, 11*(12), 1443.
56. Iannace, G., Ciaburro, G., & Trematerra, A. (2019). Fault diagnosis for UAV blades using artificial neural network. *Robotics, 8*(3), 59.
57. Ciaburro, G., Iannace, G., & Trematerra, A. (2020). Research for the presence of unmanned aerial vehicle inside closed environments with acoustic measurements. *Buildings, 10*(5), 96.
58. Iannace, G., Ciaburro, G., & Trematerra, A. (2020). Acoustical unmanned aerial vehicle detection in indoor scenarios using logistic regression model. *Building Acoustics*, 1351010X20917856.
59. Ciaburro, G. (2020). Sound event detection in underground parking garage using convolutional neural network. *Big Data and Cognitive Computing, 4*(3), 20.
60. Ciaburro, G., Iannace, G., Passaro, J., Bifulco, A., Marano, D., Guida, M.,.... & Branda, F. (2020). Artificial neural network-based models for predicting the sound absorption coefficient of electrospun poly (vinyl pyrrolidone)/silica composite. *Applied Acoustics, 169*, 107472.

Software-Defined Location Privacy Protection for Vehicular Networks

Abdelwahab Boualouache, Ridha Soua, Qiang Tang, and Thomas Engel

Abstract While the adoption of connected vehicles is growing, security and privacy concerns are still the key barriers raised by society. These concerns mandate automakers and standardization groups to propose convenient solutions for privacy preservation. One of the main proposed solutions is the use of Pseudonym-Changing Strategies (PCSs). However, ETSI has recently published a technical report which highlights the absence of standardized and efficient PCSs [1]. This alarming situation mandates an innovative shift in the way that the privacy of end-users is protected during their journey. Software Defined Networking (SDN) is emerging as a key 5G enabler to manage the network in a dynamic manner. SDN-enabled wireless networks are opening up new programmable and highly-flexible privacy-aware solutions. We exploit this paradigm to propose an innovative software-defined location privacy architecture for vehicular networks. The proposed architecture is context-aware, programmable, extensible, and able to encompass all existing and future pseudonym-changing strategies. To demonstrate the merit of our architecture, we consider a case study that involves four pseudonym-changing strategies, which we deploy over our architecture and compare with their static implementations. We also detail how the SDN controller dynamically switches between the strategies according to the context.

A. Boualouache (✉) · R. Soua · T. Engel
University of Luxembourg, Esch-sur-Alzette, AVE, 2 Avenue de l'Universite, 4365
Esch-sur-Alzette, Luxembourg
e-mail: abdelwahab.boualouache@uni.lu

R. Soua
e-mail: ridha.soua@uni.lu

T. Engel
e-mail: thomas.engel@uni.lu

Q. Tang
Luxembourg Institute of Science and Technology (LIST), 5 Avenue des Hauts-Fourneaux, 4362
Esch-sur-Alzette, Luxembourg
e-mail: qiang.tang@list.lu

© The Author(s), under exclusive license to Springer Nature Switzerland AG 2021   395
U. Ghosh et al. (eds.), *Machine Intelligence and Data Analytics for Sustainable Future Smart Cities*, Studies in Computational Intelligence 971,
https://doi.org/10.1007/978-3-030-72065-0_21

1 Introduction

As part of the vision of 5G, connected vehicles will be an important pillar of Cooperative Intelligent Transportation Systems (C-ITS), with the aim to ensure road safety, avoid traffic congestion and provide a better driving experience for users during their journey. Although the deployment stage for connected vehicles is imminent, many security and privacy issues are still unsolved. Location privacy is one of the main issues that may impede the wide acceptance of Cooperative Connected and Automated Mobility (CCAM) applications. Indeed, location tracking of vehicles may reveal every place visited by drivers. This is because there is generally a one-to-one relationship between the vehicle and its driver. The visited locations may include very personal places like hospitals, banks, insurance companies, etc. and hence can reveal sensitive information about the end-user.

On the other hand, the main wireless communication technologies for connected vehicles, such as IEEE802.11p, present several privacy concerns. Indeed, IEEE802.11p mandates that each connected vehicle should frequently send a safety message, called CAM (Cooperative Awareness Message), to ensure cooperative awareness among neighboring vehicles. These messages include sensitive information such as identifiers, positions, speeds, etc., and are sent in clear text; hence vehicles could be tracked on the basis of the information transmitted by the Vehicle to Vehicle (V2V) and Vehicle to Infrastructure (V2I) communications. To mitigate this privacy risk, the use of pseudonym schemes has received significant interest from the research community and standardization authorities. For instance, both the European standard ETSI TS 102 941 [1] and the American standard SAE J2735 [2] have adopted a pseudonym scheme. However, several studies have shown that the use of a simple pseudonym-changing is insufficient to provide unlinkability between the pseudonyms and have suggested using strategies for changing the pseudonyms. Although, there is a significant number of proposed Pseudonym-Changing Strategies (PCSs), there are no recommendations by standardization bodies for PCSs to apply. The main reasons for this can be summarized as follows: (i) several proposed strategies are strongly topology-dependent i.e. they could only be applied in a given situation or area such as signalized intersections, parking lots, and gas stations; (ii) some strategies propose the use of radio silence without considering its critical impact on the exchange of safety messages, or dynamically readjusting radio silence duration; (iii) each PCS is evaluated with individual privacy metrics that may not be suitable for another PCS. This absence of unified evaluation metrics complicates the comparison between existing strategies; (iv) The scarcity of scientific studies focusing on the non-cooperation behavior of vehicles. Selfish vehicles could significantly decrease the efficiency of PCSs; (v) most of the existing PCSs assume a strong global passive adversary. However, this assumption is not realistic, since the global presence of the adversary is difficult to achieve due to the large scale of deployment of connected vehicles, and the high cost of ensuring complete coverage; and (vi) pseudonyms could be easily used to perform Sybil attacks. This vulnerability is not taken into the account by most of the strategies proposed in the literature.

One possible solution for dealing with these various PCSs is to propose a comprehensive architecture that is able to encompass all of them and their intrinsic features. This architecture should be forward-looking in the sense that it should support future PCS solutions. Software-Defined Networking (SDN) has recently been exploited to provide a dynamic and context-aware security solution for vehicular networks [3]. In this chapter, we exploit SDN to propose a software-defined architecture for location privacy in vehicular networks. This architecture extends and leverages the concepts of SDN to PCSs and ensures the selection of the appropriate PCS according to the context of vehicles and other factors, as will be detailed later. The SDN control plane orchestrates the selection and adjusts the parameters of the selected strategy dynamically, based on information received from the data plane. The SDN strategy rules are also forwarded from the control plane to the data plane to ensure the correct execution of the selected PCS. The proposed architecture is flexible and enables the efficient integration of new proposed solutions and new functions of the PCS. The contributions of this work can be summarized as follows:

- Integration of SDN into vehicular networks, allowing new PCSs to be deployed, easily updated and dynamically reconfigured.
- Introduction of novel pseudonym-changing modules in the control plane and the definition of their different interactions to ensure a context-aware PCS.
- Introduction of novel complementary pseudonym-changing modules in the data plane and the definition of their interactions.
- Definition of SDN rules (which can be modified dynamically at the controller) to establish a set of actions that will handle the PCS.
- Definition of a Sybil attack agent to interact with the external misbehavior system controller.
- Definition of self-learning module that is able to analyze and learn from its immediate context while autonomously adapting the PCS accordingly to ensure a high level of privacy protection. This module is crucial, as it guarantees network intelligence and leads to a network that is self-privacy-preserving.

2 Pseudonym-Changing Strategies: Standardization Efforts and Open Issues

Security standardization bodies have agreed to adopt PCS to protect the location privacy of connected vehicles. However, while in the US, the Society of Automotive Engineers (SAE) suggests that vehicles change their pseudonym every five minutes [2], the European telecommunications standardization organization, ETSI, does not suggest the adoption of any PCS [1]. In the light of this, many PCSs are proposed in the literature. In [4], we presented a comprehensive survey and classification of these strategies. This paper also highlights open issues and presents recommendations, including the importance of developing a dynamic system to select the applying PCS according to the vehicular context. Recently, ETSI published a technical report

(ETSI TR 103 415) [1] that presents a pre-standardization study of PCS. This document surveys the existing categories of strategies. It also discusses and describes the suggestions of the European projects (PRESERVE, SCOOP@F, and C2C-CC) regarding PCS. The document identifies the open issues of PCSs and proposes a set of recommendations addressing these issues. In the following, we discuss the open issues highlighted in [1, 4] and the related recent advances.

- **Impact on road safety**: as shown in [4], strategies using radio silence are the most efficient solutions. However, their major drawback is their significant negative impact on safety-related applications. This was first investigated in [5], where the authors recommend that the silent period should be shorter than two seconds and that long silent periods can result in hazardous situations, since many safety messages will not be transmitted due to radio silence. The ETSI technical report [1] also discusses the problems of "missing vehicles" and "guest vehicles". Missing vehicles are those that put radio silence into effect after changing their pseudonyms; at the end of this period, these vehicles suddenly appear in the LDMs (Local Dynamic Map) of neighboring vehicles. This may generate unpredictable reactions as highlighted in [1]. In contrast, the problem of the guest vehicle is observed when a vehicle changes its pseudonym while his old pseudonym still populates the LDMs of its neighboring vehicles [6]. Subsequently, LDM messages contain two entries that correspond to the same vehicle, leading to a misinterpretation of the surrounding environment by neighboring vehicles. Unlike the missing vehicle problem, the ghost vehicle problem is not only linked to radio silence based strategies, but to PCSs in general.
- **Non-cooperative behavior**: by triggering the change of their pseudonyms at the same time slot, cooperative vehicles ensure a high level of anonymity and create confusion for the attacker. Consequently, the existence of non-cooperative vehicles will significantly hinder the efficiency of the PCS specifically under lower vehicular density. The authors of [7] study PCSs under a non-cooperative environment. They propose a game theory model and find a Nash equilibrium of the PCS under different types of games (static/dynamic, with and without complete information). Other works such as [8, 9] propose incentive mechanisms to motivate non-cooperative vehicles to participate in the PCS.
- **Attacker model**: It is not trivial to estimate the power of tracking attackers that may exist in the future deployment of vehicular networks. Attacker power can be expressed in terms of tracking capabilities (strong or weak sniffing stations, efficiency of the tracking algorithm, etc.) and the coverage area. In addition, it is critical to properly define a realistic attacker model. For this reason, most of proposed PCSs have assumed the extreme case of the attacker model (global attacker full of capabilities); however, this assumption is not realistic, because global coverage entails a significant surveillance cost. Consequently, the authors of [10] propose a mid-sized attacker whose power is in between that a local attacker and a global one. They also distinguish three tracking periods (i.e. short-term, mid-term, and long-term) and two levels of surveillance granularity (i.e. Road-level and Zone-level).

- **Evaluation metrics**: many metrics are proposed to assess the performance of PCSs. The recent study carried out by Zhao et al. [11] show that there is no single privacy metric that outperforms all others under different contexts (mobility, traffic conditions, road section, etc.). For this reason, it is recommended to combine all metrics to obtain a fair performance evaluation of a PCS.
- **Privacy model**: the privacy level depends mainly on the considered attacker model and the evaluation metrics. The authors of [7] proposed a linear model to quantify the loss of privacy after the last change of pseudonym. In this model, the privacy level of vehicles linearly decreases according to a sensitivity parameter, which characterizes the power of the adversary. However, this model has two major drawbacks: (i) it does not specify how the sensitivity parameter is measured. (ii) the linearity of this model is not justified.
- **Sybil attacks**: In this attack, vehicles use multiple identities, called Sybils, which can be exploited to create a fake traffic jam and hence to alter other vehicles' perceptions. Pseudonyms could be exploited to launch Sybil attacks. The ETSI technical report [1] gives some recommendations on thwarting Sybil attacks, such as setting the maximum number of pseudonyms that can be used simultaneously and the minimum duration for which the pseudonyms should be used. The technical report also recommends the use of misbehavior detection systems.
- **Pseudonym lock**: ETSI standards specify that the PCS could be locked on-demand for a maximum of 255s, in particular when a critical safety situation occurs. The priority levels of such a situation are respectively "0" or "1" [12]. PCS locking is also proposed by the SAE. However, the conditions when the pseudonyms are locked are not yet defined.
- **Pseudonym reuse**: Although the reuse of pseudonyms minimizes the storage capacity and facilitates the management of pseudonyms, it can decrease the level of privacy. This is why the reuse of pseudonyms is not recommended as a privacy best practice. However, the Car2car consortium considers the reuse of pseudonym while defining some KPI to increase the privacy level [4].

3 Proposed Architecture: Building Blocks

Our self-privacy-preserving architecture leverages the SDN paradigm and thus follows its main principle, which is the separation between the data and the control plane. The control plane is responsible for dynamically selecting the PCS, adjusting the parameters of strategy, and planning the strategy rules. On the other hand, the data plane translates the defined rules into actions to apply the PCS. The communications between the control plane and the data plane are secure.

3.1   Control Plane

Figure 1 shows the logical modules of the control plane in our architecture. The PCS module receives a demand from the application layer to provide the location privacy service. This module chooses the most convenient PCS to be executed based on the information received from two modules: the Mobility and Topology module and Attacker Model module. Once the strategy is selected, the PCS module invokes (i) the Parameter Settings module to request the parameters of the strategy; (ii) the Incentive Model module to request the appropriate incentive method to motivate non-cooperative vehicles; and (iii) the Privacy Metric module to request indicators and KPIs for the evaluation of PCS performance. In the following, we detail these modules.

- **Road Safety Monitoring**: this module monitors road conditions and its impact on traffic safety. Based on this assessment, the module develops appropriate SDN rules which are sent to the data plane. In addition, this module provides the necessary

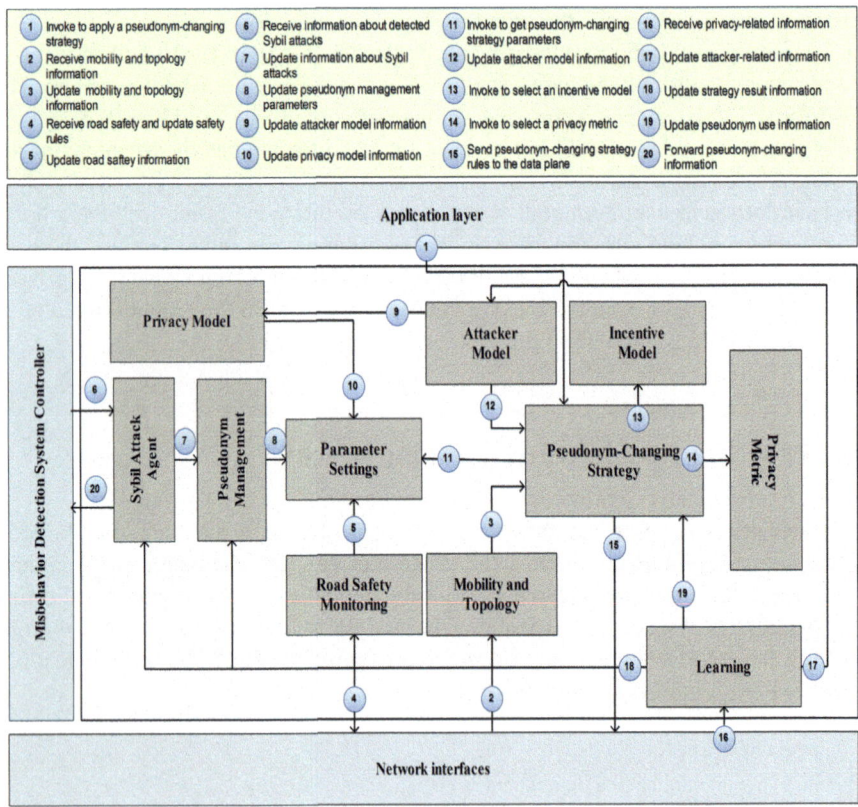

Fig. 1 The logical modules of the control plane and the interactions between them

information to the Parameter Settings module to tune the PCS parameters, such as the duration of radio silence and the lock period.

- **Misbehavior Detection System Controller**: this is an external component, which detects misbehaving attacks such as message injection, denial of service (DoS) and Sybil attacks. The SDN controller of our self-privacy-preserving architecture uses the information received from the Misbehavior Detection System Controller to update its parameters in order to limit Sybil attacks and returns information to help in detecting Sybil attacks and accurately evaluating the trust levels of vehicles.
- **Sybil Attack Agent**: this interface is used to interact with the Misbehavior Detection System Controller, receiving information from it and forwarding it to the Pseudonym Management module to adjust some PCS parameters. It also receives information from the Learning module and forwards this to the Misbehavior Detection System Controller to enhance the attack detection ratio.
- **Pseudonym Management**: this module plans the rules that orchestrate the use of pseudonyms: the reuse of pseudonyms, the frequency of changing of pseudonyms, the number of pseudonyms that can be used in parallel, etc. This module receives information from both the Sybil Attack Agent and learning modules and sends the resulting rules to the Parameter Settings module.
- **Privacy Model**: This is used to model the loss of privacy of vehicles over time. As explained in the previous section, the loss of privacy mainly depends on the strength of the attacker model. For this reason, this module receives input from the Attacker Model module. The Privacy Model provides input to the Parameter Settings module, which in return specifies the parameters of the Privacy Model.
- **Mobility and Topology**: this module monitors the mobility pattern of vehicles and the road topology in real time.
- **Parameter Settings**: this module sets the different parameters of the PCS, such as the duration of the radio silence period and the minimum duration of the use of pseudonyms. The definition of these parameters is made according to the information received from the Road Safety, Pseudonym Management, and the Privacy Model modules.
- **Attacker Model**: this module evaluates the power of the attacker. As discussed in the previous section, the attacker can be internal or external, local or mid-sized, long-term. It can perform simple syntactic linking of pseudonyms, but can also carry out more advanced semantic linking of pseudonyms. This module gets regular updates from the learning model and sends feedback to the Pseudonym-Changing Strategy module.
- **Incentive Model**: this module defines the incentive model, which is used to motivate selfish vehicles to participate in the PCS.
- **Privacy Metric**: this module defines the privacy metrics used to evaluate the PCS. It worth mentioning that the privacy metrics can be selected by the PCS to evaluate its own performance.
- **PCS Module**: this module defines the strategy to be executed based on the information received from the Mobility and Topology module and the Attacker Model module. Once the strategy is selected, this module invokes the Parameter Settings module to obtain the most appropriate parameters of the selected strategy. This

module also invokes the Privacy Metric module and the Incentive Model module to select the evaluation metric and the incentive method respectively.

- **Learning**: this module periodically receives privacy-related information from the data plane (i.e. the privacy levels of vehicles, the presence of an attacker, and the set of selfish vehicles). This information is analyzed and forwarded to the corresponding modules: (i) the Attacker Model module to adjust the attacker model being used; (ii) the PCS module to tune the strategy parameters, and the Incentive Model module, and to select an additional potential privacy metric. (ii) the Pseudonym Management module to adjust pseudonym management related parameters, and finally (iv) the Sybil Attack Agent, which forwards pseudonym-changing information to the Misbehavior Detection System Controller. The purpose is to support this controller in the accurate detection of Sybil attacks and trust assessment of vehicles.

3.2 Data Plane

The data plane is composed of the different vehicles that are involved in the PCS. Figure 2 depicts the modules of the data plane, which are responsible of the execution of the PCS. The data plane uses the vehicles' communication interfaces to collect pertinent information concerns the surrounding vehicular environment. The data plane sends mobility, safety, and privacy information to the control plane, while it receives safety and strategy rules. In the following, we describe the modules and the databases of the data plane:

- **Safety Message Management**: this module sends and receives pseudonymous safety messages. It also receives instructions from the Strategy Engine. These instructions vary according to the applied strategy. In addition, this module provides the status of the surrounding environment and the impact of the applied PCS to the Road Safety Engine, the Topology and the Mobility engine, and finally to the Strategy Engine.
- **Mobility and Topology Engine**: Equipped with a map and GPS, this module sends the mobility information of the vehicle such as position, speed, and acceleration and the topology information to the Road Safety Engine and to the control plane.
- **SDN Safety Rules**: This is a database, which contains the safety rules that are used to assess road conditions. The rules data is received from the control plane.
- **Road Safety Engine**: this module receives, stores and updates the safety rules received from the control plane. These rules are used to evaluate road safety based on the information received from the Topology and Mobility Engine and the Safety Message Management module. This module periodically sends road safety information to the control plane.
- **SDN Strategy Rules**: This is a database that contains the rules related to PCS. These rules describe where, when and how pseudonyms change. The database

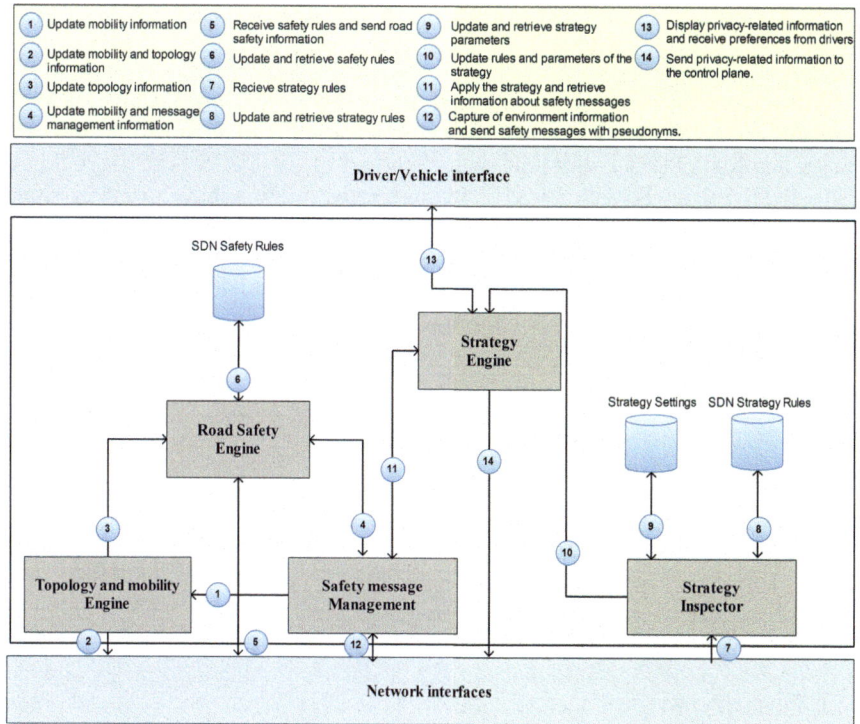

Fig. 2 The logical components of the data plane and the interactions between them

is regularly updated by the Strategy Inspector module; based on the information received from the control plane.

- **Strategy Settings**: This is a database that contains the settings of the applied strategy such as the duration of radio silence period after the changing of pseudonym. This database is also regularly updated by the Strategy Inspector module according to the information received from the control plane.
- **Strategy Inspector**: this module represents an interface, which communicates with the PCS module of the control plane. It receives information from the SDN controller(s) and stores them in two databases: the SDN Strategy rules and the Strategy Settings databases. This module also forwards these PCS rules and settings to the Strategy Engine module.
- **Strategy Engine**: this module executes the PCS according to the rules and settings received from the Strategy Inspector module. To execute the strategy, the module continuously monitors and sends instructions to the Safety Message Management module. This module provides privacy protection related information to the driver from whom it receives privacy level recommendations. This module also sends privacy-related information to the control plane.

4 Case Study

To demonstrate the merit of our proposed architecture, we conducted the following case study. As shown in Fig. 3 (1), we populated a Software-Defined Location Privacy Controller (SDLP) with four state-of-the-art PCSs: UPCS [13], TAPCS [14], PRIVANET [9] and SocialSpots [15]. In this section, we first show how these strategies are integrated into our architecture. Then, we illustrate how the SDLP performs a context-aware PCS selection. The context is mainly defined by mobility and topology, as well as the attacker model. Finally, we conduct a simulation-based study to demonstrate how our proposed architecture dynamically updates the security parameters of each strategy.

4.1 PCSs Deployment

Our proposed architecture is flexible enough to support any state-of-the-art PCS. Table 1 shows how the considered strategies are mapped to our architecture. This table has six columns: (i) Mobility and topology: specifies the topology where the strategy can be used; (ii) Parameter Setting: specifies the parameters of the strategy; (iii) Attacker model: specifies that attacker model from which the strategy provides protection; (iv) Privacy model: specifies if the strategy uses a privacy model or not; (v) Privacy metric: specifies the metric used to evaluate the strategy; (vi) Incentive model: specifies if the strategy uses an incentive model or not.

Control plane modules are activated or deactivated according to the requirements of each PCS. For example, the Incentive Model module is disabled for UPCS and TAPCS since these strategies do not propose any mechanism to motivate non-cooperative vehicles to change their pseudonyms; while the Privacy Model module is only activated for PRIVANET strategy.

Figure 3 (2) illustrates the different steps of the selection of a PCS. The SDLP first checks the information received from the Mobility and Topology module. For instance, if the vehicle is entering a signalized intersection, two PCSs could be applied to this context: UPCS and SocialSpots. To decide which of the two strategies to apply, SDLP checks information received from the Attacker Model module. If the attacker model can perform both syntactic and semantic pseudonym linking attacks, then UPCS is selected. Otherwise, if the attacker can perform only syntactic attacks, SocialSpots is selected. More details on syntactic and semantic pseudonym linking attacks can be found in [4].

Table 1 The deployments of PCSs in the self-privacy-preserving architecture

	Mobility and topology	Parameter setting	Attacker model	Privacy model	Privacy metric	Incentive model
UPCS [13]	Signalized intersection	Red traffic light duration: 30s, 60s	Global external passive and local internal passive (Semantic and syntactic linking)	No	The entropy of the annonymity set	No
SocialSpots [15]	Signalized intersection	Red traffic light turns green	(Syntactic linking)	No	The size of the anonymity set	Yes
[14]	Traffic congestion	Speed threshold	Global external passive and local internal passive (Semantic and syntactic linking)	No	The entropy of the anonymity set	No
PRIVANET [9]	Roadside Infrastructure e.g. Gas station	The capacity of RI The threshold of privacy	Global external passive and local internal passive (Semantic and syntactic linking)	Yes	The size of the anonymity set	Yes

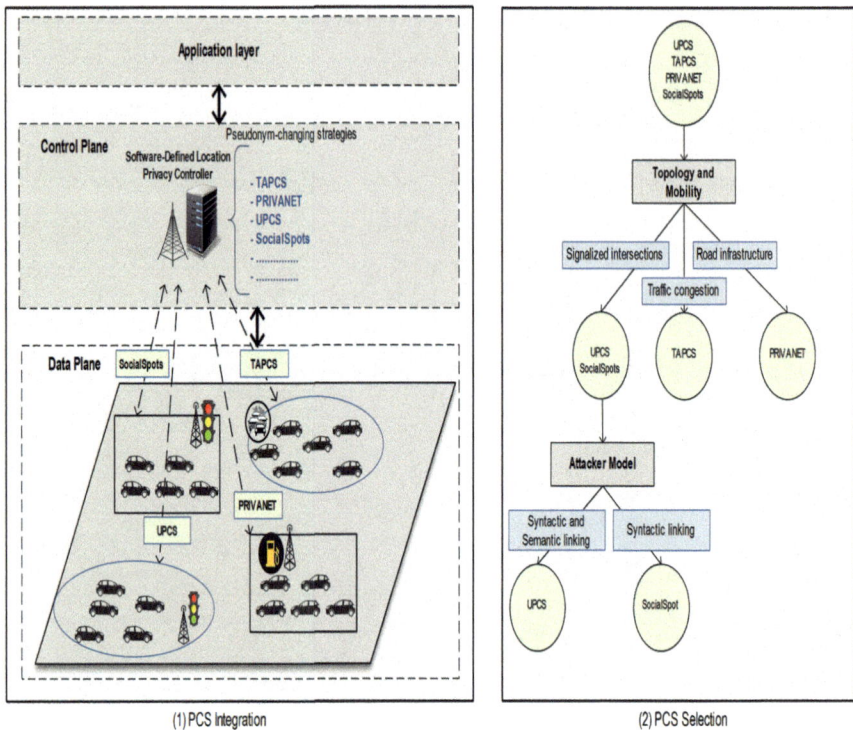

Fig. 3 The selection of pseudonym changing strategy

4.2 Simulation Setup

We carried out a simulation-based analysis to demonstrate the merit of our SDN-based and self-learning architecture and how it dynamically adapts the PCS security parameters to the context. This simulation-based analysis was performed using Veins Simulation Framework [16]. The considered scenario is similar to that proposed in [9]. Three strategies are simulated: UPCS, TAPCS, and PRIVANET. SocialSpots was excluded, as it has the same application context (signalized intersections) as UPCS.

Table 2 details the configurations of the simulated strategies. This table has four columns: (i) Changed context: specifies the context we change during the simulation; (ii) Configuration: specifies the values we assign to the context' parameters; (iii) Action: specifies the action to perform when the parameter is changed; (iv) Results: specifies the obtained results when the action is applied. To demonstrate the dynamic changing of PCS parameters according to context, three different scenarios are considered.

Table 2 The configuration of pseudonym-changing strategies

Strategy	Changed context	Configuration	Action	Results
SDN-based UPCS [13]	Road safety	10% of vehicles in dangerous situation	Pseudonym lock	Low safety risk Acceptable privacy level
		20% of vehicles are in dangerous situation	Pseudonym lock	Low safety risk. Acceptable privacy level
SDN-based TAPCS [14]	Attacker model	Simple attacker	Select privacy metric	The size of the anonymity set
		Medium attacker	Change the privacy metric	The entropy of the anonymity set
		Advanced attacker	Keep the privacy metric	The entropy of the anonymity set
SDN-based PRIVANET [9]	Privacy model	Sensitivity parameter = 0.1	Update privacy model	High privacy level
		Sensitivity parameter = 0.2	Update privacy model	Low privacy level

1. **Scenario 1**: uses UPCS strategy in a road safety context, where the number of vehicles in a dangerous situation can be 10% or 20%. The pseudonym changing in such a situation can generate traffic collisions and accidents.
2. **Scenario 2**: uses TAPCS strategy, where we study how this strategy adapts the privacy metric to the attacker model. Three configurations of the attacker model are considered: simple, medium, and advanced.
3. **Scenario 3**: uses PRIVANET focusing on the privacy model. We consider two configurations of this model by varying the sensitivity parameter value, which characterizes the power of the adversary.

4.3 Simulation Results

Figure 4 compares the static implementation UPCS (static UPCS) to its SDN-based variant (SDN-based UPCS). Two performance indicators are considere: the privacy level and safety. As shown in Fig. 4, static UPCS provides a higher level of privacy protection compared to SDN-based UPCS. However, SDN-based UPCS has a lower safety r isk than static UPCS. The reason for this, as described in Table 2, is that SDLP takes an action to lock pseudonym-changing processes of vehicles in a dangerous situation. This lock slightly decreases the privacy protection level, while reducing the safety risk.

Figure 5 makes a comparison between Static TAPCS and SDN-based TAPCS. In Static TAPCS, the entropy of anonymity set is used as a performance metric,

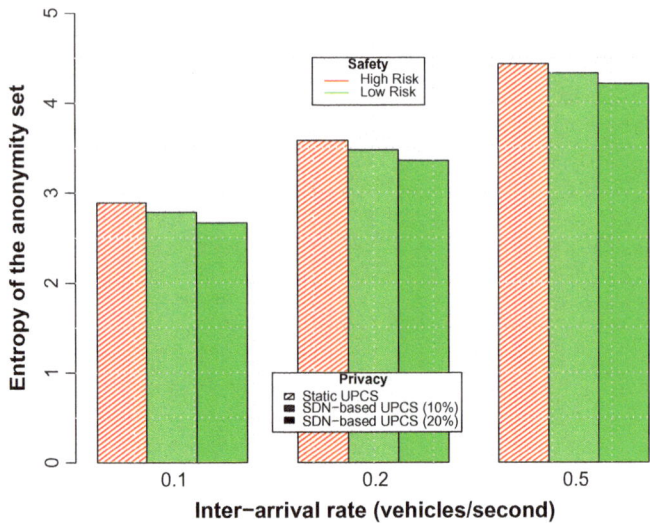

Fig. 4 Static UPCS versus SDN-based UPCS

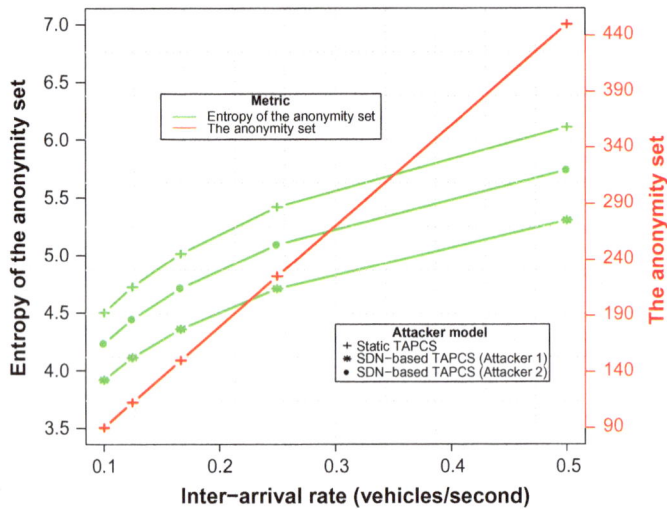

Fig. 5 Static TAPCS versus SDN-based TAPCS

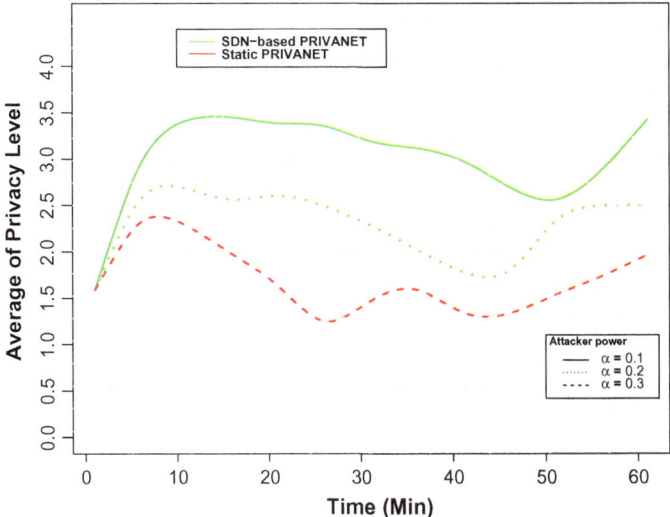

Fig. 6 Static PRIVANET versus SDN-based PRIVANET

whatever the used attacker model. However, the SDN-based TAPCS varies the performance metric according to the power of the attacker. For instance, the size of the anonymity set is chosen when the attacker is simple, while the entropy of the anonymity set is considered when the attacker is medium or advanced. This selection of the performance metrics is based on the probabilities of distinction between vehicles in the considered area. In the former case, these probabilities are equal and hence the measuring size of the anonymity set performs well. In the latter case, these probabilities are not equal; hence the need to take the entropy into account.

Finally, we compare the static implementation of PRIVANET and the SDN-based version. As illustrated in Fig. 6, the sensitivity parameter (α), which characterizes the power of the attacker, remains unchanged in Static PRIVANET and is equal to 0.3. However, for SDN-based PRIVANET, the sensitivity parameter is updated according to the information received from the data plane. The change in the power of the attacker (the sensitivity parameter) has a direct impact on the privacy level obtained by vehicles. Indeed, as illustrated in Fig. 6, the high values of the average of privacy are obtained when the sensitivity parameter equals 0.1. However, the lower values of the average of privacy are obtained when the sensitivity parameter equals 0.3.

5 Conclusion

The imminent deployment of connected vehicles requires significant attention to the security and privacy aspects. Privacy protection is a critical issue that influences the user acceptance of this technology. Pseudonym-changing strategies are considered as the key solution to overcome this acute need. However, the absence of recommended pseudonym-changing strategies (PCSs) represents an obstacle to achieving this objective. To this end, we propose an innovative architecture that exploits Software-Defined Networking (SDN), one of the key technologies for 5G networks. Our proposed architecture is flexible and self-learning and hence can encompass PCSs proposed so far in the literature and even upcoming PCSs. The selection of the appropriate PCS and it security settings are context-aware. The control plane is modular and includes the main building-blocks of PCSs which can support any future solution. As future work, we plan to carry out extensive simulations to assess the performance of the proposed architecture.

Acknowledgements This work was supported by the H2020 5G-DRIVE project (ID: 814956) and the H2020 5G-MOBIX project (ID: 825496).

References

1. ETSI TR 103 415, Intelligent Transport Systems (ITS); security; pre-standardization study on pseudonym change management. *ETSI Standards* (2018).
2. J2945/1, On-board system requirements for V2V safety communications. *SAE Standards*(2016).
3. Garg, S., Kaur, K., Kaddoum, G., Ahmed, S. H., & Jayakody, D. N. K. (2019). SDN-based secure and privacy-preserving scheme for vehicular networks: A 5g perspective. *IEEE Transactions on Vehicular Technology, 68*(9), 8421–8434.
4. Boualouache, A., Senouci, S.-M., & Moussaoui, S. (2017). A survey on pseudonym changing strategies for vehicular ad-hoc networks. *IEEE Communications Surveys & Tutorials, 20*(1), 770–790.
5. Lefevre, S., Petit, J., Bajcsy, R., Laugier, C., & Kargl, F. (2013). Impact of v2x privacy strategies on intersection collision avoidance systems. In *Vehicular Networking Conference (VNC), 2013 IEEE*, Dec 2013, pp. 71–78.
6. Jemaa, I. B., Kaiser, A., & Lonc, B. (2017). Study of the impact of pseudonym change mechanisms on vehicular safety. In *IEEE Vehicular Networking Conference (VNC). IEEE*, pp. 259–262.
7. Freudiger, J., Manshaei, M. H., Hubaux, J.-P., & Parkes, D. C. (2013). Non-cooperative location privacy. *Dependable and Secure Computing, IEEE Transactions on, 10*(2), 84–98.
8. Ying, B., Makrakis, D., & Hou, Z. (2015). Motivation for protecting selfish vehicles' location privacy in vehicular networks. *Vehicular Technology, IEEE Transactions on, 64*(12), 5631–5641.
9. Boualouache, A., Senouci, S.-M., & Moussaoui, S. (2019). PRIVANET: An efficient pseudonym changing and management framework for vehicular ad-hoc networks. *IEEE Transactions on Intelligent Transportation Systems*.
10. Petit, J., Broekhuis, D., Feiri, M., & Kargl, F. (2015). Connected vehicles: Surveillance threat and mitigation. *Black Hat Europe*, p. 11.

11. Zhao, Y., & Wagner, I. (2018). On the strength of privacy metrics for vehicular communication. *IEEE Transactions on Mobile Computing, 18*(2), 390–403.
12. ETSI TS 101 539-1 (V1.1.1), "Intelligent Transport Systems (ITS); V2X applications; part 1:Road Hazard Signalling (RHS) application requirements specification," *ETSI standards* (2013).
13. Boualouache, A., & Moussaoui, S. (2017). Urban pseudonym changing strategy for location privacy in VANETs. *International Journal of Ad Hoc and Ubiquitous Computing, 24*(1–2), 49–64.
14. Boualouache, A., & Moussaoui, S. (2017). TAPCS: Traffic-aware pseudonym changing strategy for VANETs. *Peer-to-Peer Networking and Applications, 10*(4), 1008–1020.
15. Lu, R., Lin, X., Luan, T. H., Liang, X., & Shen, X. (2011). Pseudonym changing at social spots: An effective strategy for location privacy in VANETs. *IEEE Transactions on Vehicular Technology, 61*(1), 86–96.
16. Sommer, C., German, R., & Dressler, F. (2011). Bidirectionally coupled network and road traffic simulation for improved IVC analysis. *IEEE Transactions on Mobile Computing, 10*(1), 3–15. Jan.

Printed by Printforce, the Netherlands